AF442811

RECENT ADVANCES IN EMBRYOLOGY

RECENT ADVANCES IN EMBRYOLOGY

Vol. 4
Avian Embryology

A.P. Diwan
&
N.K. Dhakad

ANMOL PUBLICATIONS PVT LTD
New Delhi-110 002

ANMOL PUBLICATIONS PVT LTD
4374 / 4B, Ansari Road
Daryaganj, New Delhi-110 002

Recent Advances in Embryology
© Reserved
First Edition 1994
ISBN 81-7041-974-3 (Set)

PRINTED IN INDIA

Published by J.L. Kumar for Anmol Publications Pvt. Ltd., New Delhi
and Printed at Efficient Offset Printers, Delhi.

Preface

Embryology is a branch of biology which has a most immediate bearing on the problem of life. It has become an immense field whose confines are difficult to delineate. The focus has been and remain on the embryos, on the gradual emergence of form and structure from what appears to be a very modest beginning. The essence of embryonic development is change transition from one stage to another. Embryo is a fleeting stage, a continum along the axis of time. Embryology is a bank division of biology. It stems the anatomy, the histology and the physiology of adult. To understand it well is to aid in the comprehension of the other biological disciplines.

Though, the authors have taken special care to present a current account, yet they are fully aware about their limitations, and the readers may come across the mistakes of various types. For all types of mistakes they extend their due apology.

The authors have freely consulted other standard books and research papers while preparing the manuscript, so the authors claim no originality of the work.

The authors express their thanks to their friends and colleagues whose continuous inspirations have invited them to bring out this text.

The authors are also thankful to Shri J.L Kumar of M/s Anmol Publications Pvt. Ltd. for bringing out a handsome print of this set in a comparatively short time.

Constructive criticisms and suggestions for improvement of the book will thankfully acknowledged.

Authors

Contents

Reproduction in Birds

Most anyone who eats breakfast with any regularity is familiar with the key to avian reproduction, the egg, and knows that it comes in a variety of sizes ranging from very small to extra large. Unlike the other classes of vertebrates, birds exhibit only egg laying (oviparity) as a reproductive strategy; they cannot retain young and nurture them until birth as mammals do (viviparity), and they do not retain eggs within the body until they hatch, as some reptiles do (ovoviviparity). Yet, within this seeming constraint, birds show remarkable variability. For example, while most eggs hatch with young that are entirely dependent on parental care, some young hatch and are totally self-sufficient and immediately able to feed themselves.

Associated with this variation in egg characteristics are specific traits of parental care, growth rates, fledging times, and so forth. In this chapter, we shall discuss the general mechanics of reproduction, with emphasis both on the constraints faced by various types of birds and on the flexibilities they may have. We begin with a brief look at the basic anatomy and physiology of egg production.

Production of An Egg

Reproduction serves two important purposes. Obviously, it is the way an organism reproduces itself so that some of its genes survive in subsequent generations. Of perhaps equal importance, though, is the fact that reproduction allows an organism the chance to produce, through sexual recombination, offspring with varying genetic traits that may affect survival in the future. The occurrence of sex is costly: if males were not necessary, all birds could produce eggs, potentially doubling the production of young. However, the lack of variability resulting from a single sex system would be maladaptive; the cost of a two-sex system are more than compensated for by the production or young with a variety of genetic characteristics.

Although a two-sex system is adaptive, the system does not necessarily represent a balanced division of reproductive effort. Only females produce eggs, and these eggs require a much greater investment of energy than the sperm produced by the males. While males may try to compensate for this imbalance in a variety of ways, this asymmetry in initial costs of reproduction cannot be ignored.

We have briefly mentioned the morphology of the avian reproductive tract in earlier chapters. There we noted that it was seasonally variable in size in order to better accommodate flight, and was associated with the urinary tract. In the female, the apparent selection for reduction in weight has resulted in one functional reproductive tract in what originated embryologically as paired tracts. In all birds, the left side of the reproductive tract is developed. In most species, the right side is only rudimentary, although in some birds of prey and in kiwis a mature right ovary is present and functional.

The Female Reproductive System
The female reproductive system consists of an ovary and an oviduct (Figure 1.1). The ovary is situated in the center of the body cavity, anterior and ventral to the kidney. During the breeding season it may enlarge up to 50 times with the development of several mature ova (singular, ovum). The ovary contains hundreds of thousands of oocytes, but only a very few of these develop during each breeding period. The oocytes that do develop move from the outer layer of the ovary to the center or medulla, where they become surrounded by concentric layers of vascularized ovarian tissue forming a follicle. The follicle protects and nourishes the oocyte during

the development and passes to it the lipids and proteins synthesized in the liver (*vitellogenesis*). These materials form the large inner mass that we recognize as the yolk of an egg. At first, follicular growth is slow and it may take months to years for the oocyte to grow only a few millimeters in diameter. Then, over a period of several weeks there is rapid deposition of protein in the yolk, and finally, a short period of extremely rapid growth occurs during the 7-11 days prior to ovulation, when lipids synthesized in the liver are added to the yolk. During the latter period in the chicken, the follicle enlarges from an initial 8 mm to 37 mm in diameter and increases in weight from 0.08 g to 15-18g. The oocyte undergoes one meiotic division during   development, the second meiotic division is initiated at fertilization, so that only half of the parental complement of DNA remains.

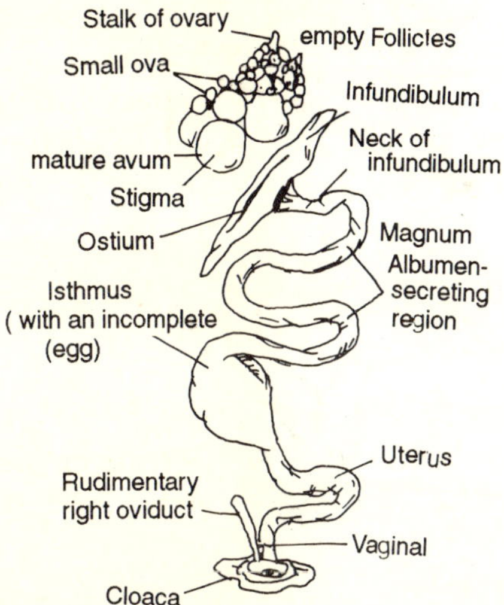

Fig. 1.1. The reproductive organs of a female chicken.

The development of the follicle and the oocyte is regulated by Follicle-Stimulating Hormone (FSH) secreted by the anterior pituitary gland. When the follicle reaches maturity, the outer layers break down, and, under the influence of another anterior pituitary hormone, Luteinizing Hormone (LH), the follicle

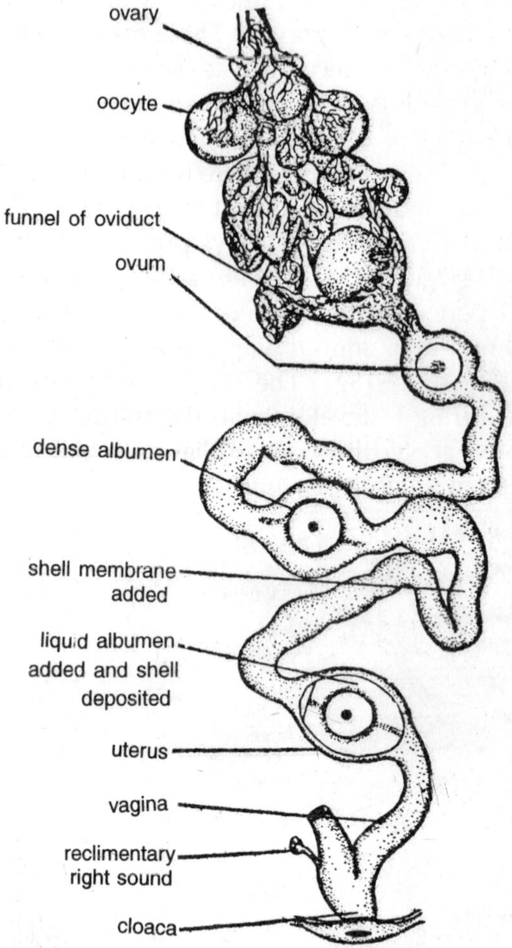

Fig. 1.2.   Egg Transport in hen.

ruptures, releasing the mature ovum into the body cavity. This is the process known as *ovulation*. The ovum does not float aimlessly in the body cavity because the waving fibria of the anterior end of the oviduct, called the *infundibulum*, draw it into the mouth of the oviduct.

The role of the oviduct is to complete the construction of the egg by providing a watery albumen layer to buffer the yolk and a protective shell and to move the egg to the cloaca. Most birds lay an egg a day, and ovulation often occurs just after laying so that only one egg occurs in the lower part of the oviduct at a time. The oviduct (also called *Mullerian tube*) is a long, winding tube that can be divided into five segments, each with a different function. Ciliated cells in the infundibulum, or mouth, direct the ovum into the oviduct by an active process that works only at the time of ovulation. Usually fertilization occurs here, within 15 minutes of ovulation. The ovum then moves into the magnum, a thicker part of the oviduct. The ovum then moves into the magnum, a thicker part of the oviduct. Fertilization must occur prior to the ovum's arrival, before more layers are added to the yolk that could impede sperm entry. The magnum retains the ovum for 2 to 3 hours while several layers of albumen, or egg white, are added. Once this is done, the egg moves to the isthmus, where the two keratin shell membranes are applied, a process taking 1 to 5 hours. The final construction of the egg is completed in the uterus or shell gland, where a watery albumen layer is added, and the egg is sealed with a calcareous shell and a cuticle. Pigments that are added to the shell during the 20-hours shell-forming process are secreted by uterine glands. The colours are derived from bile and blood (hemoglobin) pigments, and the patterns are a result of the movement of the egg in the shell gland. The vagina is the last region of the oviduct. It lubricates the passage of the egg by mucous secretions and its muscular walls move the egg to the cloaca. This region is also important for the collection and storage of sperm (see below):

The above process of egg formation requires a significant energy expenditure by the female. For example, a female House Sparrow (*Passer domesticus*) lays a clutch of four to five eggs, whose energy content is 14.4 kcal. If the total time for production of these eggs is about 7.7 days, and her efficiency of converting assimilated energy into egg material is about

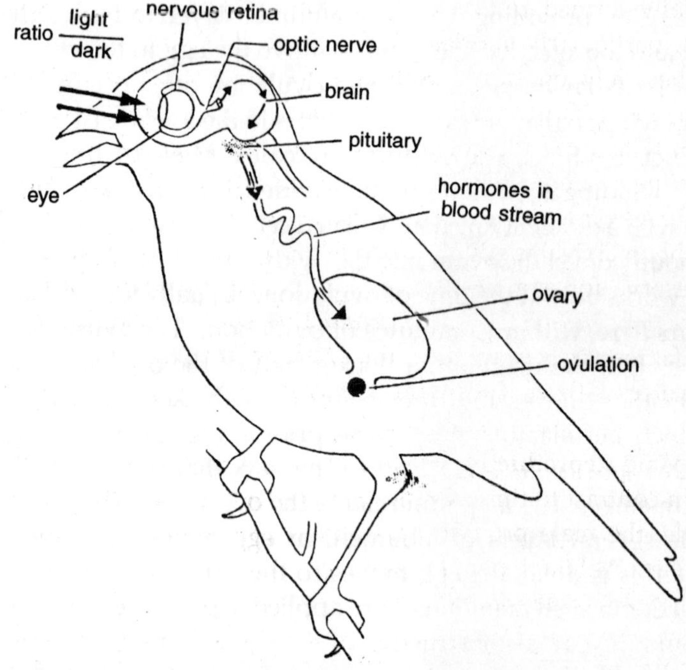

Fig. 1.3.   Steps in stimulation of ovulation by light in a bird.

77%, then the energy requirement for egg laying would be 2.4 kcal per bird-day. This cost represents a 26.6% elevation of her basal metabolism. In addition, many physiological changes occur in the female that enable her to rapidly synthesize and deposit nutrients in the yolk. A laying bird typically has a blood glucose concentration that is about twice normal. Blood lipid and blood calcium levels also increase dramatically. However, even the increased levels of blood calcium cannot provide enough resource for the production of eggshell. Approximately 2000 mg of calcium are needed to form the shell of a chicken egg. This represents about 100 mg per hour for the 20 hours the egg is present in the shell gland. A 2-kg hen has a total of only 20-30 mg of calcium circulating in her plasma. Thus, a laying hen must secrete approximately four times more calcium into the eggshell than is found in the blood. This is accomplished by mobilizing calcium from specially formed reservoirs of intramedullary bone. Many birds, particularly species such as waterfowl that lay large clutches, rely on stored nutrients to successfully complete laying in as short a time as possible. Female Lesser Snow Geese (Chen caerulescens) accumulate fat reserves on their wintering ground in the southern United States and then migrate to the Arctic to breed before the snow has melted. Their weight on arrival is still 20% above normal. Females use their fat reserve for egg production and to sustain them during incubation. In fact, their reproductive success depends on their fat reserve because the number of eggs produced is closely correlated with their weight on arrival.

The Male Reproductive System

In contrast to this enormous energetic undertaking in the female, the male provides only the spermatozoa to fertilize each egg. Although of little energetic value, each spermatozoan is critical for the genetic information it contains, as noted earlier. The reproductive system of the male bird consists of a pair of testes, ducts that carry the spermatozoa to the cloaca, and in some species, an ejaculatory groove and phallus in the cloaca. Bird do not have the complement of accessory glands that contribute the liquid portion of the semen in mammals. Instead, seminal fluid is formed in the tubules and ducts in the testis.

The testes are located ventral to the kidneys and just posterior to the adrenal glands (Figure 1.3). These bean-shaped organs change in size as much as 300 times from non-breeding to breeding condition. It has been estimated that the testes of a duck enlarge to almost 10% of its body weight at the height of the breeding season. Generally, the left testis is larger than the right, although both are functional. Within them, specialized germ cells called *spermatogonia* undergo meiosis and cellular transformations to produce the spermatozoa, whose only job is to find and fertilize an ovum. Other cells in the testis secrete androgens, which influence both development of secondary sexual characteristics associated with the male and courtship behaviour.

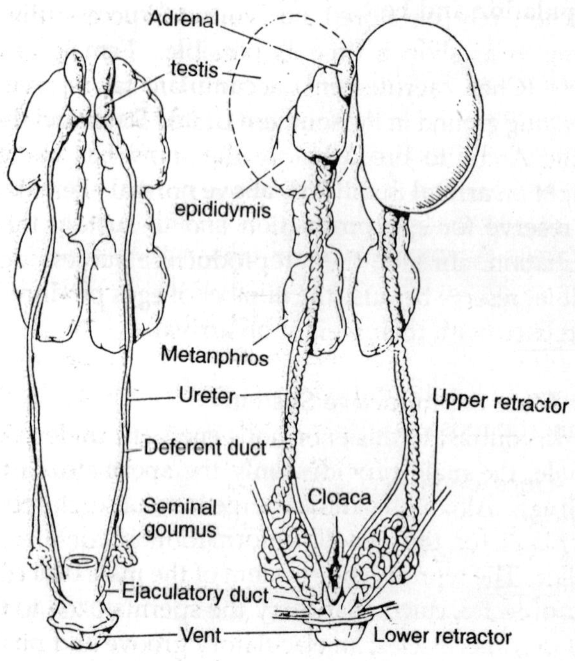

Fig. 1.4.   Urogenital organs of a House Sparrow (*Passer domesticus*) during the breeding (*right*) and non-breeding (*left*) seasons, illustrating the seasonal changes in the size of the sex organs.

Spermatozoa move from the tests to the cloaca via the highly coiled *ducts deferens (or vasa deferentia)*. This organ also increases greatly in size with breeding condition; in passerines, the sperm storage sac at the posterior end of the ductus is about 100 times larger during breeding than during nonbreeding. Spermatozoa that have moved down the ductus remain in this storage sac until the time of coition, when they are transferred to the female. Spermatozoa are sensitive to high temperatures in all animals. In birds, this problem is often solved by displacing the sperm storage sac away from the body core in a cloacal protuberance. Here, temperatures may be as much as 4° C lower than the core, which aids the sperm's survival.

Copulation and Fertilization

Copulation in birds consists of the transfer of semen from the cloaca of the male to the cloaca of the female. In most male birds, only a very small, erectile phallus occurs, and coition involves nothing more than close apposition of cloacas of the two sexes. In a few groups such as ratites, tinamous, and ducks and geese, a fairly large, erectile, and grooved penis occurs to aid in copulation. In the ostrich this organ may be as long as 20 cm. Perhaps because of the relative inefficiency of sperm transfer in most birds, the semen contains a high concentration of spermatozoa. It has been reported that as many as 8.2 billion sperm per ejaculate are produced by some domestic roosters.

Once deposited within the female's cloaca, the semen may be stored in a sac within the vagina. Although fertilization may occur just several hours after copulation, in many species peak fertility occurs several days after copulation. Because of their ability to store viable sperm, females can remain fertile for periods of several weeks following copulation. One of the events that occurs at the time of fertilization is sex determination. The sex determination system of birds is the reverse of that found in other vertebrates. Female birds produce two types of eggs, ones with a male sex chromosome and ones without. All sperm contain the male sex chromosome. The egg with a male sex chromosome will produce a male when fertilized; the egg without the male sex chromosome will produce a female when fertilized. Thus, the female gamete is the one that determines the sex of the offspring.

The sex organs of both male and female also contain several endocrine glands in the interstitial tissue surrounding the germ cells that produce sex hormones that affect both development and behaviour. We shall discuss these further when we look at the timing and synchronization of breeding events.

EGG LAYING AND INCUBATION

Egg Laying

Once a female has reached the egg-laying stage, an egg is laid each day in most species until the clutch is complete. Larger birds such as geese, swans, herons, hawks and owls may lay eggs at about two-day intervals, while some species lay in even longer intervals. Eagles and condors exhibit four to five days interval between eggs, and five to seven days may elapse between the first and second eggs of some seabirds. Most species lay their eggs in the morning, presumably because the shell has formed and formed and hardened overnight when the hen is not active, but there is a great deal of variability in the time of laying both within and between species.

The number of eggs in a clutch is also quite variable both within and between species. Most species within a particular region exhibit a typical clutch size, although this varies somewhat with age and time of nesting. Young birds in general lay smaller clutches than older birds, and first clutches during . the breeding season are generally larger than second clutches. For example, the *Great Tit (Parus major)*has an average clutch size of ten eggs in April nestings, and only seven eggs in June nestings.

The mechanisms that determine when the bird will stop laying and start incubating are not fully known. Some species are determinate layers; that is, a certain number of follicles mature within the ovary each spring, and once these have been laid, the clutch is complete, regardless of the number of eggs actually in the nest. Other species are indeterminate layers, because they have the capacity to keep laying eggs well beyond the number in a typical clutch. Normally; these species must use visual or tactile cues in concert with hormonal adjustments to stop laying. If eggs are removed from the nest, though, they will lay eggs for long periods. Unfortunately, so much variation occurs among birds that one cannot clearly classify all species as either determinate or indeterminate layers.

Incubation

When the clutch is completed, incubation of the eggs commences. The heat required to raise egg temperature to the optimal levels for embryonic development is usually provided by the parents through incubation. Little heat is produced by the embryo, except at the end of incubation. Only in rare cases is the incubation provided by a non-parent, or even, as in the case of the megapodes, by decaying vegetation. Developing embryos are quite sensitive to fluctuating temperatures, so the parent bird must control the thermal environment of the egg through various behavioural and physiological adjustments.   Optimum temperatures for development in 37 species averages 34°C. Temperatures above 43°C for just 1 hour are lethal for embryonic Heerman's Gulls (*larus heermanni*), and development of most avian embryos ceases below 25°C. Several studies have shown how the amount of incubation by the parents is directly related to ambient temperatures. In general, incubating birds adjust the heat delivery to their eggs by varying the time spent in direct contact with them, not by raising their own heat production to warm the eggs. In order to maintain egg temperature at about 35°C, parental attentiveness (i.e., sitting on the eggs) increases with decreasing air temperature below 25°C and with increasing air temperatures above 35°C. At air temperatures between 25o and 35°C, parents allow the eggs to passively heat or cool. The result of these adjustments is a surprisingly uniform thermal environment for the developing embryo (Fig. 1.4). This is more easily managed when both parents share incubation responsibilities, so that while one is on the nest, the other can leave to forage and rebuild its own energy stores. The eggs of single sex incubators are much more likely to vary in temperature over the course of the day when parent must leave the nest to forage. The amount of time spent incubating also varies with the stage of development. Parental attentiveness (as measured by the per cent of time spent incubating) increases during the first one third of the incubation period in *Herring Gulls (Larus argentatus)*. In this case, external heat supplied by the parent plus the heat generated by the embryo itself help maintain the embryo temperature between 37° and 38° C (Figure 1.4). Only in larger birds do embryos generate more heat through growth late in development than they lose through evaporation, and consequently, less incubation is required at this time. In similar eggs, such as those of the House Wren *(Troglodutes aedon)*, heat produced by the embryo does little to elevate egg temperature.

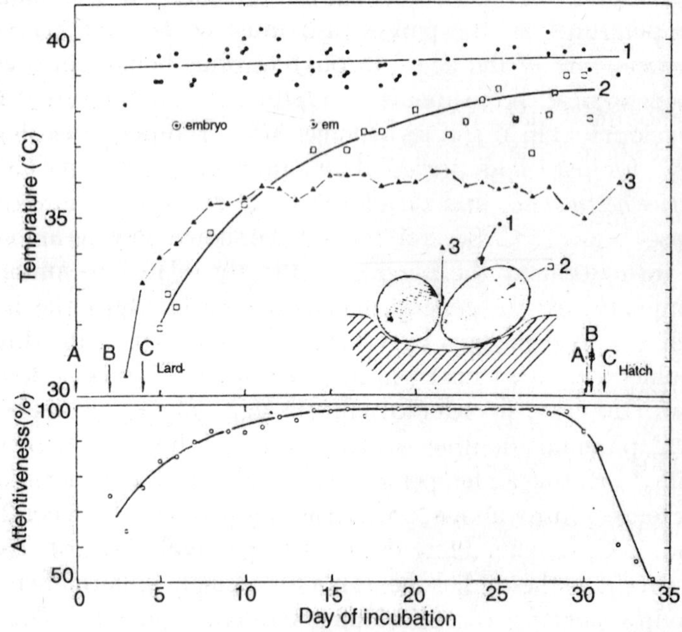

Fig. 1.5.   Nest and egg temperatures (top) and parental attentiveness (bottom) during incubation in the *Herring Gull (Larus argentatus)*. Sites of temperature measurements are indicated in the diagram.

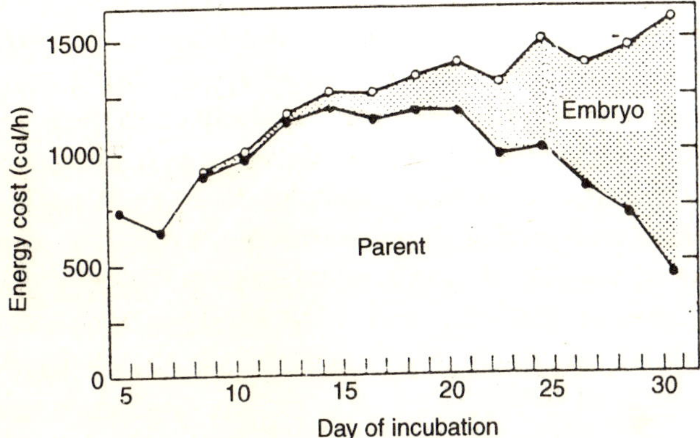

Fig. 1.6. Contributions of parental and embryonic heat to the regulation of egg temperature during incubation of Herring Gulls (*Larus argentatus*).

Heat is probably more threatening to embryonic development than chilling temperatures, and overheating by direct solar radiation of the nest can occur in any climate. Some species have developed means of regulating eggs temperature under hot conditions. In cases where external temperatures are warm but no extreme, parent bird may not incubate but will simply shade or perhaps fan the eggs. In more extreme cases, though, the contact between eggs and incubating bird can be used to cool the eggs, or at least to prevent them from reaching a lethal temperature. Heat from the egg is transferred to the parent who dissipates it through radiative, convective, or evaporative means. Studies with the Double-banded Courser (*Rhinoptilus africanus*), a ground nester of the *Kalahari Desert of Africa*, showed that parent birds shaded but did not incubate eggs when temperatures were between 30° and 36° C, but employed constant incubation for maintaining egg temperature at air temperatures above 36°C. Similarly, White-winged Doves (*Zenaida asiatica*) nest in open desert and despite intense solar radiation and air temperatures up to 45°C maintain egg temperatures at 39.2°C by constant incubation (Figure 1.6). Interestingly, these doves accomplish this without resorting to panting or gular flutter; the means by which they dissipate heat during incubation is not known. In extreme case, a few open-habit, ground-nesting species bring water to the eggs by wetting their feathers. This both cools the nest and provides a moister environment for development. The Egyptian Plover (*Pluvianus aegyptius*) incubates i:s eggs at night but covers them with sand as the air warms in the morning. During the heat of the day it drops water onto the sand from its soaked ventral feathers, and evaporation keeps the egg temperatures at about 37.5°C, compared to nearby sand temperatures of 46°C.

In cold habitats, birds must incubate continuously to maintain the large thermal gradient between egg temperature and air temperature. Cold climate creates energetic problems for incubating adults because of the restriction of their foraging time. Several adjustments are made by species that regularly nest under these circumstances. Some species of hummingbirds, which are small, single-sex incubators, employ temporary hypothermia at night when ambient temperatures fall below 0°C. This, of course, slows down embryo development. Other small-bodied birds circumvent the potential energy drain of thermoregulation at night by utilizing the nest microclimate to buffer the decline in air temperature. The air temperature in nests placed in cavities, caves, or even within dense conifers, which can greatly diminish the heat lost by radiation and convection from the surface of an incubating bird, may be as much as 10°C warmer than air outside the nest. In other single-sex incubators, such as the *Great Horned Owl (Bubo virginianus)*, which nests during the middle of winter in the temperate zone, the male brings food to the female both during incubation and during the early nesting stages. Sometimes, however, the parent is forced to abandon its eggs in order to forage at great distances from the nest, leaving the eggs to chill in the nest. Some species of procellariiforms, for example, neglect their eggs for hours or even days while they are foraging. Their embryos tend to be very resistent to chilling, but the effect of this intermittent incubation is reduced hatchability and a greatly prolonged incubation period.

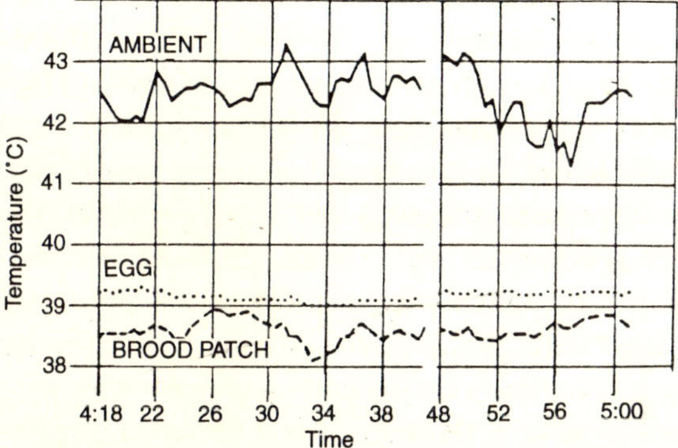

Fig. 1.7. Variation in the air, brood patch, and egg temperatures measured in an exposed nest of the desert-dwelling White-winged Dove (*Zenaida asiatica*). In this species the brood patch may be used to either heat or cool the egg.

Incubation depends on intimate contact between the incubating birds and the eggs, so that the heat generated by the adult is passed to the eggs. This interaction can be viewed as though the eggs were simply an additional appendage and the parent and eggs were a single unit. The site of heat transfer between parents and eggs is the incubation or brood patch which is usually found on the lower breast and abdomen of the bird. The area is generally characterized by a lack of feathers, edema leading to flabbiness of the superficial skin and thickening of the epidermis, and an increase in the number and size of blood vessels and the musculature around these vessels. All of these modifications increase the efficiency of heat transfer between egg and parent by allowing closer contact of the skin with the egg and by increasing the amount of heat present at the surface of the well-vascularized brood patch.

In most species, the brood patch develops through the hormonal influence of prolactin, which is secreted by the anterior pituitary gland, and estrogen, which is secreted by the ovary.

Prolactin causes defeathering of the region, and estrogen produces epidemal thickening and vascularization. In many water-fowl the brood patch is actually plucked by the bird, and the downy feathers are used to insulate the nest. In pigeons, the brood patch occurs in a region without feathers (a trait possible in part because all pigeons and doves have clutches of only one or two eggs). In some species, the brood patch consists of distinct regions that are egg-sized and arranged as the clutch is arranged. For example, gulls have three distinct brood patches to match their typical three-egg clutches. In many species both sexes develop brood patches to aid in incubation; among these are the grebes, albatrosses and relatives, pigeons, woodpeckers, shore-birds and cranes. Only the female has a brood patch in the galliforms and owls; only the male has a brood patch in phalaropes, jacanas, and some sandpipers. In these latter cases, occurrence of a brood patch reflects the type of mating system in which the female defends the territory and the male incubates. Among the hawks and passerines, there is great variation in the occurrence of brood patches, with many species having patches in both sexes and many with only the female incubating. Much of this variation may also reflect mating and parental care characteristics. Only the pelecaniform seabirds, some of the auklets, and the Bank Swallow (*Riparia riparia*)do not develop a brood patch. Many of these are burrow nesters, and the insulative properties of the burrow may preclude the need for a brood patch. At least some of the pelecaniform species use their very large, webbed feet as the heat exchange organ in place of an abdominal brood patch.

Through the interaction of incubation and the thermal environment provided by the nest, the parents attempt to provide the nest, the parents attempt to provide the proper developmental environment for the eggs. At the same time, through both properties of the nest and various behavioural patterns, the parents also attempt to keep predators away from the eggs and the nestings.

Egg Structure and Embryonic Development

Once an egg has been laid, the parents are unable to provide any nutrients to the young bird until the egg hatches. How is the egg constructed so that it can provide the proper medium for development of a young bird? What can or (in most cases) must the parents do to aid the proper development of the embryo within the egg until it hatches? How much variation occurs in the processes of egg development and the final product, a baby bird?

As we attempt to answer these questions, we must remember that a bird egg starts out as a single cell composed of all the appropriate nutrients needed for development but with very little structure. The process of growth within the egg consists largely of the incorporation of these nutrients into the embryo.

Egg Structure

Since the whole purpose of an egg is the production of a young bird, let us start our look at egg structure with the embryo. When the egg is laid, the embryo is a tiny spot called the germinal disc, which sits on top of the yolk mass. The embryo remains at this location even when the egg is moved, because the yolk floats freely within the egg. The yolk of a chicken egg has a definite structural organization. A whitish layer of yolk within the center of the mass is highly proteinaceous. Yellow yolk is stratified in concentric layers around this core and is composed largely of lipoprotein and proteins that are the main nutrient source of the embryo. The yolk mass makes up 20% - 65% of the egg, depending in part on the type of development (altricial or precocial).

Surrounding the yolk are the "whites," or albumen. Albumen makes up about 65% of the egg mass in precocial species and is composed entirely of protein and water. It is arranged into three compartments. Immediately surrounding the yolk is the chalaziferous layer, a thick, viscous layer that forms twisted fibrous strands (chalazae) that anchor the yolk to the poles in the long axis of the egg. Next is an inner, thin, watery layer of albumen around the yolk, and then a middle layer that is much thicker and more viscous, and finally an outer, thin, watery layer. The differences in the various layers lies only in the amount of water or fibrous ovomucin protein they contain. The yolk is suspended in these layers of albumen yet anchored so that, as the egg rotates, the embryo stays on top of the yolk. The albumen is also important for several other reasons; it provides an aqueous environment for development, it retards desiccation, it has some antibacterial properties, and it provides an additional nutrient source for the embryo.

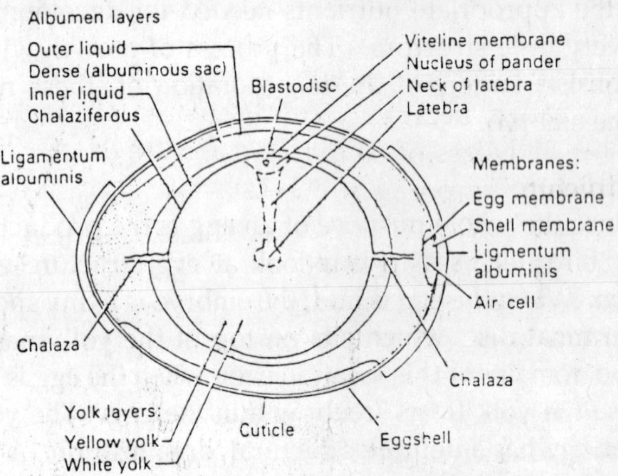

Fig. 1.8.   Structures of the egg of the domestic fowl, *Gallus domesticus*.

Surrounding the albumen are two fibrous shell membranes made of another protein, keratin. The inner membrane rests on the surface of the albumen and holds this watery layer of the egg together. The outer layer is about three times thicker and has elements of the external shell anchored in it. The two layers separate at the blunt end of the egg, where they create an air space that enlarges during development, as yolk is absorbed by the embryo, and provides the first air to the baby bird prior to hatching.

The calcareous shell is the external barrier, and it serves a variety of functions that require some compromises in structure. For example, the shell should be hard enough to keep the egg from breaking easily under the weight of the incubating parent, but not too hard for the hatchling to break through. It must be porous enough for diffusion of oxygen into the egg and carbon dioxide out of it, but not so porous that the embryo desiccates or that bacteria can enter the egg. To accomplish all this, an organic (protein) framework serves as a skeleton for deposition of inorganic minerals, 98% of which are crystalline calcite (C_aCO_3). The mineral is arranged in vertical columns, between which are minute air spaces that open to the surface as oval or circular pores (Figure 1.8). There are thousands of these minutes pores over the surface of the shell; estimates range from 6000–17,000 in a chicken egg. Apparently these pores function in gas exchange, because recent studies have shown that the number and sizes of pores varies with altitude and climatic conditions. The pore size and density greatly influence the rate of gas exchange (O_2, CO_2, and water vapor) across the eggshell and are a measure of the shell's conductance or diffusivity. Eggs laid under humid or hypoxic (low O_2) conditions, such as in burrows, or under vegetation (megapodes) often have elevated conductance in order to facilitate movement of O_2 to the embryo (and CO_2 from the embryo). The greater potential for water loss from the egg is not realized in this case because of the higher humidity surrounding the egg.

The outermost covering of the egg is the cuticle, which imparts the characteristic surface texture of the egg. The adaptive significance of the various outer coverings of eggs (glossy, greasy, chalky, ridged, powdery) is not known. The cuticle has been most thoroughly examined in chickens, in which it is a thin, continuous layer of glycoprotein over the entire shell, including the pores. It imparts water repellent properties to the egg surface, and impedes both water loss through the shell and bacterial entry. All of the above structures results in an eggs that is, at the time of laying, about 65% water, 12% protein, 10% lipid and 11% mineral.

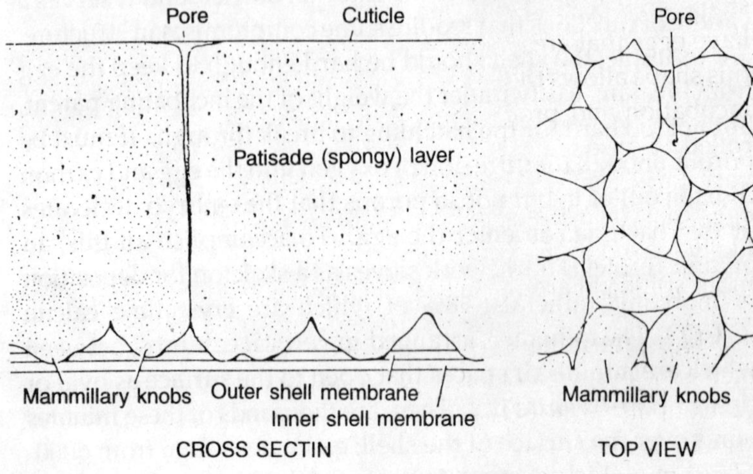

Fig. 1.9.   Structure of an avian eggshell and location of the pores.

Egg size and shape varies with the type of bird that lays the egg. The largest egg known to man was that of the Elephant Bird (*Aepyornis maxima*) of Madagascar, which had a capacity of more than 9% and measured 34 by 24 cm. The smallest egg known is that of a hummingbird, the Jamaican Emerald (*Melisuga minima*); the egg is only 1 cm by 0.65 cm and its mass is 0.5 g or about 1,50,000 of an Elephant Bird egg. Many factors influence the size of bird eggs. Generally, larger birds lay larger eggs, but the size of the egg relative to its parent decreases with increasing size of the bird (Table 1.1). For example, an ostrich lays an egg that is about 1.8% of its body weight, while a wren lays one that approaches 14% of its weight. Many exceptions to this general trend occur, though. For example, the kiwi lays an egg that is about 18% of its body weight, rather than the 3% expected for a bird of that size. Birds lay larger eggs as they get older, and the eggs of precocial species are usually larger (10%-15% of the female's body weight) than those of similar sized altricial species (5% of the female's weight).

Generally, eggs are semielliptical, with one pole slightly flattened and the other more pointed, but some variation in shape does occur. Birds that nest on the ground or cliffs often have eggs that are quite pointed and pear-shaped (pyriform); this shape allows the eggs to be packed tightly together during incubation and produces a very tight circle when the egg is rolled. Consequently, when the egg is moved, it does not go very far. Some birds, particularly some cavity nesters, lay eggs that are nearly spherical. The shape of the egg is most likely related to the pelvic structure, the deeper the pelvis, the more spherical are the eggs.

Eggs are quite variable in texture due to differences in their cuticle layer, as discussed above. Egg colour is also highly variable. While it has been suggested that these colours are merely a means of excreting metabolic waste products, it is more likely that their primary function is a protective one—to camouflage the eggs and to shield the developing embryos from incident UV radiation. Support for the latter view is the fact that egg colour of hole-nesting species, which typically experience lower predation and little to no incident solar radiation, is usually white. The colour and pattern of markings on the eggs of the Common Murre (*Uria aalge*) is extremely variable and may aid the parent in locating its single egg within the densely populated nesting colony on rocky ledges. *Oniki* (1985) proposed that the brightly coloured eggs of the tinamou allowed the parents to find all the eggs after they had been protectively camouflaged with leaf litter. However, the main benefit of egg colour is protection against predation, accomplished by camouflage. For example, blue eggs in dark nests that are placed in isolated areas receiving partial sun seem to imitate the spots of light on green leaves in a forest. Buff-coloured eggs occur in birds that lay them in leaf little or on other dull but well-lighted substrates. Spotted white eggs occur in thinly formed nests in poorly lighted sparse foliage, where the egg tends to vanish against its speckled background (*Oniki* 1985).

TABLE 1.1

Egg Weight as a Proportion of Female Body Weight

Species	Adult female body weight (g)	Egg weight (g)	Egg weight/ body weight (%)
Ostrich	90,000	1600	1.8
Emperor Penguin	30,000	450	1.5
Mute Swan	9000	340	3.8
Snowy Owl	2000	83	4.1
Peregrine Falcon	1100	52	4.7
Mallard	1000	54	5.4
Herring Gull	895	82	9.2
Puffin	500	65	13.0
Robin	100	8	8.0
House Sparrow	30	3	10.0
House Were	9	1.3	13.7
Vervain Hummingbird	2	0.2	10.0

Note: Both egg weights and female body weights are approximate.

Embryonic Development

The development of the embryo begins with cell divisions almost immediately after fertilization. The second meiotic division of the ovum occurs after penetration of the sperm. The male and female pronuclei fuse to form the zygote nucleus, and the first cleavage division occurs three to five hours after fertilization while the egg is in the magnum and the inner layer of albumen is being added. The second cleavage division coincides with the laying down of the shell membranes in the isthmus. By the time the egg reaches the uterus, it has reached the 16-cell stage, and when it is laid, the embryo consists of a double-layered blastula oriented at right angles to the long axis of the egg. In most cases, embryonic development is suspended after an egg is laid, then resumes with the regular application of heat to the egg through incubation. It is beyond the scope of this book to examine the many details of embryonic development, but we do want to look at general patterns of development and the requirements for them.

Early in its development, the embryo becomes enveloped in the amniotic layer, which forms a fluid-filled chamber around it and protects it. Next, cells grow downward and envelop the yolk, forming a yolk sac; veins develop on the surface of the yolk sac which transport nutrients from the yolk to the embryo. Shortly before hatching the yolk sac is drawn into the body so that stored nutrients are still available after hatching. In many species, hatchlings can live for several days on stored reserves from the yolk. Finally, in the early development of the embryo a third membrane develops, the allantois (Figure 1.8 also called the *chorioallmtois*). This grows out as a pouch from the hindgut to line the inner surface of the shell membrane, where it serves as both a bladder, or waste depository, and as a respiratory organ. A rich supply of blood vessels develops in the allantois in close proximity to the shell and facilitates gas exchange. The allantois also serves as a depository for nitrogenous wastes, generally in the form of uric acid salts. By excreting uric acid into the allantois, the avian embryo can avoid the potential toxicity of accumulated nitrogenous waste and conserve water at the same time. (See the discussion of water conservation by uric acid excretion in Chapter 2.) The uric acid salts form crystals, which are simply stored outside the embryo until hatching.

The optimal development of an embryo requires the appropriate gaseous environment, an external source of heat, and the proper egg position, including in most cases, turning of the egg. The gaseous environment is controlled by gas exchanged through eggshell pores. The outward diffusion of CO_2 and inward diffusion of O_2 is of primary importance movement of water vapor from the saturated interior of the egg to the less humid microclimate of the nest must also be regulated. Variation in the construction of the eggshell between species enables all avian embryos to be exposed to roughly the same conditions. In fact, experiments on domestic fowl have shown that levels of oxygen less than 15% or greater that 40% (normal is 21% of atmospheric air) or CO_2 levels greater than

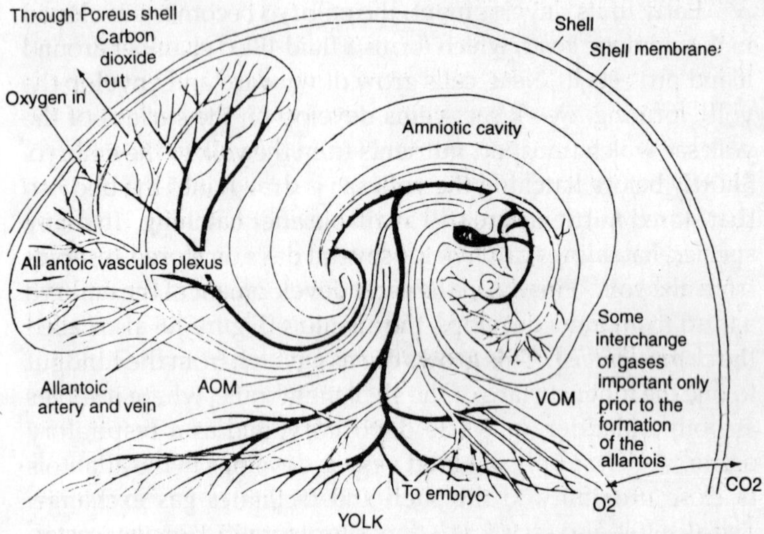

Fig. 1.10. Embryonic membranes and circulation. The amnion surrounds the embryo which floats in the watery amniotic secretions. The allantois contacts the shell membrane and through its vasculature provides a route for gas exchange. It also serves as a depository of uric acid waste. Vitelline vessels transport nutrients from the yolk sac to the embryo.

1% (normal is 0.03%) greatly retard embryonic development. As a consequence, despite large variations in egg mass, incubation period, climate, geography, and so forth, avian embryos complete their development with similar water losses (about 16% of the initial egg mass), similar amounts of oxygen consumed per g of embryo, similar gas pressure gradients across the shell, and similar O_2 and CO_2 content in the air cell before hatching (Carey 1980b). Hatchability depends on the consistency of these parameters during development.

Development also requires parental (or other) input of external heat energy in order to maintain egg temperature at an optimal 37°-38°C (see discussion of incubation above). We have pointed out that temperature tolerances of embryonic development are rather narrow, and it should be noted that the optimal developmental temperature is very close to the lethal temperature for avian embryos (above 43°C). In addition, tolerance to temperature extremes declines with development; embryos close to hatching cannot stand the extremes that a freshly laid egg can.

Not only must parents provide the heat for incubation but they must manipulate the eggs so that each receives uniform heat. A temperature gradient of as much as 5.6°C was measured between the central and peripheral eggs in the clutch of the Mallard (*Anas platyrhynchos*; Turning the eggs is also important to prevent adhesion of the membranes to the shell early in development. Repositioning of eggs also aids the development of its equilibrium position. At first, the yolk mass is free to revolve in the shell and the embryo remains uppermost because it floats on a lighter portion of the yolk. Later, however, the extraembryonic membranes fuse with the shell membranes, and the embryo position becomes fixed in the shell with the head oriented toward the blunt pole and the air space. The egg then becomes asymmetric in weight, and will always assume a certain orientation, embryo uppermost, when it is rolled in the nest.

The time required for embryonic development is variable among birds, even among birds of similar size and taxonomic affiliation. Generally, large birds lay larger eggs and have longer development times than small birds. However, other factors affect the length of incubation and embryonic development. Young birds at hatching are not equivalent developmentally. Woodpeckers, for example, hatch at an early stage, compared to small passerines, and although their development time would appear short, they are not as mature as other altricial nestlings with longer development. The same is true when comparing precocial and altricial young. Precocial young have much longer development times than altricial young do, but their organ systems are much better developed, and they can essentially feed and take care of themselves soon after hatching (see the section on nestling development below). It has also been suggested that birds exhibit some ecological adjustment of embryonic development, such that open nesting birds have shorter development times than hole nesters. This may occur because of the higher predation pressures on open nesters, and the need to hurry the developmental as well as the subsequent nestling growth process.

Hatching

Hatching is obviously a critical stage in the development of a baby bird, for it involves escape from the confinement of the shell and conversion from the embryonic to adult form of such physiological processes as breathing, excretion, and so forth. The shift between functional systems must be done fairly quickly.

Hatching success is highly dependent upon the proper position of the embryo within the egg. During the days before hatching, the embryo assumes what is known as the "tuck" position, with the head between the right wing and the body and with the bill pointed toward the blunt end of the egg. From this position, the first act of hatching involves puncturing the air sac and the initiation of lung breathing by the embryo. This stage occurs a day or two before actual hatching. Pipping, or shell breakage, also occurs from this position. The first pip may occur ten hours or more before hatching as the result of a random movement of the embryo. At some point, this movement becomes much more active and involves strong thrusts of the beak into the shell accompanied by propulsive movements of the entire body by pushing with the feet. In the process, the embryo rotates within the shell causing a ring of cracks near the blunt end. This process is aided by specialized structures in the full-term embryo: a horny knob or egg tooth on the upper mandible, and hatching muscles on the back of the head that are responsible for the vigorous back thrust of the head during pipping (Figure 1.11). Both structures either fall off or regress shortly after hatching. Usually one rotation within the egg will sever the cap of the egg, although two or three rotations have been observed in some birds. Eventually, the embryo will break off the cap and the thrusting motion will push the baby bird out of the shell, but this may take several hours to even several days in albatross chicks.

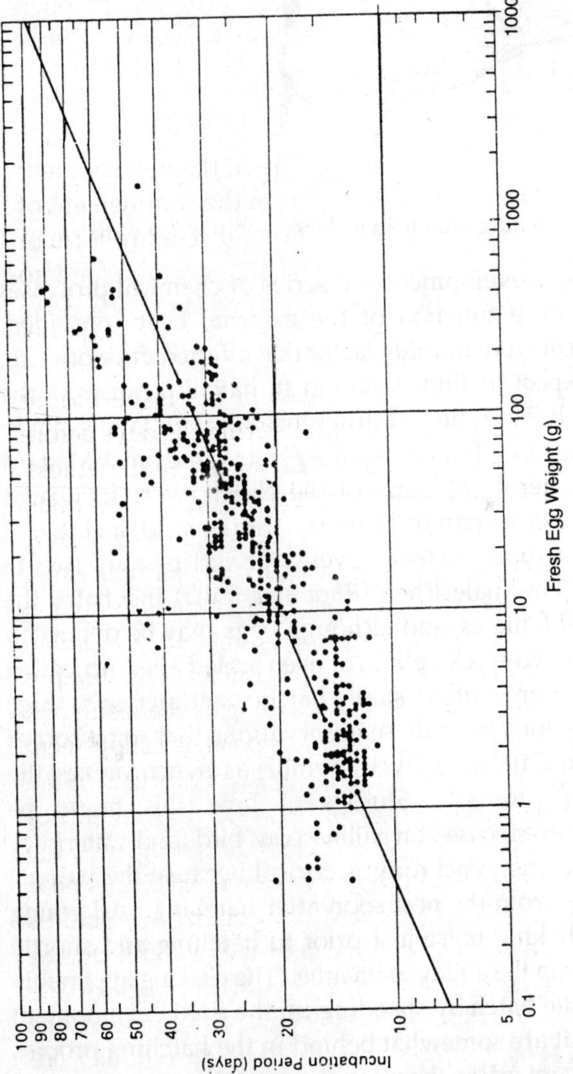

Fig. 1.11. The relationship between incubation period and egg weight in birds.

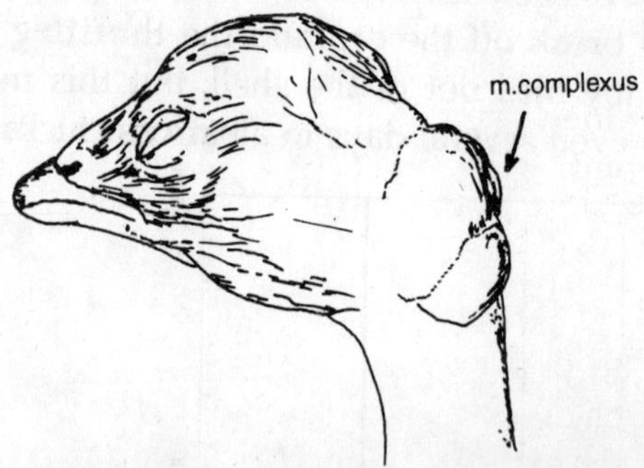

Fig. 1.12.   The hatching muscle in a chicken on the day of hatching.

Embryonic development is a series of chemical processes whose rates are a function of the external heat provided. However, if heat were the only factor that affected development, one would expect to find variation in hatching times of the eggs in a clutch that reflected differences in their laying times. This is not the case. For example, eggs taken from a Mallard nest, shortly after completion of the clutch, were incubated and hatched over a span of 16 hours. In the wild, a clutch of Mallard eggs usually hatches over a period of only two to eight hours. The male Rhea (*Rhea americana*) incubates the eggs of several females, and although eggs may be deposited in a nest over a two-week period, or even added after incubation is started, the entire clutch usually hatches within two to three hours. It is obvious from these observations that some sort of behavioural modification by the embryos synchronizes the final hatching process. Studies of how this might be accomplished have focused on gallinaceous birds and waterfowl, where the hatching synchrony is critical because the parents lead the young from the nest soon after hatching. All young birds emit a clicking noise just prior to hatching and shortly after they develop the ability to breathe. The clicking apparently synchronizes the hatch by speeding up the final development of nestlings that are somewhat behind in the hatching process and perhaps by retarding the movements of the more advanced individuals Thus, through acoustic communication between the embryos, hatching of the entire clutch occurs in a minimal period of time.

Stage of Development at Hatching

Up to this point, we have looked at a general pattern of development, but most everyone knows that baby chickens are different from baby robins in how they behave and in the amount and type of care they require. While baby chicks or ducts may be good pets for children, one would not want to give a child a clutch of recently hatched robins or bluebirds as pets! Obviously, the extent of development during incubation varies among birds.

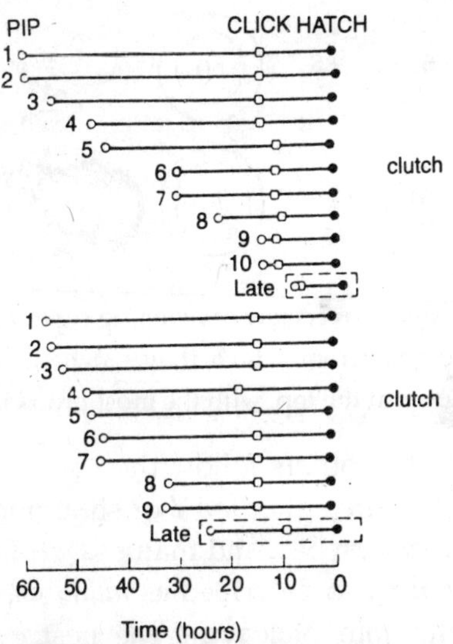

Fig. 1.13. Events which may determine hatching synchrony in two clutches of the Bobwhite Quail (*Colinus virginianus*). Eggs are ordered by the occurrence of pipping, but note how the clicking and then actual hatching are synchronized.

Recently, hatched birds have been classified into several developmental categories. *Precocial* young are those that hatch with their eyes open, are downy-covered, and have the ability to leave the nest in a matter of a few hours or a day or two. Some precocial chicks actually grow up independently of

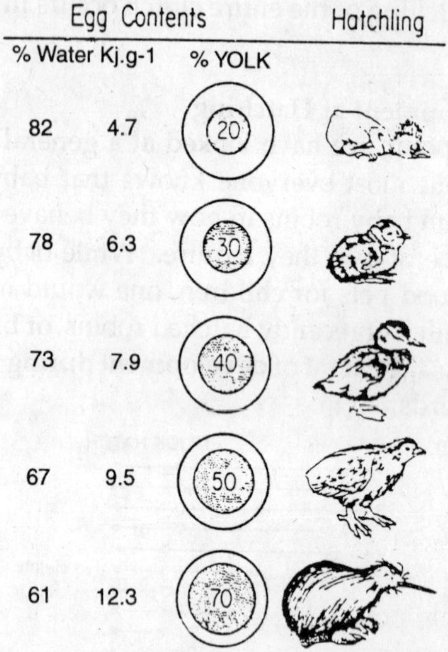

Egg Contents			Hatchling
% Water	Kj.g-1	% YOLK	
82	4.7	20	
78	6.3	30	
73	7.9	40	
67	9.5	50	
61	12.3	70	

Fig. 1.14.   A simple overview of hatching types and the composition of the eggs from which they hatched. An altricial young is shown at the top, with the most precocial types at the bottom.

their parents, but others follow their parents (walking and or swimming) and are either fed / or shown food by the parents. Ducks, chickens, grebes, and many sandpipers have precocial young; they are also described as *nidifugous*, or nest fugitives, for their behaviour of leaving the nest. *Altricial* birds are those at the other extreme; they hatch with their eyes closed, have little or no feathering, and require great amounts of care and feeding. Most passerines and other small birds fit into this category; these young are also termed *nidicolous*, or nest dwellers, because of their prolonged occupancy of the nest.

Within these groups two other categories are regularly recognized. *Semi-precocial* birds hatch with their eyes open and have a downy covering, but, although able to walk, they

Mode		Down	eyes	Mobility	Parental nourishment	Parental attendance	Examples
Precocial	1	○	○	○	○	○	Magapodes
	2	○	○	○	○	●	Ducks, shore birds
	3	○	○	○	○	●	Quail, grouse
	4	○	○	○	◐	●	Grebes rails,
Semi-Precocial		○	○	◐	●	●	Gulls terns
Semi-altricial	1	○	○	●	●	●	Herons howks
	2	○	●	●	●	●	Owis
Altricial		●	●	●	●	●	Passerines

Fig. 1.15. Variation in the characteristics of precocial, altricial, and two intermediate types of nestlings. Open circles denote precocial characteristics; closed circles denote altricial traits.

do not leave the nest and are generally fed by their parents. Included among these are gulls and terns. *Semi-altricial* birds are somewhat less developed than the above; they are down-covered but are not able to leave the nest. The semi-altricial nestlings of hawks and herons hatch with their eyes open, while those of owls have their eyes closed.

The difference in the product that hatches reflects differences in egg composition between precocial and altricial birds; eggs of precocial young often have larger yolks than those of altricial young, and much of the yolk remains at hatching as an internalized yolk sac. More important, however, is how growth is apportioned to the various organs of the body during embryonic development of precocial and altricial young. Altricial nestlings hatch with still relatively immature organ systems. Only the digestive tract is well-developed at hatching, enabling them to maximize their assimilation of food. In contrast, the precocial hatchling is a miniature adult with relatively mature organ systems that allow even a new hatchling to stand immediately, to run from predators, and to seek its own food.

NESTLING DEVELOPMENT

Growth of Altricial vs. Precocial Nestlings

Most young birds grow rapidly at first, but their growth rate slows as they approach adult size. The pattern for most birds is one of a sigmoidal-shaped growth curve. Although the shape of the curve and the actual daily growth rates may vary greatly between species, this is a consistent pattern of growth in all birds.

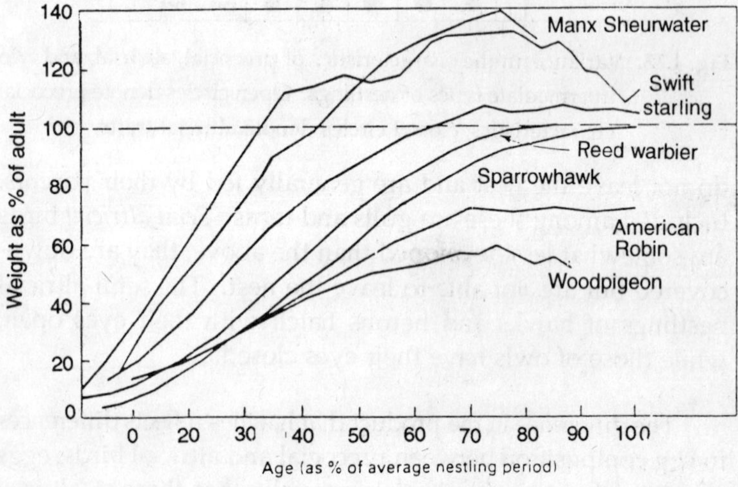

Fig. 1.16.   A comparison of growth curves of young of several species.

The gradual diversion of food energy from primarily growth to maintenance and activity during development is the basis for the sigmoidal shape of the growth curve. To illustrate this point, we shall describe some characteristic features of the nestling development of altricial young and compare their development with that of precocial young.

After hatching, altricial nestlings are totally dependent on their parents for food and for maintaining their body temperature. During the first third of the nestling period, the young move about very little but begin to have better control of their head and neck; feathers emerge and begin to grow but the body is still largely devoid of insulation. The nestlings have little control of their body temperature when exposed to cold at this early stage. Because most of the energy ingested goes into building tissue, they young can achieve very high growth rates. During the next third of the nestling period, feathers cover more of the bare skin and the young begin to shiver and to move around in the nest. With better insulation and more heat production from the musculature, the young now exhibit better control of body temperature during cold exposure. They are alert to external cues and exhibit appropriate innate behaviours (e.g., begging from the parent, crouching in fear). During the final third of nestling development, feather development is completed, temperature and sensory responses become acute, and the young exercise in the nest, preparing themselves for flight. More intensive foraging efforts by the parents for these large and voracious nestlings means that the young must expend more energy regulating their own body temperature. Consequently, the ingested energy channeled into growth decreases and the growth rate slows. Increased activity of older nestlings is another source of energy diversion away from growth, but most nestlings are close to adult size at this time.

Although the growth curve of precocial nestlings is sigmoidal, the growth rate is only about one-third that of altricial nestlings. As we have discussed, precocial young can walk, see, and feed themselves soon after hatching. They also are able to thermoregulate with limits. This means that they must apportion more of their daily energy to maintenance and less to growth; the further development of their relatively advanced organ systems leaves even less energy for overall growth. Finally, these young must move around to feed and to escape predation, and some energy must be diverted away from growth and into activity costs.

The marked differences in developmental rates between altricial and precocial nestlings suggest two major options in reproductive strategies among birds. Those with altricial young can produce offspring rather rapidly due to their high growth

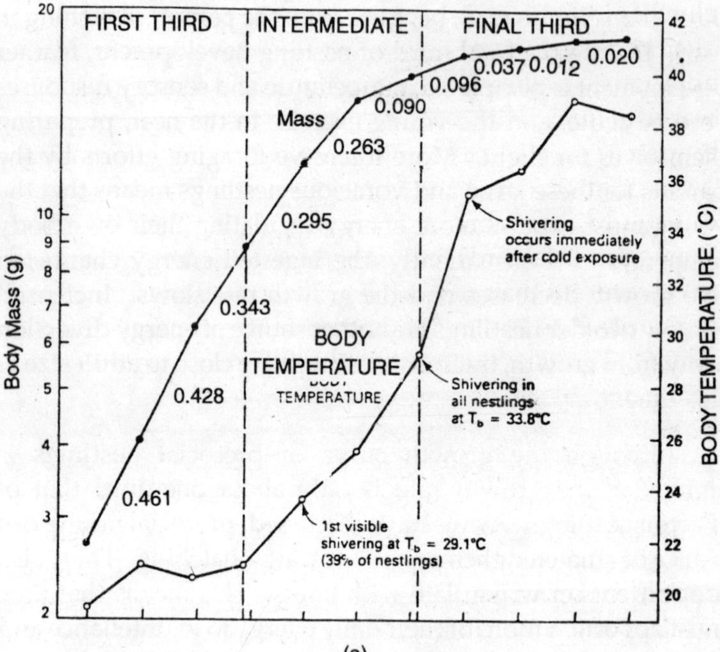

Fig. 1.17.   Growth (body mass), development of temperature regulation, and feather growth in nestling mountain White-crowned Sparrows (*Zonotrichia leucophrys oriantha*). Body temperatures are means of 20-30 nestlings after they were removed from the nest and exposed to air temperatures of 5-6°C for ten minutes. Numbers between the data points on the growth curve are daily growth rates.

rates, but these young require extensive amounts of care, including brooding to keep them warm, and must be provisioned with high energy foods. This strategy also carries a higher risk of predation, and the parent risks losing all of its reproductive investment during the nesting period.

In contrast, precociai young are at risk from predation only during the incubation period, because nestlings can usually avoid predators from time of hatching on, and rarely are whole clutches lost. Precocial young do not require as much care as altricial young, since they often need only to be guided to appropriate feeding locations and protected (often just through warning) from predators. The primary cost in such a seemingly easy strategy is a longer embryonic and nestling development period than that of altricial young; as a consequence, it is more difficult to rear more than one brood per breeding season. Young precocial birds require access to prey that is fairly easy to capture, but their reduced growth rates also mean that they can survive on somewhat lower-quality foods than altricial young. The above factors both tend to release the parents from some of the constraints of care that are related to the number of young, so that birds with precocial young tend to lay larger numbers of eggs in a clutch.

Factors Affecting Growth Rate

After hatching, the development of most baby birds is dependent on various features of parental care and a set of external and internal constraints that limit growth; diet and food availability, predator pressure, climate, internal allocation of energy in the young, even biochemical processes all have an impact on growth rate. Generally, small birds grow more quickly than large ones, young in open nests grow faster than those in cavities, temperate zone birds mature faster than their tropical counterparts, species at high altitudes mature faster than those at sea level, and young fed insects develop more quickly than those fed a fruit diet.

Explanations for the intraspecific variation in growth rates among young birds focused on several of these constraints. British ornithologist *David Lack* (1968) regarded growth rate as a balance between predation pressure selecting for rapid growth and food supply selecting for slower growth. Lack proposed that young birds exposed to high risk of predation should grow as fast as possible in order to outgrow this possible mortality factor. The Lack hypothesis is supported by the slower growth rates found in hole or cavity nesters (low predation pressure) compared to that in open nesting birds (high predation pressure) in similar areas. In addition, Lack hypothesized that growth rate of the young is adjusted to the food the parents can provide. A slower growth rate in hole nesters means that less food is required per young per day, which in turn implies that parents can therefore raise larger broods of young. In fact, hole nesters do have larger broods. Other groups exhibiting slow growth include those that live on predator-free islands and those whose young can run away from predators, all of which support the Lack hypothesis.

American ornithologist *Robert Ricklefs* (1969) was not convinced by Lack's arguments and retested the hypothesis that growth rates were optimized by mortality patterns of each species. A clear correlation of growth rate and mortality was absent in the many species be examined. Ricklefs further countered that food requirements do not increase proportional to increases in growth rate. For example, doubling the growth rate of the slow-growing Leach's Storm-petrel (*Oceanodroma leucorhoa*) would increase the energy required for growth only 5%.

The Lack hypothesis also does not explain why most tropical birds with high nest predation have slower growth rates than their temperate counterparts, with comparatively lower nest predation. Lower metabolic rates of tropical species and the lower nutritional quality of their diet could be responsible. In fact, tropical birds that feed their young fruits, which may be of poor nutritional quality, tend to have slow-growing offspring.

In contrast, nestlings fed highly nutritious food, such as insects, grow more rapidly. The protein deficient diet of young petrels fed an oily regurgitant by their parents has been proposed as one of the reasons for their slow growth. However, parental choice of highly nutritional items for their young should generally rule out the nutrition hypothesis for all but a few species.

Organismal constraints on growth rate may be set by the young bird's ability to assimilate the ingested food. The size of the digestive tract is proportionally larger and is more mature in altricial young than in precocial young. This disparity of development between the two groups leads to the conclusion that altricial young can assimilate more energy and thus can grow faster than precocial young. *Perrins'* (1976) feeding experiment on Blue Tits (*Parus caeruleus*) also supports this hypothesis. To demonstrate that the slower growth of late broods of this species was related to the high concentration of oak leaf tannins in their prey, Perrins compared the growth rates of young reared on plain mealworms with that of young fed mealworms contaminated with tannic acid. The uncontaminated young gained weight faster and exhibited more intense begging activity and interest in food.

Ricklefs (1979) has proposed that two types of physiological constraints limit growth rates of birds. Biochemical and molecular constraints limit the extent to which functionally mature tissue can continue to grow and proliferate. Thus, growth rates are determined by a balance between the mature and embryonic functions of tissues. Generally, growth and differentiation are two competing and mutually exclusive processes in tissue development. That is, maturation of nerve and muscle tissue into a more refined locomotory system cannot be accomplished while muscles are growing in size and developing neuronal connections. This is the reason that growth slows as tissue matures and explains why, in altricial young, we see a rapid growth phase first, followed by a maturation phase of slower growth.

TABLE 1.2

Growth rates of neotropical birds in relation to the nestling diet

Diet	Range	Mean $\pm$ SD	(n)
		Growth constant	
Fruit	0.098-0.460	0.262 ± 0.151	(5)
Mixed fruit-insect, mostly fruit	0.280-0.464	0.375 ± 0.076	(6)
Mixed fruit-insect, mostly insect	0.199-0.536	0.379 ± 0.102	(9)
Insect	0.236-0.524	0.357 ± 0.078	(13)
Nectar-insect	0.256-0.382	0.317 ± 0.055	(3)
Seed	0.472-0.520	0.496 ± 0.034	(2)

The second physiological constraint is a consequence of maturation of the tissues. Mature organ systems have higher maintenance costs, and a greater percentage of the assimilated energy must be diverted away from growth into maintenance. Similarly, on an organismal level, as the young mature, more of their assimilated energy is diverted into thermoregulation and activity, leaving less for growth.

Fledging

The growth and development process generally stops when the offspring reaches adult size. In most species, at or near this point the process of fledging occurs when the young bird leaves the nest, learns to fly, and fends for itself. Parental care often is terminated shortly after the young leave the nest; parents may refuse to feed a begging youngster, forcing it to look for food itself. Eventually the young learn to follow the adults and search where they are feeding, until finally they become completely independent. In some species, such as blackbirds and sparrows, large flocks of young birds form at the end of the breeding season; they roost together and may migrate together, separately from their parents. However, in some owls, the fledgling period is prolonged for months, as the young learn where and how to find their food under limited light conditions. Birds receiving extended parental care probably benefit either from avoiding food shortages while still inexperienced, or from learning how to forage.

THE TIMING OF REPRODUCTIVE EVENTS

In an evolutionary sense, birds should breed at the time when they can produce the most offspring that can survive to breed in the future. In the temperate zone, this is usually during the spring or summer when temperatures are moderate and food is abundant. Even in the tropics, the wet-dry seasonal cycles affect nesting seasons. Rather than focus here on the ultimate, evolutionary factors affecting the timing of nesting (many of which we shall discuss later), we shall address the proximate factors that determine breeding time. Given that evolution mandates that June, for instance, is the time to nest, how do birds synchronize the many changes necessary to initiate the breeding process and advance from one stage of reproductive behaviour to the next?

The synchronization of the stages in the reproductive cycle is complex. Initiation and termination of one stage are requisite for proceeding to the next: arrival of spring migrants is followed by establishment of territory which leads to courtship of the female, ovulation, completion of the clutch, incubation, and so forth. This sequencing is rather rigid in birds; that is, individuals at one hormonal stage in the cycle may be unresponsive to external stimuli associated with another stage. For example, during the nest-building stage, adults may be completely unresponsive to the begging stimuli of nestlings placed in their nest.

Photoperiod Control of Reproduction

In the temperate zone, the external cue that most often initiates the development of the reproductive organs and start of the breeding process is an increasing photoperiod, that is, an increase in the amount of light each day. Factors such as rainfall, social conditions, or food supply may also affect the initiation of breeding, but often these serve as final cues for the fine tuning of the reproductive response. In some species, however, these factors may be the primary cues used to initiate reproduction. For example, in the Red-billed Quelea Finch

(*Quelea quelea*), rainfall and green vegetation seem to be the proximate inductive factors. These birds migrate across Africa following the rains and stop to breed wherever and whenever the females can store enough energy reserve to produce a clutch of eggs.

To use daylength to regulate reproductive function, birds must have photoreceptors and some means of measuring time. The eyes are the most logical photoreceptor, but it has been demonstrated by many investigators that the eyes are not essential, at least in testicular development. In a classic experiment, *Benoit* and *Ott* (1944) used a quartz rod to direct light of various wavelengths directly to the brain of blinded mallards and found that the maximal testicular growth occurred with red light directed toward the hypothalamus. In fact, the hypothalamus was 100 times more sensitive to photostimulation than was the retina. Other researchers have used luminescent beads or optic fibers to deliver light to selective areas within the brain and have found that stimulation of the ventromedial hypothalamus with light produced the greatest response of gonadal (testis) growth.

The existence of a biological clock for measuring daylength, or even longer periods of time up to a year, has been documented for both plants and animals, beginning with Bunning's experiments in the 1930s. The structures that comprise the clock are the pineal, the suprachiasmatic nucleus (located in the hypothalamus above the optic chiasm), and perhaps the retina as well. The biological clock is responsible for maintaining the internal (endogenous) daily (circadian) rhythmicities of hormonal cycles, body temperature cycles, metabolic activity, locomotor activity, and so forth. In order to prevent the daily rhythms from becoming out of phase with the real world, the clock is "entrained" to an external cue (*zeitgeber*), such as light. In the absence of light, the clock may be entertained to sound, or a number of other peripheral cues that act as time cues. Thus, sparrows entertained to a light-dark cycle will continue a particular rhythm of body temperature, for example, when maintained in constant darkness. Without the light cue each day, however, the temperature rhythm assumes the periodicity of the biological clock (usually somewhat longer than 24 hours) and becomes asynchronous with real time. *Binkley et al.* (1971) found that removal of the pineal gland (a small appendage of neural tissue near the dorsal surface of the brain) caused the activity of sparrow kept in constant darkness to be sporadic with no periodicity at all. The pineal secretes a hormone called *melatonin*, which can affect the secretion or action of other brain hormones. Even in culture and in darkness, the pineal gland releases melatonin rhythmically, with peaks with peaks occurring about every 24 hours.

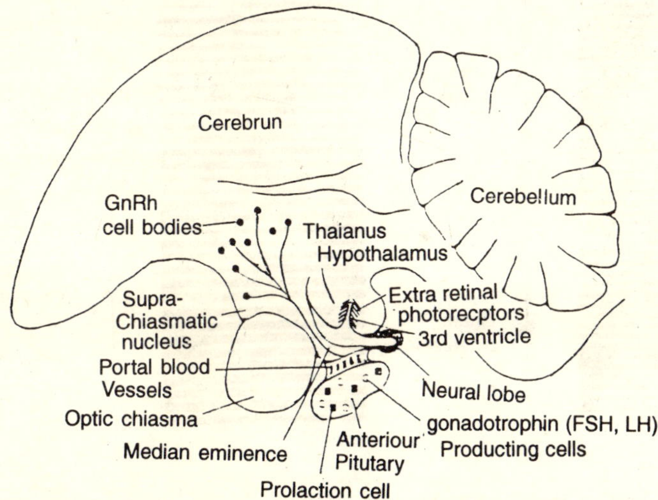

Fig. 1.18. Sagittal section of the avian brain showing the structure of the hypothalamus and pituitary, external area of photo-sensitivity (*hatch lines*), and location of neurons which manufacture GHRH and other releasing factors for pituitary hormones. Releasing factors are carried to the anterior pituitary by the portal blood vessels.

In addition to the endogenous circadian rhythms already mentioned, there is a circadian rhythm of photosensitivity. A particular physiologic event may be induced when light is received during the photosensitive period. Based on this, it is easy to see how changing daylength might initiate the entire cascading series of events in the reproductive cycle. At present there are two theories that explain how light induces reproduction.

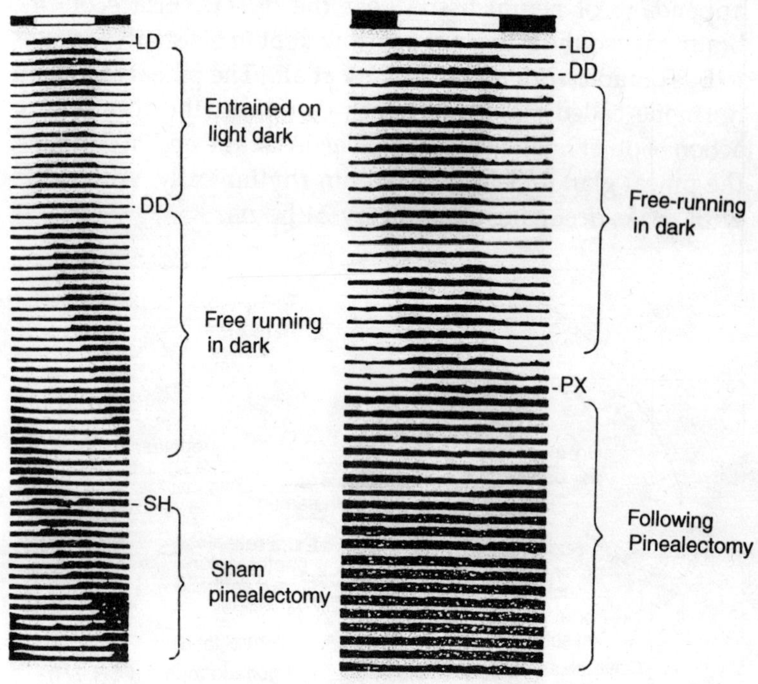

Fig. 1.19. Circadian rhythms of body temperature in House Sparrows (*Passer domesticus*). The figure at left shows that a circadian rhythm of body temperature is maintained by a bird transferred from a 12-hour photoperiod (LD) to constant darkness (at DD). The cycle begins to drift in the absence of a light cue. At SH a sham pinealectomy was performed, and the rhythm continues to free-run. The figure at right shows the same pattern until the pineal gland is removed at point PX, after which circadian rhythmicity is lost.

The *external coincidence* model states that gonadotrophin release, for example, occurs when light is coincident with the

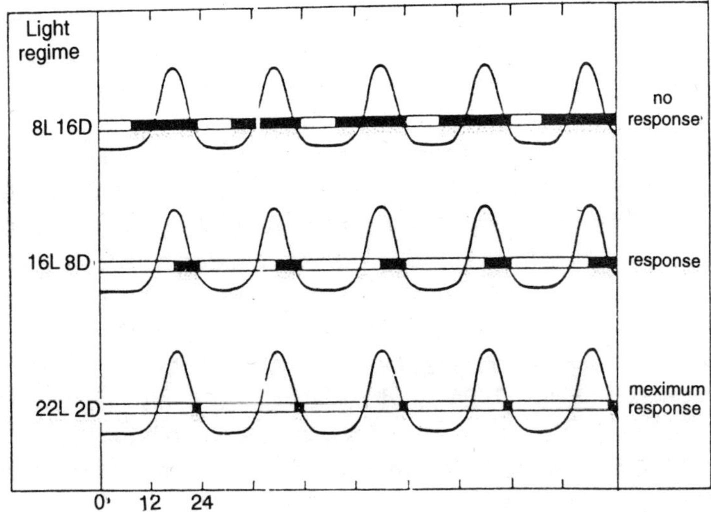

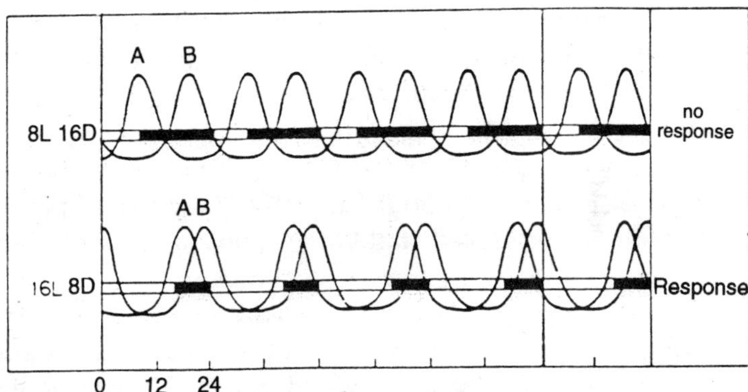

Fig. 1-20. The external coincidence model *(top)* shows a sinusoidal curve of entrained circadian rhythm superimposed with the light-dark cycle *(bar)*. The maximum response occurs when it is light during the peak photosensitivity period (curve is above the light-dark bar). The internal coincidence model *(bottom)* shows two oscillating circadian rhythms, each entrained independently by the light-dark cycle. In this case, *A* is entrained by lights on, *B* by lights off. A change in the photoperiod from 8L to 16L changes the temporal relationship coincidence between the peaks, bringing them closer together. The coincidence between the peaks is the basis for the photoperiodic response.

photoinducible phase of the daily cycle. *Hammer* (1963) exposed photosensitive male House Finches (*Carpodacus mexicanus*) to light-dark cycles of varying length to test this model. The duration of light in Hamner's experiments was always six hours, but the total length of the cycles ranged from 12 to 72 hours. At the end of the experiment, birds kept under 12, 36-, and 60-hour regimes had large testes, while birds kept under 24 (the control), 48-, and 72-hours regimes had tests of the normal immature size. The daily photoinducible phase continued to oscillate in birds kept in darkness for as long as 66 hours (the 72-hour regime). The reason that gonadal growth occurred in the 12, 36-, and 60-hour treatments was that the six-hour light period occurred during the photoinducible phase. When the light occurred only during the first half of the cycle, there was no stimulation.

The *internal coincidence model* assumes that changes in photoperiod wiil cause changes in the temporal relationship of two or more circadian rhythms. A specific temporal relationship between the two rhythms induces the physiological process. For example, *Meier* and *Ferrell* (1978) have found that injection of prolactin 12 hours after the daily rise of plasma corticosterone promoted fattening, gonadal growth, and northward-oriented migratory restlessness in White-throated Sparrows (*Zontrichia albicollis*), all characteristic of a spring-breeding bird. In contrast, injection of prolactin 8 hours after the rise in corticosterone in these birds resulted in no fattening or catabolism of fan no gonadal growth, and no migratory restlessness, which is more typical of a midsummer-breeding bird. The two coincidence models are not mutually exclusive, but are both useful in understanding the initiation of reproductive events, because some of the processes seem to be better explained by one model and other processes by the other model.

The amount of daylength necessary to stimulate gonadal growth differs greatly among species, but, in general, the daylength required to photostimulate is about the same as that experienced at the breeding site. For example, there is a direct relationship between critical daylength and latitude of breeding in swans. Bewick's Swan (*Cygnus columbianus*), which breeds at 65° N latitude, requires 16.5 hours of light for photostimulation; the Mute Swan (*Cygnus clor*), which breeds at 55° N latitude, requires 14.5 hours of light; and the Black Swan (*Cygnus atratus*), which breeds at 28° S latitude, requires only 10.7 hours of light. Of course, the consequence of this is that species are then limited in selection of breeding sites to those that have the critical daylength.

Hormonal Control Reproduction

Stimulus of the ventromedial hypothalamus by long photoperiods (i.e., light received during the photoinducible phase) results in the release of gonadotrophin-releasing hormone (GnRH) and prolactin releasing and inhibiting factors from neurons in the preoptic or supraoptic areas of the hypothalamus. Only tiny amounts of GnRH are released and move the short distance to the pituitary, where they stimulate the release of large amounts of the gonad-otrophic hormones, luteinizing hormone (LH) and follicle-stimulating hormone (FSH) into the general circulation. FSH stimulates the development of the gametogenic aspect of the gonads and promotes the production of mature ova and sperm. It also promotes the growth of the steroidogenic interstitial tissue of the gonads, but LH is responsible for stimulating the synthesis of the steroids (androgens and estrogens) and their release into the blood. In the male, the interstitial tissue (*Leydig cells)* secretes testosterone; in the female, the granulosa and thecal cells that surround the ovum secrete progesterone and estrogens, respectively. These steroids circulate through the blood to the secondary sex organs (e.g., the oviduct of the female or the vas deferens of the male) and are responsible for the growth, vascularity, and secretory nature of these structures during reproduction. They also circulate back to the brain, where they affect the further production of GnRH in a dynamic manner, and also affect other behaviour such as courtship, nest building, singing, and territoriality (Figure 1.20). Because these changes occur rather rapidly following a period of sexual quiescence, this stage is called the *acceleration phase* of reproduction.

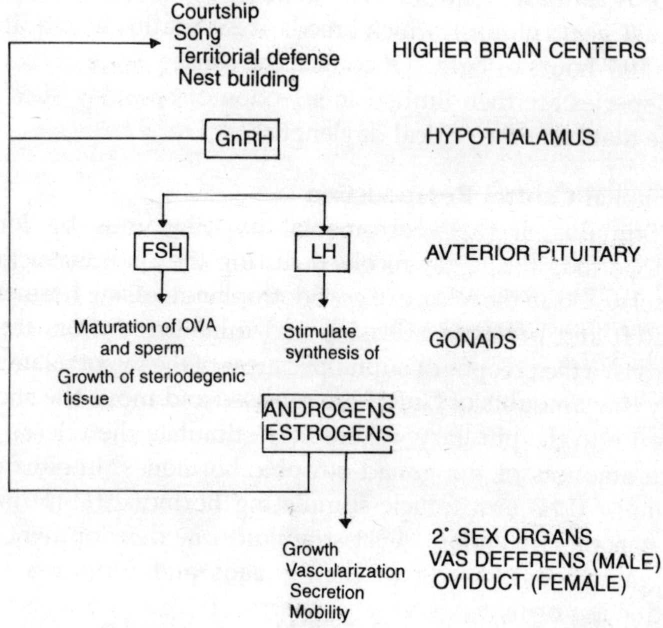

Fig. 1.21.   Effects of hormones (*boxes*) on the brain, gonads, and secondary sex organs during early gonadal development (acceleration phase of gonadal growth).

Following a period of gonadal growth, the ovarian follicles mature and begin to secrete less estrogen and more progesterone. Progesterone has a positive effect on LH secretion and causes plasma LH levels to rise, culminating in a preovulatory spike that causes ovulation. The wave of maturing follicles causes plasma progesterone levels to rise each day and LH levels to rise with it, each speaking about 4-6 hours before ovulation. Birds generally lay their eggs at the same time each day, which is a reflection of the circadian rhythm of ovulation.

However, in some birds, notably the domesticated fowl, it takes longer than 24 hours from ovulation to oviposition, and consequently, eggs are laid later each day, until eventually, the timing of the LH surge does not fall within the sensitive period (4-11 hours following the beginning of the dark cycle), causing the bird to miss an ovulation or skip a day's egg production. However, we must remember that daily egg production in all birds depends on the hen's ability to accumulate or mobilize her own resources rapidly enough to produce eggs.

Expulsion of the egg (oviposition) involves the relaxation of abdominal muscles and muscular contraction of the shell gland, the vagin and the cloaca. Two hormones of the posterior pituitary gland induce oviposition in the laying hen, oxytocin and arginine vasotocin (avian ADH). Both hormones cause contraction of smooth muscle in both birds and in mammals. In chickens, stimulation of the preoptic area of the brain will cause premature expulsion of an egg, via release of posterior pituitary hormones.

The ovulation-oviposition cycle continues until the clutch is complete. How this is determined is not fully understood. High levels of progesterone may eventually inhibit ovulation by negative feedback on GnRH release from the hypothalamus. Visual and tactile cues are also important in the termination of egg laying, but how birds perceive that their clutch is complete remains a mystery. During egg laying, levels of prolactin (secreted by the anterior pituitary) rise and this hormone may also have an negative effect on GnRH release and thereby suppress further ovulation.

Once the clutch is completed, incubation procedes. Most females incubate with greater intensity as the clutch increases in size, perhaps as a result of increasing levels of prolactin during laying. Brood patch development also progresses during laying, as rising levels of prolactin and estrogen and progesterone secreted by the ovary promote defeathering and vascularization of the ventral abdominal area. In phalaropes, in which the male incubates, testosterone and prolactin stimulate the production of a brood patch. Contact with the eggs promotes further increases in plasma prolactin, which in turn promotes more intense incubation efforts. When high levels of prolactin suppress the release of gonadotrophic hormones, the ovaries regress in size and the unovulated, yolky follicles are resorbed.

The role of prolactin in avian reproductive behaviour beyond this point depends on the species. In birds with altricial young, prolactin levels rise somewhat throughout incubation, peak shortly after hatching, and then decline. In birds with precocial young, such as the turkey, prolactin levels drop rather sharply after hatching, presumably because less parental care is required in this species. High levels of prolactin are maintained in adult pigeons and doves for the first week after the young hatch; in these species prolactin stimulates the production of a special crop secretion, "pigeon's milk," which is fed to the young.

Termination of Reproduction

To maximize its fitness, a species should have evolved control mechanisms to terminate reproductive activity when it is no longer energetically feasible to produce young and in time to complete its preparation for winter. In many species, the reproductive effort is terminated by a period of *photorefractoriness*, or insensitivity to long photoperiods. Often this occurs when daylengths are still long and, perhaps, even before the summer solstice. But the ecological advantage of such a control is that it allows the young time to grow and mature during a period when food is abundant, and it allows the parents time to complete the postnuptial molt and to fatten before migrating or overwintering. Through a variety of experiments with male White-crowned Sparrows (*Zonotrichia leucophrys*), *Farner* and *Gwinner* (1980) concluded that it is not possible to induce any of these latter events without previous exposure to long days; thus, it appears that both reproductive and postreproductive events are driven by the same photoinducible stimulus. In species that breed only once per year, individuals that have fledged young enter what is called an *absolute refractory phase*, in which they are insensitive to a long photoperiod. This phase is usually short but is followed by a *relative refractory phase* during which individuals remain sexually inactive although they may be responsive to long photoperiods (demonstrated experimentally). The latter state may last for several months, usually through the winter in temperate species, after which the cycle is repeated with the natural increase in photoperiod. Many birds must go though a photorefractory period to be photostimulated again.

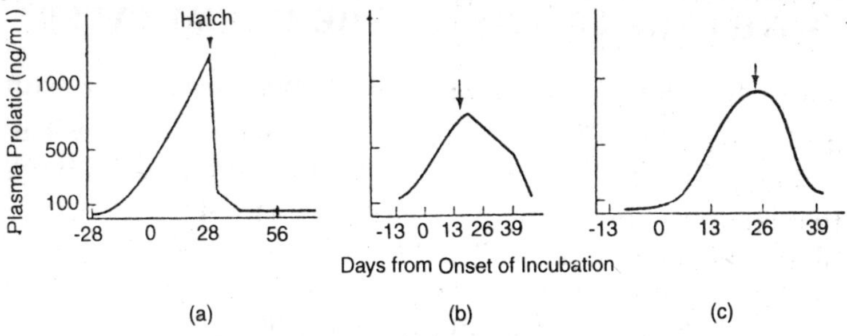

Fig. 1.22.   General patterns of variation in prolactin levels during incubation in a bird with precocial young (a) one with altricial young (b) and one with altricial young that are fed crop milk (c).

We have implied that photorefractoriness occurs in all birds following fledging of young. In fact, some species can produce several broods in a season. These birds start breeding in early spring and delay photorefractoriness until late summer or autumn. In multiple brooded species, external stimuli related to the development of the young or the adequacy of the environment for rearing young must reinitiate the reproductive process. The gonads do not regress completely in these species, and this shortens the period between nestings. Pigeons and doves continue to produce young as long as climatic conditions allow, irrespective of photoperiod. This group also is known to omit the refractory phase in the reproductive cycle.

There are exceptions to the normal pattern of termination of reproduction. In the reproductive effort is prematurely terminated by nest or mate loss early in the cycle, many species can renest. The speed with which renesting occurs depends on a variety of factors, including the stage of the cycle in which the interruption occurred, the physiological state of the female, and the availability of mates.

COORDINATION OF THE REPRODUCTIVE CYCLE WITH OTHER EVENTS OF THE ANNUAL CYCLE

All of the above physiological phenomena associated with breeding are energetically expensive, yet they are necessary to produce offspring. For temperate birds, and especially those that migrate, termination of breeding may partly be explained by the demands of the events in the annual cycle that follow reproduction. Some optimal balance of time and energy must be developed to complete the annual cycle.

Following the breeding season birds undergo a postnuptial molt, premigratory or prewinter fattening, and migration (in some). All of these physiological events must be squeezed in between fledging young and the first frosts of winter. Most birds must undergo their annual molt at this time simply because energy demands at other times of the year are too high. For example, high costs of thermoregulation in the winter coupled with diminished food resources preclude having enough energy to molt then. Costs associated with breeding and spring migration obviously rule out a complete molt at these times. This leaves only the period immediately prior to breeding, and some species do undergo partial body molts at this time to acquire their brightly coloured breeding plumage. However, an energy-demanding complete molt prior to breeding would deplete the energy reserve that is required for spring migration and for a successful reproductive season. These constraints leave the postbreeding period as the best time to undergo a complete molt. The advantage of a postnuptial molt is that the new set of feathers provides residents with better insulation for winter and provides migrants with new flight feathers to speed their travel south.

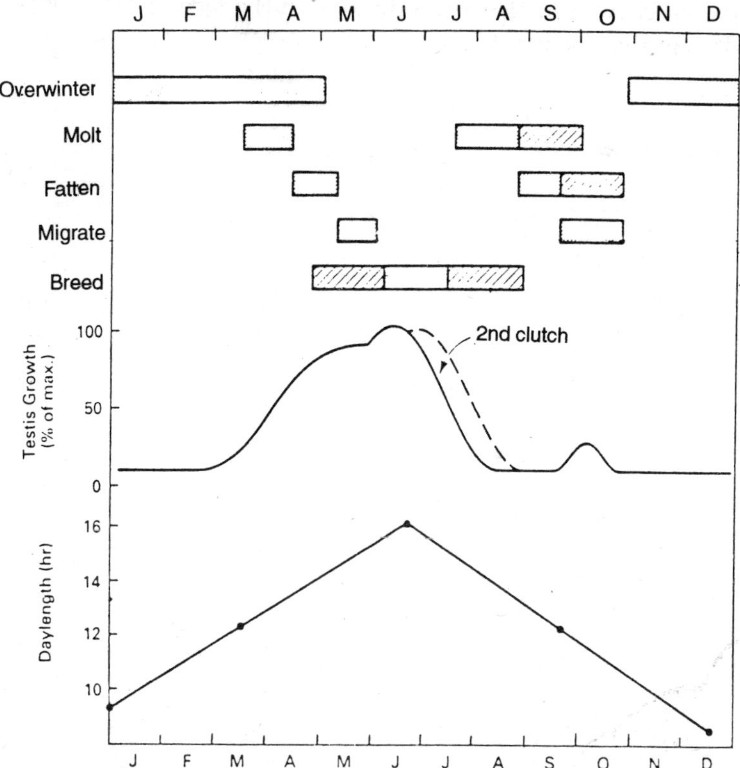

Fig. 1.23. The annual cycle of a typical temperate zone passerine. Dark bars represent the schedule for a migrant, hatched bars for a resident. Note that events are nonoverlapping, because of the way in which energy must be allocated by the bird. The hump at the peak of testis growth results when migrants arrive at the breeding ground. A smaller regrowth of the testis may occur in the fall when the absolute refractory period is terminated.

The cost of the autumn migration is a further constraint limiting the reproductive period of migrants. Migrants must accumulate extensive fat reserves before leaving the breeding site, and this cannot be accomplished until both reproductive efforts and molting are completed. Birds may use various environmental cues, such as weather or food supply, to judge the efficacy of further reproduction in the face of the energy-demanding events that follow.

2

Variation in Reproduction

The one previous chapter have focused on the mechanistic aspects of reproduction, especially the physiological and behavioural adaptations needed to produce eggs and young. Physiological and behavioural factors also serve as constraints on the number of offspring birds can produce. Incubation, the energy needed for egg production, or the time required to find food for altricial young all serve to put limits on the maximum number of offspring a pair of birds could raise at a time. If we look at the actual reproductive characteristics of birds, however, we see a great deal of variation, both between and within species. For example, a House Wren (*Troglodytes aedon*) that lives in Panama usually lays two or three eggs per clutch, whereas one breeding in Canada may lay up to seven eggs. A duck nesting in the Arctic may lay a clutch of eight to ten eggs, whereas a seabird in the same region may lay only one egg.

These examples suggest that many birds are responding to evolutionary and ecological factors that result in reproductive efforts smaller than purely physiological or behavioural constraints might allow. In other words, white physiology or behaviour might set maximal limits to reproduction, many birds seem to be producing young at some lower rate. Evolutionary biologists tend to think of this reduction as an adaptive compromise that balances a variety of factors other than just the absolute number of young that can be produced at a particular time. This chapter examines the variation that occurs in birds to produce optimal number of young rather than the number that may be physiologically possible.

Throughout this chapter we shall assume that reproductive characteristics are the products of natural selection. For this to be the case, reproductive traits must be inherited through genetic means and natural selection must favour those individuals with particular reproductive traits. For example, for clutch size to be an evolved trait, offspring of a female must have the genetic propensity to lay clutches of the same general size as their mother. In this way, an initial population of females might exist with the genes to lay a variety of clutch sizes. Natural selection will then favour those females that lay a particular sized clutch, while females laying either more or fewer eggs will not reproduce as well (see below for more details on this). If clutch size is not a heritable trait, natural selection cannot operate in this manner. To date, studies have indicated that virtually all reproductive traits, while showing some variation due to age or breeding condition, have basic genetic components that are passed from generation to generation. Therefore, all reproductive traits are considered to be evolved traits.

As the conditions that promote various options in reproductive behaviour are examined, it must be remembered that this variation occurs because a subset of individuals with a particular set of genetic traits produce more young under those ecological conditions than individuals with other genetic traits. That is the essence of natural selection. In many cases, we shall try to explain a particular situation by asking questions about possible options: Does a female that lays two eggs produce more surviving young than a female that lays four eggs ? While the answer is the result of selection favouring either two- or four- egg clutches, many scientists like to approach such questions by asking themselves what the best strategy for a female would be under particular circumstances. Under this approach, one might ask: What is the best egg-laying strategy for a female House Wren in Panama? When is sharing a male mate a better strategy than monogamy? When is helping my parents a better strategy than attempting to breed myself? Whereas the conditions in which a particular strategy is most effective may be identified by looking at the ecological and evolutionary conditions in which the behaviour is found, it must always be remembered that birds do not actually have conscious strategies in these reproductive behaviour. Rather, they show a variety of behaviour and natural selection chooses those that work best in those conditions. Nevertheless, the concept of strategies is sometimes a good way to look at how birds response to the variety of conditions in which they attempt to breed.

BIRD POPULATIONS AND THEIR REPRODUCTIVE CORRELATES

Population Fluctuations

In studying bird populations it is usually impossible to actually count the total population of a particular species. Rather, the number of birds per unit area, density, is studied. As a general rule, if a habitat does not change significantly from year to year, the density of birds in that habitat at a particular time of year will also be very consistent. This stable population level for a species within a habitat is generally identified as the carrying capacity of that habitat, as it suggests that at that point the habitat is sufficiently full of the species that no more individuals can be supported by the habitat.

Recent work with tropical birds suggests exceedingly stable populations, while long-term studies with birds in temperate climates show some fluctuations around an average population level. This consistency in density suggests that some factors must be at work regulating these populations, and the fact that the tropical and temperate zones differ in the stability of populations suggests that climate must play a role in this regulation. If one follows these same populations through the year, one sees seasonal variation due to the production of young. This seasonal variation is also generally greater in the temperate zone than it is in the tropics.

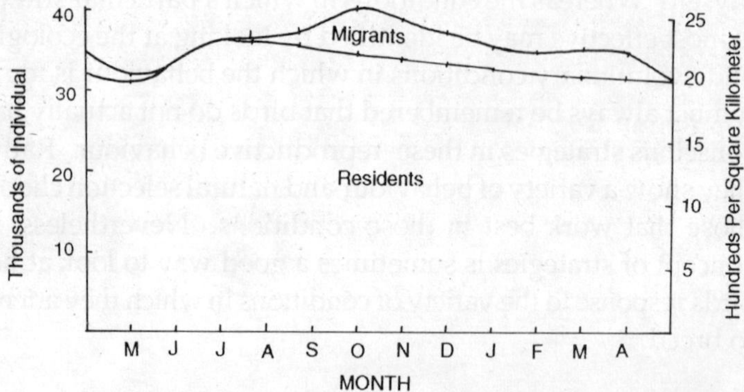

Fig. 2.1.   Variation in resident bird density on Barro Colorado Island, Panama, through the year. Note how stable populations are when compared in similar measurements made in a temperate forest.

Under undisturbed conditions, most populations of a species show fairly limited variation in size throughout their ranges. Some populations fluctuate greatly on a local scale. While these may seem like unstable populations, if these species' populations are considered on a larger scale, they probably do not vary as greatly as one might think.   Truly irruptive populations probably occur only in very seasonal habitats such as some deserts, where drastic fluctuations in food supply and other factors occur over periods of many years, and many species are unable to migrate to other locations.

Other noteworthy changes in populations reflect changing conditions or the expansion of a species into a new habitat. Habitat modification in the New World has certainly increased the populations of such species as the American Robin (*Turdus migratorius*), which does well in suburban habitats. Introduced species such as the European Starling (*Sturnus vulgaris*) or Ring-necked Pheasant (*Phasianus colchicus*) have also shown great population increases in some areas. In general, however, these populations have eventually stabilized as they increased to match the available habitat (thereby reaching carrying capacity). Thus, observed over long enough time periods or on the proper regional scale, even the seasonal fluctuations of most bird populations suggest some form of regulation rather than drastic population expansions and crashes.

Population Regulation

Factors that can regulate bird populations come in two general categories. *Density-independent factors* are those that operate, as the name implies, independently of the density of the species. Climatic factors such as cold or rain are the thief density-independent factors; the lowest temperature in winter is obviously not a function of the number of sparrows living in a habitat, although it certainly may affect that number. Trying to understand the population levels of species affected by density-independent factors is largely a matter of trying to understand how each species deals with fluctuations in its general environment.

Density-dependent factors, on the other hand, affect individuals in a variable way that depends upon population density. When populations are low, these factors may be relatively unimportant, but as populations increase they have a greater effect. Food is perhaps the most obviously density-dependent factor, but other factors that show density-dependency may include competition, predation, roosting or nesting sites, disease, and so forth.

Factors from these two general categories obviously interact, for in the absence of any density-independent limitation populations would grow until food or some other factor was limited by density, as which time density-dependent limitation would occur. On the other hand, density-independent factors may reduce populations with such regularity that they rarely reach densities great enough to show density-independent limitation. Often, the overall affect of density-independent factors such as extreme climatic conditions is a function of a species density. A cold, harsh winter may cause more deaths when populations are high than when bird populations are already low for some other reason.

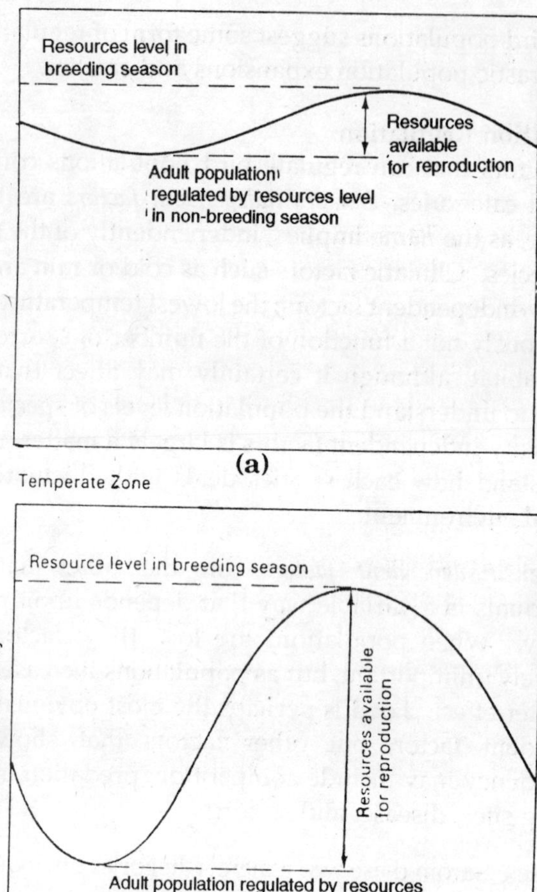

Fig. 2.2.   A general comparison of resource and or population variation in a stable tropical habitat *(top)* and the temperate zone *(bottom)*. The amount of variation between maximum and minimum levels accounts for much of the difference in reproductive traits found in the birds frequenting these regions.

Several correlations occur between the patterns of change in the size of populations and the relative importance of density-dependent or density-independent factors, and these relationships greatly affect the general patterns of reproduction found in birds. Populations that vary little from year to year, or even within the year, tend to suffer little from density-independent factors (often because they live in stable, mild climates) but are subject to relatively high levels of density-dependent limitation. The tropical populations discussed earlier rarely are subjected to cold or other extreme climatic conditions, thus they fluctuate little and their carrying capacity is determined primarily by density-related factors. Temperate zone populations are reduced nearly every winter by factors associated with harsh climatic conditions (a density-independent factor); each spring, populations are below the carrying capacity of the habitat such that density-dependent factors are relatively less important through the breeding season.

As a general approximation, this difference between the effects of density-independent factors in tropical and temperate habitats explains some of the differences in reproductive characteristics in these different regions. Certainly, if one compares a standard temperate zone population typified by rather great seasonal fluctuations in density with a more stable tropical zone population, it should not be surprising to see that temperate zone reproduction is characterized by greater production of young than in the tropics. Temperate zone species have the potential to increase populations severalfold during the breeding season through such means as large clutches or multiple breeding attempts. Because total densities rarely differ from near the carrying capacity of the habitat, tropical species are less likely to show such characteristics of high productivity.

Mortality factors also must be examined in understanding reproductive characteristics in different environments. A bird in a tropical habitat is not exposed to harsh winter conditions that may consist its life. Rather, once this bird has reached adulthood it has a high probability of a rather long life-span. During this life, though, it must deal with a variety of limitations caused by the number of conspecifics with which it must live and the diversity of other species that live in this stable climate, many of which may serve as predators, competitors, or parasites of the bird. While the bird may live a long time under these conditions, it may find gathering enough extra food for the young difficult, which may limit the number of young. Natural selection may also work against behaviour that involves risking its future life-span with small chances for reward. Such risk-benefit trade-offs may limit how often or where a bird nests. The result may be an individual that breeds infrequently and produces only a few young at a time, although this is enough to match the low mortality rates that exist under these conditions.

In contrast, a bird living under conditions with substantial density-independent population control faces a decidedly different set of circumstances. Even if it is successful in reaching adulthood and breeding condition, it generally does not live a long time. When it is time to breed, though, an almost unlimited supply of food may be available because populations of nearly all species are well below carrying capacity due to the effects of winter. This means that the bird can find enough food to feed many young, and the chance that it may not live to breed again means that it should take more risks to maximize each breeding attempt. The general result is a temperate zone species that may breed often and have large clutches; this, too, tends to result in similar numbers of birds of this species at a particular time of the year each year, but the life history traits of this temperate zone bird are very different from its tropical relative.

CLUTCH SIZE VARIATION

Clutch size has been one of the most intensively studied of the reproductive traits or birds for a variety of reasons. It is a discrete entity, involving only the number of eggs in a nest, and is generally presumed to come from the parent(s) that constructed that nest. Since many species nest only once a year, if that nest is successful, it is the best estimate available of annual reproduction. Clutch size also varies greatly, both with in and between species, which allows for studies on the effects of environmental factors in the evolution of clutch size. Finally, clutch size has been important because it can be easily manipulated (by adding or removing eggs) in ways that other reproductive traits cannot.

To understand the factors (other than physiological constraints) that affect variation in clutch size, we must first look at how various factors are balanced to produce an optimal clutch size for a region. With this knowledge, the variation that occurs between species is clearer, which also aids the understanding of regional variation within species. This section ends with a look at some adaptations of species found living in variable environments.

The Optimal Clutch Size for a Regional Population

It can generally be said that the optimal clutch size is that which leaves the most offspring in subsequent generations. For a small population living in a particular habitat type, it seems that those individuals that produce the most offspring (within any physiological or behavioural constraints) will most affect the gene pool of subsequent generations. The key here, as in any situation, is that the offspring must survive long enough to breed. If the production of many nestlings results in poor-quality offspring, few of these offspring may survive long enough to breed and the trait for a large clutch size is therefore not passed on to future generations. Thus, it appears that in any habitat there is some balance struck between producing the maximum number of young and producing young of a high enough quality that they have a chance of surviving to breed.

The quality of offspring produced is strongly related to how much food they receive, particularly in altricial birds. If well fed, young birds can develop properly and reach adult size rapidly, both of which may increase the bird's chances of surviving into the future. If we think of parent birds as having a maximum amount of food that they can bring to a brood, adding an egg to a clutch divides this amount by some increment. This may affect the growth rate of the young. Studies with the Great Tit (*Parus major*) have shown that clutch size tends to be inversely related to the weight of the fledglings, individuals with large clutches produce lighter young. In some cases, these lightweight young do not develop properly and do not even fledge. In order cases they may be able to leave the nest, but studies with several species (including the Great Tit) have shown that, in general, smaller offspring have lower chances of surviving to reach breeding age. This suggests that birds producing relatively large clutches actually do not leave as many breeding offspring in subsequent generations as those somewhat fewer, but larger, young.

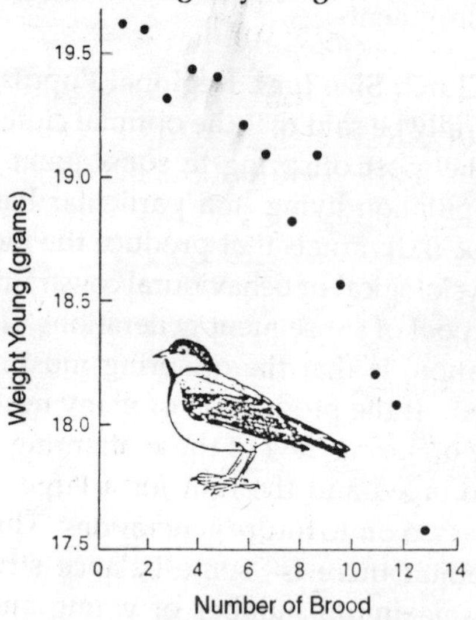

Fig. 2.3.   The relationship between weight of young and brood- size in the Great Tit (*Parus major*).

Given that the quality (health) of the young is an important factor, one should consider the effects of producing just one, very healthy offspring on an individual's potential contribution to the gene pool. Despite this offspring's having an excellent chance of surviving to breed in the future, parents that produce two or more offspring that are healthy enough to survive to

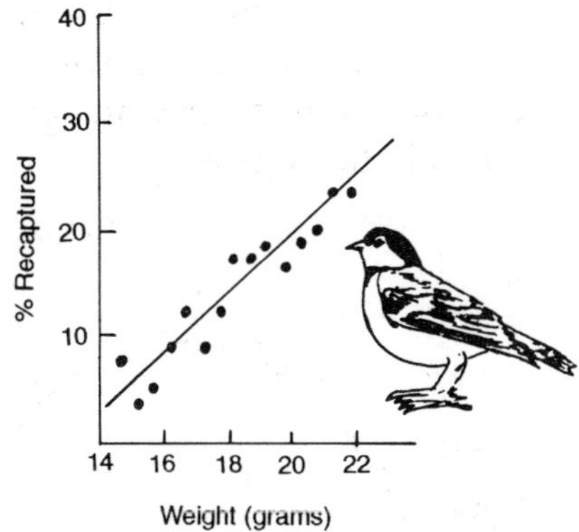

Fig. 2.4.   The relationship between weight when 15 days old and recapture rate in the Great Tit. Recapture rate is a measure of survival.

breed will contribute more genes to subsequent generations than the parents of a single, super young. To the extent that clutch size is generally determined, a clutch of greater than one will be favoured.

The average clutch size for a region, then, is the balance that is struck between producing the largest number of young possible and ensuring that these young are healthy enough to have a good chance of surviving to breed in the future. A clutch size that is too large actually may leave few surviving offspring, while one that is too small will be outnumbered genetically by those individuals producing more young. We assume that a particular balance between these two factors occurs for each species in each region to lead to the evolution of a particular clutch size for that population. Such a mechanism does not mean that a specific clutch size is fixed within the population. Rather, factors related to age or experience, food supply (both due to climatic variation or population fluctuations), or seasonal variation may result in some variation around an average clutch size. In species living in environments that vary greatly between years, clutch size variation may be adaptive. For example, particularly dry years may favour individuals with smaller clutches than average, while wet years favour those with larger clutches, resulting in a variable clutch size within the species. In species living under relatively predictable conditions, however, large amounts of clutch size variation should be selected against.

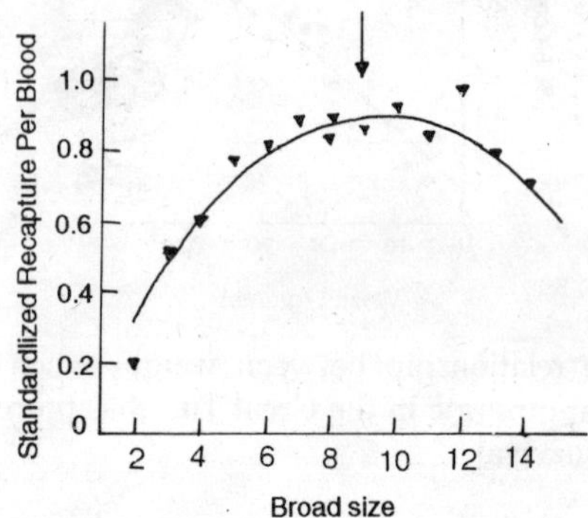

Fig. 2.5.   The relationship between survivorship (which is related to fledging weight and brood size in the Great Tit. This suggests an optimal clutch of nine to ten eggs, close to that actually observed.

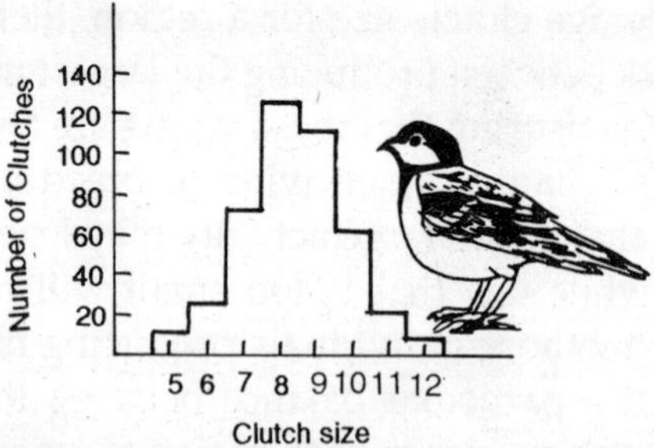

Fig. 2.6.   Distribution of clutch size in the Great Tit in Wytham Wood, England.

Clutch Size Variation between Species

If we assume that the optimal clutch size for each species is determined by this balance between the quantity and quality of offspring produced, the differences between species in clutch sizes should be mostly related to the factors that determine quality of offspring. since in most cases this is food supply, it is not surprising that the availability of food for the young seems to be the chief determinant of interspecific clutch size variation. Species feeding on foods that are rare or widely distributed will be able to raise fewer young than species feeding on abundant foods. In general, large foods are rarer than small foods, so we might expect large birds to have smaller clutches for this reason. Additionally, birds with young that feed themselves (precocial birds) should have more young than birds where the parents must provide all the food, all other things being equal.

The clutch sizes of a variety of avian groups are shown in Table 13-1. It is not surprising that seabirds and raptors have relatively small clutches compared to those of small insectivores or birds with precocial young. While our model of optima clutch size helps explain all these differences, in a few cases other constraints appear to be operating. For example, small shorebirds (sandpipers and the like) produce precocial young, but the requirements of a large egg coupled with the small body of the incubating parent apparently limit clutch size greatly. A large pheasant or goose does not face this limitation. In many of the species with very small clutches, actual physiological or behavioural limitations may be operating as increasing the clutch by even one egg may result in severe losses among all the young. For example, swifts living in the same general area as Great Tits lay only two or three eggs, and the addition of a single egg may cut the number of fledged offspring. Swallows, however, which have very similar general food habits and share many habitats with swifts, have appreciably larger clutches. Most seabirds lay only a single egg and adding a second may result in the loss of both young.

Predation pressures may play some role in interspecific differences in clutch size, particularly when explaining differences between hole-nesting and open-nesting birds. It appears that those species nesting in relatively predator-free locations such as cavities lay larger clutches than similar species that use open nests. Predator pressures may also affect clutch size limitations among open nesting birds, particularly among ground-nesting birds with large clutches of precocial young. The longer it takes to lay and incubate eggs, the greater the chance of a predator finding the nest. At some point, the chances of predation may outweigh the benefit of a larger clutch. Within the same habitat when looking both between and within species, it is hard to measure the predation factor, except in the case of a predator finding the nest as noted above. (The role of predation in intraspecific variation on a geographic scale is discussed below.)

TABLE 2.1

Examples of Typical Clutch sizes Within Different Types of Birds

Group	Number of Eggs
Penguins	1-2
Loons	2
Grebes	3-6
Procellariiformes	1
Pelicans and cormorants	1-4
Herons and egrets	3-5
Geese and swans	3-6
Ducks	7-12
Hawks, eagles, and vultures	1-5
Grouse, pheasants, quail, and ptarmigan	5-18
Cranes	2
Rails, coots and allies	5-12
Shorebirds (sandpipers and allies)	3-4
Gulls and terns	2-3
Auks	1-2
Pigeons and doves	2
Owls	2-7
Nighthawks and nightjars	2
Swifts	1-5
Hummingbirds	2
Woodpecker	3-6
Antbirds	2-3

Manakins	2
Tyrannid flycatchers	2-6
Swallows	3-7
Birds of paradise	2-3
Chickadees and titmice	4-15
Wrens	2-11
Old World warblers	2-10
White-eyes	2-5
Sunbirds and honeyeaters	1-3
Orioles and blackbirds (Icterinae)	2-6
Goldfinches and allies	3-6
Estrildid finches	4-10

This leaves a variety of factors that may affect the evolution of clutch size when all species are considered. Yet, the effects of body size, population parameters, food supply, predator pressures, and the developmental type of the young help explain the general patterns discussed above.

Variation in Clutch Size Within Species

We have already pointed out that the House Wren may lay two or three eggs in the tropics and up to seven in the temperate zone. Geographic variation in mean clutch sizes has been found in numerous species along a variety of gradients. Much work has been done looking at latitudinal gradients in clutch size, where it has been shown that most temperate birds lay larger clutches than their tropical counterparts. Similar patterns have been shown along longitudinal gradients from coastal areas to the interior of continents, or with latitude (with lowland birds having smaller clutches than montane species). Many populations on offshore islands have smaller clutches than their mainland counterparts, while species in more seasonal habitats in the tropics have larger clutches than those of nearby, less seasonal habitats.

A variety of explanations have been offered over the years to explain these patterns. Some have suggested that physiological limits cause these patterns, but experiments in many cases have shown that this is not the case. Because small clutches

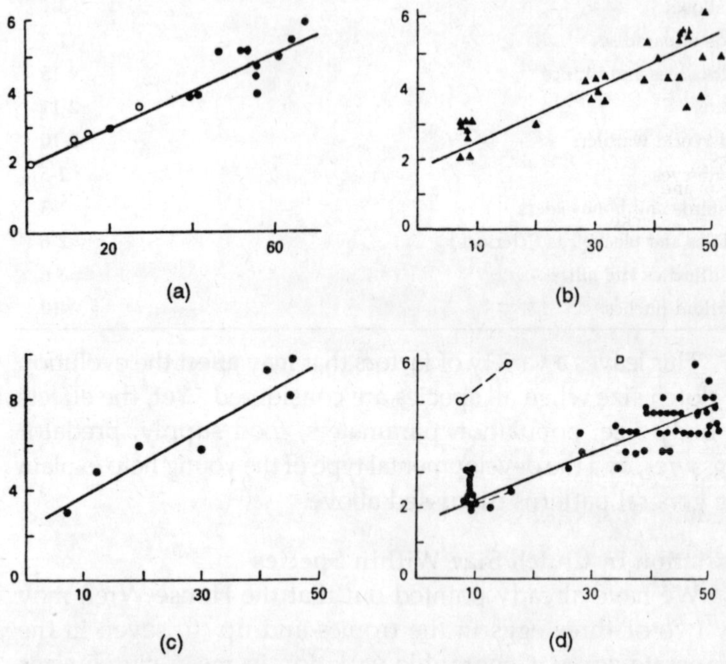

Fig. 2.7.   The relationship between clutch size (ordinate) and latitude (*abscissa*). (a) shows members of the *Old World genus Emberiza*; (b) shows members of the *New World Icterinae*; (c) shows members of the duck genus *Oxyura;* and (d) shows members of the *New World Thraupinae and Parulinae* (open points represent atypical hole-nesting species).

often occur in more stable environments, it has been suggested that the birds are purposely "balancing mortality." While it may in fact be true that reproductive rates tend to match mortality rates (see below), this pattern is more a result than a cause of clutch size patterns. Modern evolutionary theory provides no mechanism by which a bird may choose to balance mortality (early arguments favouring this hypothesis require some form or group selection). Another explanation suggested that daylength was the critical factor, as a temperate bird has much longer each day to feed its young than its tropical counterpart. While this extra time certainly does not hurt, it does not explain altitudinal or longitudinal patterns. Additionally, nocturnal owls show the same latitudinal patterns as other species, even though their foraging time is at its minimum during the breeding season in the North Temperate Zone. The "spring bloom" hypothesis suggested that the increased productivity of the temperate zone allowed much larger clutches; while this does appear to help explain some of the variation, changes in productivity seem to be only loosely related to clutch size variation. Finally, the predation hypothesis suggested that species in the tropics showed smaller clutches because tropical environments have larger numbers of nest predators. Large clutches (and hence large broods) might attract nest predators more than small broods because the parents must feed large broods more often, and thereby assist the predator in locating the nest. Many studies have suggested that nest predation rates are much higher in the tropics than in the predator-poor temperate zone. On the other hand, species such as hole nesters that are relatively immune to nest predation still show latitudinal gradients in clutch size, as do the predators themselves.

While none of the above hypotheses can really explain geographic variation in clutch size, it is obvious that many of them have some validity and that some sort of combination of factors is at work in determining clutch size. Although a number of models have been proposed that depend on combining factors, the author's favorite is that of Martin Cody (Figure 13-9). He developed a model using the *principle of allocation* (the idea that a bird has a limited amount of energy that it must allocate in an optimal way to produce offspring). He proposed that clutch size is determined by the interaction of three factors, one related to predator avoidance, one of the food supply, and the third the allotment of energy to clutch size. Predator avoidance takes into account a bird's chances of loosing its nest and any chance the bird itself might take while nesting. In temperate birds, the allocation of energy to this factor is generally low, as predation hypothesis mentioned above apply here.

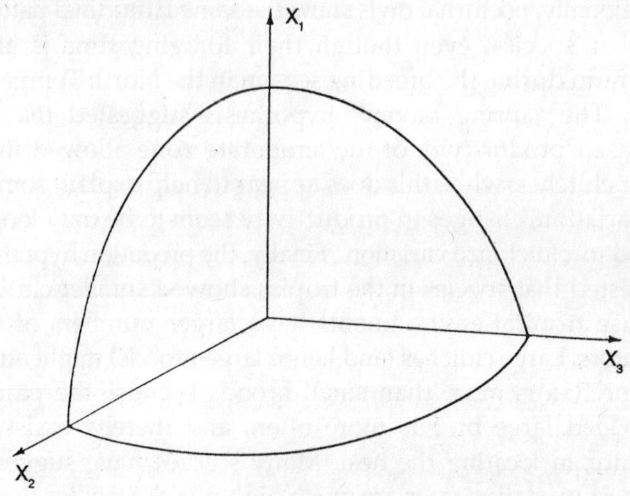

Fig. 2.8.   The three-dimensional model for the evolution of clutch size. A limited amount of energy must be allocated along the three axis related to avoiding predation, finding food, and clutch size. Given a set amount of energy, movement along one axis affects the position along another axis.

The food factor in this model takes into account the amount of food available to a species for egg formation and the feeding of the young. Tropical environments have stable populations and a high species diversity. Because of this, competition is high and food is relatively hard to find. With the spring bloom, decimated populations following winter, and relatively low species diversity, temperate species should have to spend less energy finding a similar amount of food.

The dependent variable in this model is then the allocation of energy to clutch size. If we examine two species with similar amounts of total available energy, a tropical species might need to spend more of its energy in predator avoidance (including building a better nest) and more in food gathering than a temperate species, leaving less energy for investment in the clutch size. Thus, it would produce a smaller clutch than a species in the temperate zone, where less energy had to be allocated to predator avoidance and food gathering.

Although this model was designed to explain latitudinal gradients in clutch size, it is general enough to explain clutch size variation in almost any situation in which stable environmental conditions are compared with less stable conditions. Energy investments in relatively stable situations where populations are near carrying capacity all the time can be compared to situations where populations vary greatly. As it turns out, all of the gradients described earlier (latitudinal, coastal-interior, longitudinal, altitudinal and island-coastal) are gradients of environmental stability. Where populations vary little from carrying capacity, more energy must be allocated to the predation and food axes. Where they vary greatly, more energy is left to allocate to large clutches.

The beauty of this model is that one can see that a change in allocation of one factor affects at least one of the other factors (usually clutch size). Let us consider hole-nesting birds, which we discussed earlier. Using this model, one would expect that hole nesters need to allocate less energy to predator avoidance and thus could put more energy into clutch size. Data show that hole nesters have larger clutches than open nesters within a region, but they also show the same latitudinal variation in clutch size, most likely due to changing food supply. This model can also make predictions about clutch size when comparing birds with precocial young with those having altricial young; birds with precocial offspring do not face the energetic cost of food gathering that birds with altricial young face, so we might expect the former to have generally larger clutches.

It should be noted that this model suggests that larger clutches should appear in populations living in less stable environmental conditions, which would mean that the production of young would tend to balance mortality. This pattern is to be expected from the varying conditions involved, and not because the birds are actually attempting to consciously regulate their populations.

Bet-hedging Strategies

Seasonally variable environments also often show annual unpredictability with regard to rainfall, temperature, and so forth. Anyone who has lived for several years in the temperate zone knows that one summer season might continue for months, while another might occur for just a few days. Such variability in climate may also affect the evolution of clutch size, as what may be optimal clutch in one year may be too big or too small in another. While in many cases this may select only for higher intraspecific variation in clutch size, in other cases mechanisms have evolved that allow an annual "fine-turning" of the production of young under such unpredictable conditions. These are often termed *bet-hedging strategies*. They are adaptive in unstable conditions because several weeks elapse between the time when a breeding bird lays its clutch and the time when it must find enough food to feed the offspring. To the extent that clutch size laid is an evolutionarily adapted "bet" about environmental conditions, these strategies give the bird a chance to adjust the effective brood size at the last minute to produce the maximum number of viable offspring.

In most cases, bet hedging involves what is known as *brood reduction*. Reduction is necessary because it is impossible for a bird to lay a late egg when it realizes that conditions will be good for raising young. Rather, the bird might initially lay an extra egg, then try to come up with some way to quickly dispose of the young bird should conditions suggest that the brood be reduced. This removal should result in a smaller brood, hopefully one in which each nestling can receive adequate food. In good years, however, this extra young might receive enough food to survive and have good chances of success.

If a clutch hatches with one or two too many eggs for resource conditions but with all young of the same size and condition, the young all might suffer by dividing the food evenly. The result might be the production of a complete clutch of low quality young. Species that have evolved brood reduction mechanism usually set up a situation in which competition within the nest occurs if food is limited and at least one of the young is at a competitive disadvantage. In this situation, the weakest young quickly dies and the brood is reduced.

Such unevenness can be accomplished in several ways. Some species lay eggs that vary in size, thus resulting in young that vary in size and, presumably, competitive ability. Eggs of the Common Grackle (*Quiscalus quiscula*) may vary in size by as much as 45% of the mean egg weight. This has been shown to be correlated with potential survival. Some species begin incubation before the clutch is finished. In many raptors, incubation begins with laying resulting in young that hatch at at intervals corresponding to those at which the eggs were laid. Such pronounced inequality may be especially adaptive in upper trophic level carnivores, where food supply may be variable and high-quality young must be produced to have a good chance of survival. Many small birds begin incubation with the penultimate (next-to-last) egg, which means that the last egg hatches a day late. This difference may be enough to result in its starvation if food is in short supply. In some species, a two or three day interval occurs between the initiation of incubation and the laying of the last egg.

Such mechanism of brood reduction would not work for all species. Obviously, brood reduction by this means would not be particularly advantageous for birds with precocial young (although it has been suggested that precocial birds start incubating before the last egg to reduce the chances of nest predation before at least some eggs have hatched). It also would not work for species in which the young have slow growth rates or can store fat, for these attributes would tend to reduce the intense competition that causes the starvation of a young bird.

A few species that are exposed to unpredictable, short-term variation in resources (such as swifts, swiftlets and swallows that are unable to capture flying insects during rainy weather) have adaptations to help ensure the short-term survival of the young. These include fat storage and variable growth rates, which reduce the effects of days with little food, and the development of early thermoregulation, which allows both parents to forage away from the young. Although these techniques are good for short-term survival, they obviously work against brood reduction.

In most small birds, this intrabrood competition results in the rapid loss of the weakest nesting. In large species like raptors or herons, even the youngest of the offspring may survive for some time before dying. In some cases, it appears that the demise of the youngest bird may not be solely due to lack of food, but it may be the result of attacks from one or more older siblings (called *siblicide* or *fratricide*). It has been suggested that parent birds can either encourage or discourage such intractions among their young, depending in part upon resource conditions. While some have suggested that the occurrence of siblicide is an aberrant occurrence reflecting unusual food limitations, the frequency of apparent brood reduction mechanisms suggests that it really may be an adaptive way to produce young of acceptable quality in a variable environment.

THE TIMING OF BREEDING

The determination of the optimal time to breed should be the result of natural selection, with those individuals that breed at the best time producing more young that survive to breed themselves than individuals breeding at other times. Maximization of an individual's fitness would then be an ultimate determinant of the breeding season, perhaps exhibited through some genetically determined mechanism related to photoperiod or other predictable stimulus. As mentioned earlier, while such external stimuli may get a bird into breeding condition at what is evolutionarily the proper time, most species also rely on short-term stimuli that tell them conditions are favourable for breeding. These stimuli may be related to weather conditions, food supply, the physiological condition of the female, and so forth. There is considerable variation among species in the interaction between these long-term and short-term factors that affect breeding seasons. In most species, readiness for breeding is determined by photoperiod and short-term factors may affecting breeding times by only a few days. In contrast, some of the seed-eating birds that exhibit mutualistic interactions with their food supply seem to respond almost completely to the proximate stimulus of their food supply. Because of this, species such as the Pinon Jay (*Gymnorhinus cyanocephalus*) or Clark's Nutcracker (*Nucifraga columbiana*) may breed at almost any time of the year when their food supply is superabundant.

Although it is generally believed that breeding should occur at that time when the most young will be produced for long-term survival, what that particular time is may vary between species, even within the same habitat. Much of this variation is related to variation in the abundance of the foods these species use, and some of it may reflect differences in food limitation related to either feeding the young or ensuring the success of the fledglings once they are on their own. In a typical temperate woodland, the Great Horned Owl (*Bubo virginianus*) may lay its eggs in January or February, which means that the nestlings must be fed during very cold periods when food may be scarce, but which also means the young fledge early in the summer when food is abundant and presumably easier for inexperienced foragers to capture. The predatory Loggerhead Shrike (*Lanius ludovicianus*) also nests very early, often by mid-February. Other owls breed somewhat later in the spring, along with multi-brooded species such as the American Robin (*Turdus migratorius*) or Mourning Dove (*Zenaida macroura*). Most of the migrants from the tropics arrive in May and breed in late May or June, often producing only one brood before leaving for the tropics again in August. A few species, such as the American Goldfinch (*Carduelis tristis*) or Blue Grosbeak (*Guiraca caerulea*) may not breed until mid-summer, presumably because their foods are more abundant at that time. An extreme case in this regard is the Eleanora's Falcon (*Falco eleanora*) of the Canary Islands, which breeds in the fall when migrant birds constitute its major food. The Harris Hawk *(Parabuteo unicinctus)* of the southwestern United States and the tropics is known to breed both in spring and fall.

Among temperate zene breeders, it has been observed that nests are initiated early enough in the spring that they are often lost to could conditions, yet breeding may cease easy in the summer, at a time when the food supply may be reaching a peak. Suggestions as to why these patterns occur have come from studies on the House Sparrow (*Passer domesticus*) in the rather extreme environment of Alberta, Canada. These studies found that nests were often initiated in April, despite low chances of nesting success, but that nesting stopped in early July, when fledging rates were high. This was explained by looking at the survival of young from various parts of the breeding season. Even though parents had a small chance of fledging young when nesting in April, any young that did survive had a very high probability of surviving to breed the new year. Young that were produced later in the spring or early summer, when the probability of fledging success was high, had little chance of surviving to breed. This appeared to be due to dominance interactions in winter feeding flocks. Young from April or May were dominant to young from June or July, which apparently favoured the survival of the former over that of the latter. Similar patterns may be at work in favouring the early breeding attempts of other species where either dominance interactions or foraging experience may be important in affecting winter survival of the young.

The climatic constraints that are dominant in determining the breeding seasons of temperate birds are not as important to many tropical species, especially those found in rainforests. Not surprisingly, studies have suggested longer breeding periods among tropical birds. Nevertheless, many tropical habitats are seasonal due to rainfall fluctuations, such that breeding occurs during times when resources are most favourable.

A special case regarding the control of breeding season occurs in the Greater Antilles of the West Indies. These islands are the winter home for great numbers of small, insect-eating warblers (Parulinae); in most habitats, these winter residents outnumber resident insectivores manyfold. Under normal conditions; resident insectivores of the Greater Antilles do not breed until late May or June, the time when these potential competitors are absent and spring rains have ended the January-April dry season. Should these rains not occur, the Resident species will not breed during May-July, and they also will not breed after the winter residents have returned, even though heavy autumn rains may occur. In the Lesser Antilles, however, where few winter residents occur, resident birds show much longer breeding seasons, often starting in March or April and extending through the summer. Whether or not a similar high density of winter resident affects the breeding season of mainland insectivores is more difficult to determine, although some evidence suggests that this may be the case.

The vast majority of bird species breed annually, but some exceptions do occur. Many large species, such as condors and albatrosses, will not breed in a year following a successful breeding attempt. In most of these cases, young are produced only every other year. An intermediate case between annual and biannual breeding sometimes occurs in the Magnificent Frigatebird (*Fregata magnificens*). In this species it takes about 13 months to fledge an offspring. Although monogamous through most of the nesting period, the male frigatebird may desert its mate when the young is about a year old. At this time, the male then mates with a different female, while the previous female remains with the offspring and fails to reproduce in that year.

A few species seem to live in environments where either no environmental cues affect the breeding season or these cues occur in something less than one year cycles. The Sooty Tern *(Sterna fusca)* on Ascension Island breeds with about a nine-to-ten month periodicity. In many of the tropical seabirds, breeding seasons are not highly synchronized, such that members of a species may be breeding over long, overlapping periods. The use of predator-free nest sites would favour this spread in reproductive effort, as would the competition for food around colony sites.

In addition to variation in the length of the breeding season along the temperate-tropical gradient, one also finds variation in breeding times associated with changes in longitude from the coast to the interior of a continent, with altitude, and with patterns of temperature or rainfall. In many ways, these patterns of climatic predictability affect the evolution of the breeding season in much the same way as they affect clutch size (see above), with stable and/or more moderate conditions allowing longer breeding seasons.

MATING SYSTEMS

Sexual reproduction requires the interaction of a male and a female, but minimally only for the few second that it takes for successful copulation. The fact that over 90% of bird species show monogamy, in which a pair-bond lasts for much if not all of the breeding process, suggests strongly that both members of the pair are needed for successful reproduction in most species. We must remember that natural selection operates on individuals, not pairs, so monogamy must be in the best interests of both the male and the female for both to remain together. Should either sex have an opportunity to increase its fitness by behaviour other than monogamy, this behaviour should be selected for.

That monogamy is necessary for successful reproduction in many species has been shown in a variety of ways. Loss or removal of one of the parents almost always results in loss of some or all of the young, or in loss of weight in the young. Either of these outcomes lowers the potential fitness of both parents. Given that both parents have invested time and energy in pairing and preliminary nesting activities, deserting a mate should occur only if it is adaptive. To measure this adaptiveness, one must know the effect of this desertion on the number of young the remaining parent might produce and the chances the deserting parent has of successfully finding a new mate with which a second breeding could be accomplished. In most cases, the survivorship of nestlings with only one parent or the chances of finding second mates must be too low to select against monogamous mating systems.

Situations do exist, however, in which one parent could successfully raise the young, thereby allowing the other parent to attempt to mate with other individuals. These conditions have been described by *Emlen* and *Oring* (1977) as the *environmental potential for polygamy* (EPP) because they define conditions in which the constraints favouring monogamy break down and a polygamous mating system could result. In almost all cases, these EPPs are related to food, because that is most often the factor that limits reproductive success; however, they can also involve nest sites or display grounds. Because there are two sexes, EPPs occur in several combinations. *Polygyny* occurs when one male monopolizes the mating of several females. If one female mates with several males, it is *polyandry*. Multiple mates may occur at once, which results in simultaneous polygyny or polyandry, or they may occur in succession, which results in serial or sequential polygyny or polyandry. A variety of other combinations are possible if one examines mating behaviour over a longer period of time.

Because the female in producing the eggs invests much more energy in reproduction than the male, it is generally more difficult to find situations where it is adaptive for the male to take care of a female's eggs than it is to find situations favouring a male that can attract multiple females. Therefore, polygyny is the dominant non-monogamous mating system while polyandry is quite rare.

Before we look at the evolution of polygyny, it must be pointed out that not all monogamous matings are as clean-cut as they might seem. Recent work has shown the regular occurrence of what are called *extra-pair copulations* (known as EPCs or sometimes EBCs for *extra-bond copulations*) within what are basically monogamous species. In most cases, these EPCs seem to occur when males that have already mated with a female search for other females with which to copulate, while maintaining a pair-bond with the initial mate. In some cases, this behaviour may result in forced copulations. Obviously, a male whose mate is copulating with other males may end up caring for young he has not fathered, so natural selection should favour behaviours that prevent or reduce the number of such EPCs. In most cases, this is done through territorial behaviour, or mate guarding behaviour when pairs are not territorial. While this suggests that males control the situation, there is evidence in a few species that females that are paired with young males may in fact allow older males to copulate with them. In this way, a young female may receive genetic material from an other, perhaps higher-quality male, while still receiving assistance in nesting duties from a younger male.

The Evolution of Polygyny

Most models explaining the evolution of polygyny begin with monogamous, territorial birds, primarily because this seems to be the dominant, ancestral system in birds. Some special resource condition or behavioural trait that serves as the EPP is then added, which results in a single male being able to mate with several females.

The classic model for the evolution of polygyny is known as the *polygyny threshold model* developed by *Orians, Verner,* and *Willson* in the 1960s. They worked primarily with the Red-winged Blackbird *(Agelaius phoeniceus)*, which has become to mating system studies what the Great Tit is to foraging studies.   To visualize their model, consider a group of monogamous, territorial pairs living in an environment that is not uniform. Some of the males are able to defend territories that are very rich in resources, while other males defend poor-quality territories. As females arrive at the breeding grounds, the first female should recognize the best-quality territory and attempt to make with the male in that territory. The second female should mate with the male in the second best territory, and so forth. At some point along this gradient of diminishing territory quality, in arriving female may reach a point where choosing a male on a low quality territory provides little or no chance for success. On the other hand, if she could mate with one of the males holding the best territories, she might actually be able to raise more young by herself than she could with a male mate on a low quality territory.

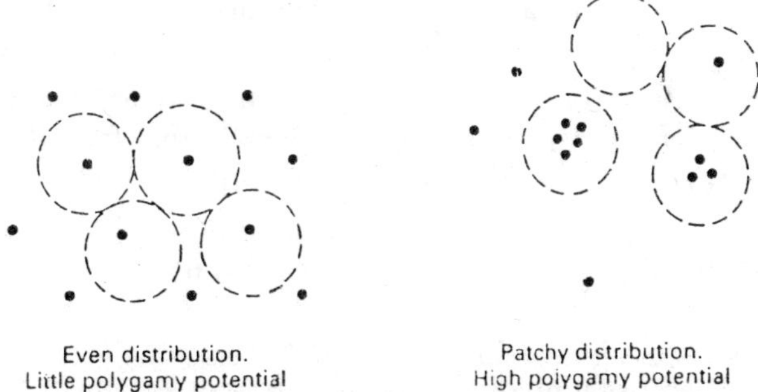

Fig. 2.9.   Schematic diagram suggesting how resource patchiness affects the environmental potential for polygamy. Circles define defended territories while dots denote resources

The point where a female would have greater fitness by raising young by herself on a high quality territory than by being monogamous on a low quality territory is termed the *polygyny threshol.* For polygynous behaviour to evolve from monogamy, behavioural shifts need to occur such that males can mate with multiple females and these females allow other females to live on their mate's territory. Because males already have invested much energy in gaining the best quality territories, the adaptive value of polygyny is easy to see, for a male can potentially increase his reproductive success by several times without greatly increasing his effort. For polygyny to evolve, the behaviours that reduce aggression between female so that they can subdivide territories must be adaptive. Because all females within these high-quality territories produce young successfully, the evolution of these behavioural shifts appears to occur rapidly.

This model for the evolution of polygyny requires, at least initially, a gradient in territory quality, with the best territories rich enough that a single female can successfully raise young. While this seems simple enough, it does not occur very often. Most of the polygynous species occur in marshes, grasslands, or early second-growth vegetation. These are all very productive habitats, and they are structurally very simple. Food is concentrated in just a few feet of vegetation or, in the case of marshes, at the water's surface. This makes it easier for the female to forage effectively. Although mature forests or other habitats may be as productive overall, the fact that this productivity is spread over many vertical feet apparently means that the female cannot forage fast enough to raise young by herself, and hence the EPP is not present.

The above form of polygyny is known as *resource defense polygyny* because it revolves around a male defending a resource with which he can attract multiple females. Another form of resource defense polygyny occurs in some colonial breeders, in which dominant males may control multiple nests or nest sites and thus are able to mate with multiple females. Most colonial species are monogamous, however, presumably because either the male serves an important role in the reproductive effort or the synchrony that is advantageous in colonial breeders results in few unmated females during the peak reproductive period. In a few species, a male is able to defend two separate territories and thereby mate with two different females. For some unknown reason, nearly all of the latter cases of polygyny have been recorded in European species, such as the Pied Flycatcher.

In birds with precocial young, the fact that it is not necessary for the adults to bring food to the young might serve as an EPP, and, indeed, species with precocial young show higher rates of polygyny. Monogamy still occurs, however, in cases where both parents may be needed to watch for predators, when the male may be needed to defend a territory for its mate and young, or when both are needed to ensure adequate incubation of the eggs. The latter case is particularly important among species nesting in the Far North, such as water-fowl and sandpipers. In many geese that are monogamous for life, the female lays the eggs, then feeds for several days while the male incubates. Once the female has recovered from the energetic stress of laying the clutch, both birds alternate in incubation, which keeps the eggs warm while allowing both to feed. With only a single parent, either that bird would starve or the eggs would chill and not develop properly or they would die. Both birds stay with the young throughout development. In contrast, most temperate or Arctic ducks are monogamous through the egg laying period, presumably because the male of the pair is needed for fertilization of the eggs or he needs to ensure his paternity of the young, and perhaps, to defend a bit of a territory for the female and young.   In most ducks, however, once the clutch is laid the male leaves the breeding grounds.

For many species of birds with precocial young, the above circumstances do not occur and polygyny develops. In some cases this may be resource defense polygyny, in which males defending the most resource-rich areas mate with the most females (also termed *territorial harem polygyny*). In many cases, however, because of the independence of the female and the young, any territorial constraints associated with this mating system have broken down. In such cases, it would be optimal for a female to mate with the highest-quality male available, then move to the best nesting habitat available. Each male would have the chance to mate with as many females as possible (rather than the maximum number he could fit on his territory). The result here is termed *male dominance polygyny*, because the most dominant males usually mate with the most females. Some have termed such a system *promiscuity*, because the interaction between male and female is so brief, but studies have shown that females are doing a great deal of selection with regard to male characteristics, such that the laxity implied by promiscuity is not present.

Species with male dominance polygyny can be arranged spatially in a variety of ways. In some cases, particularly in fairly complex habitats where males are somewhat protected from predators, the males may be scattered throughout the area. In these cases, a female must wander about, evaluating males, until she makes a choice. In many cases, however, males congregate in display groups known as *leks*. Leks may occur for several reasons. When displaying many, males either adopt conspicuous colours or behaviours, which makes them more susceptible to predation, particularly when these displays occur in open environments. By displaying in groups, these males have more eyes to spot predators, much as foraging flocks do. Grouping may also occur to attract females with the idea that a female who is going to choose among a group of males would be more likely to be attracted by an already assembled group than to wander about evaluating males singly. This would allow the female the chance to judge males while they were together and, in fact, to see the dominance hierarchies that male-male interactions have developed. While the adaptive advantage for one of the dominant males in a group seems to be obvious, it is less understandable why a subordinate bird would remain in a lek. The explanation seems to revolve around the fact that larger leks attract more females, such that the long-term chance of success for a subordinant male may be better in a larger lek.

In most leks, males actually defend tiny display territories. Following the system used earlier these may be considered *Type D territories*. Generally, the most dominant birds have territories within the center of the lek, with subordinant birds on the periphery. Females usually choose the dominants, often a few of the males in a lek participate in nearly all of the copulations. Young males seem to work their way toward the middle of the lek with age.

In the various polygynous systems, each individual should adopt a strategy appropriate for its social status and the resource conditions involved. In general, all females in male dominance polygyny will get the chance to mate with dominant males, then go off and raise the young by themselves. In resource defense polygyny, a female must evaluate options depending upon the number of females mated with a dominant male and the quality of the territory of an unmated male. As more females mate with the dominant males, the situation may develop in which a newly arriving female would be better off in a poor-quality territory than in an overcrowded polygynous territory. Male strategies also may vary, particularly with age and dominance status. In resource defense systems, many young males do not defend territories, as they can acquire only poor territories with little chance of attracting females. The best strategy seems to be to wait a year before even trying (often keeping a female like plumage in the meantime). In male dominance systems, young males may join a lek to establish a position in the dominance order. This decision, though, should depend on the advantages of joining a large lek (which may attract more females) versus the disadvantages (it may take longer to become dominant). In some cases, a better option might be to join a smaller lek that may be visited by fewer females but where breeding status may be attained earlier in life.

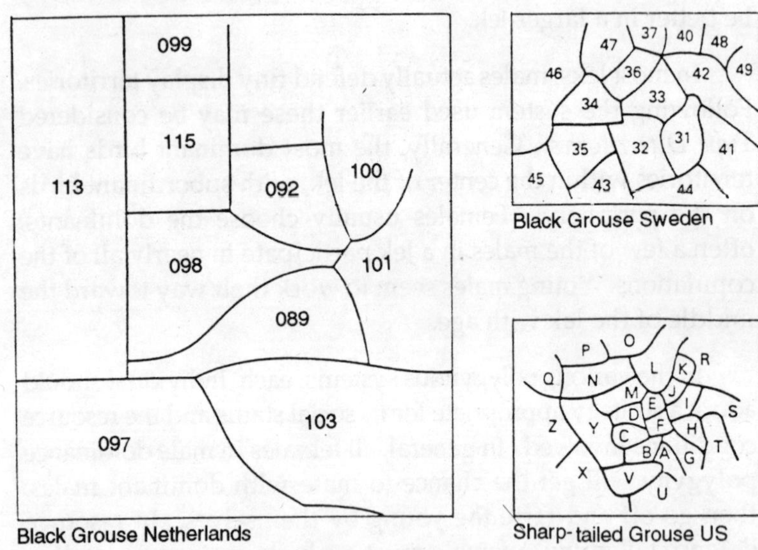

Fig. 2.10. Maps of territories on leks in different grouse populations. All maps are to the same scale; territory owners are identified with numbers or capitals.

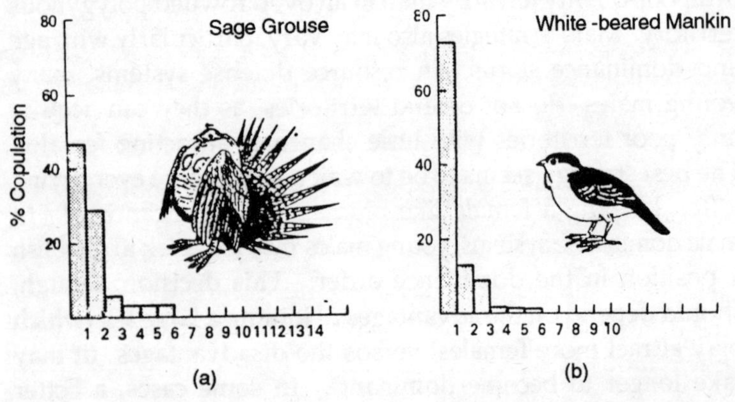

Fig. 2.11. Proportions of total copulations accomplished by males in order of dominance rank in the Sage Grouse (*Centrocercus urophasianus*) and the White-bearded Manakin (*Manacus manacus*). The most dominant male is number 1. Note how a few males dominate the copulations in these species.

Male dominance polygyny is not confined solely to birds with precocial young. Some tropical frugivores are involved in mutualistic interactions with fruits that are more nutritious than most fruits. In these cases, the female may be able to collect enough food to feed the young; these species also have the small clutch typical or tropical birds. Among the New World species with male dominance polygyny are manakins and cotingas, while the Old World is known for its birds of paradise and bowerbirds. Whereas these New World forms are generally small, the Old World forms are large, perhaps because they are able to use large foods made available by the absence of monkeys in their Australian and New Guinea homes. Most nectarivores, such as hummingbirds and sunbirds, are polygynous, presumably due to the fairly easy availability of high-quality foods in the flowers on which they feed.

Sexual Selection in Polygynous Systems

It was pointed out how male-male competition among monogamous birds might favour brightly coloured signals or unusual displays as a part of intrasexual interactions. The intensity of selection for these markers or displays is limited, however, by the amount of male parental care, as such markers might be maladaptive if they attract predators to the nest.

With the crossing of the polygyny threshold, male-male competition becomes more pronounced due to the greater possible rewards of such competition, and also due to the fact that male roles at the nest often are reduced or absent. In a resource defense polygynous system, a dominant male may be able to increase his fitness many times over that of a subordinant male. If extravagant colours or displays aid in achieving this dominance, they will be quickly passed on to the dominant bird's many offspring. Among male dominance polygynous systems, a dominant bird may be able to achieve an almost unlimited number of matings, which further selects for any advantageous colours or displays. With the reduction of nest-related behaviours acting as a control on the expression of these sexual traits, the sky almost becomes the limit. Predation tends to serve as the chief factor limiting sexual dimorphism, although in many cases this does not seem to have had much effect. Among the birds of paradise, incredibly bizarre plumages and behaviour occur, including some males that hang upside down showing their brightly coloured bellies. Some of the New World manakins combine bright colours with group displays, where two or three males move in synchrony to attract a female. When examining a bird book, one can safely predict that those species with really bizarre male plumages or mating rituals are most likely polygynous. The greater the difference betweenmale and female, the more likely it is that a non-monogamous mating system is involved. Because male-male competition often involves fighting, polygynous males are also often much larger than their female mates.

The costs of such gaudy plumages are paid in higher predation rates. Many open country species with polygynous systems, such as prairie grouse, cannot afford to be bright red all the time, so they trade bright colours in the feathers for brightly coloured, inflatable air sacs and spectacular display dances. In this way, when they go off to feed they can look like a typical cryptic grassland bird, yet when they are competing for females they can be quite attractive. Forest birds are apparently somewhat less vulnerable to intensive predation pressure, as they have bright colours more often.

An interesting compromise between the strong selective pressures favouring sexual displays among the males and the costs of predation when males are brightly coloured may have been struck in some of the bowerbirds. These tropical frugivores of Australia and New Guinea include many species with male dominance polygyny. Some species have the more typical polygynous situation, with ornately coloured males that display to attract females. Some of these modify their display locations by building structures known as *bowers*, which are an aid in the display. Other bowerbirds are quite dull-coloured, but they build extravagant bowers that are highly decorated with coloured leaves, fruits, or any other coloured material available. In these cases, it appears that the bower has taken over the role of gaudy plumages as a signal of male quality. This still allows females to evaluate and choose among the males, but it reduces predation rates on the males when they are not involved in sexual displays. Recent work with some of these bowerbirds has shown that different populations seem to favour displays dominated by different colours, perhaps as a result of cultural preferences. These preferences were tested with some ingenious experiments by *Diamond* (1982), who used coloured poker chips as an experimental tool. In addition to finding out the colour preferences of the birds, he also found that male bowerbirds spend a lot of time stealing coloured objects from one another's bowers.

Fig. 2.12. A female Buff (*center*) observing the display of two males with coloured collars and a white-collared Buff (*lower right*).

Because young males often do not have enough status to achieve successful matings in polygynous systems, in many cases they also avoid paying the costs of the male plumage by not acquiring it until later in life. This is known as *delayed maturation* and it occurs in many polygynous species. In some cases, young males may use their femalelike appearances to sneak in occasional copulations while they are visiting male territories. Because they look like females, territories males may not exclude them, giving them the chance to mate. These sneaky strategies are generally considered to be secondary strategies that develop after the adaptations associated with the primary strategy (for example, a male evolves a femalelike appearance due to the pressures of polygyny, then is able to evolve behaviors that aid it in sneaking copulations).

Perhaps the most striking case of a sneaky strategy occurs in the Ruff (*Philomachus pugnax*), a European shorebird known for the amount of individual variation in the coloured plumage of the display collars of dominant males. These coloured Ruff males defend small territories within leks, as do most lekking species, but there also exist Ruff males. These white-collared Ruffs behave subordinantly to the coloured Ruffs, which allows them to remain on coloured Ruff territories. When a female arrives at a lek, the coloured Ruffs are sometimes so busy displaying and defending their territories that the white Ruff male is able to copulate with the female.

Although the evolution of polygyny from monogamy may require some behavioural adaptations, recent research suggests that members of traditional pairs, and particularly males are much more flexible in their behaviour than previously thought. Given the opportunity, males will readily participate in extra-pair copulations or accept a second mate on a territory. Polygynous matings have been recorded at low frequencies in over 60 North American species that are generally considered monogamous. Many species may already possess the behavioural attributes needed for polygyny, but the EPP for most may not exist, either in terms of habitat quality or the regular availability of second mates.

Avian Polyandry

The tremendous investment of the female in the production of eggs usually means that it is the female who has the most to lose by leaving a monogamous pair. As a result, polygyny is the dominant non-monogamous mating system. Only two to three dozen species of birds in the world show mating systems that can be classified as polyandry. Most of these species are among the sandpipers and have evolved this behaviuor because of traits peculiar to this group and their northern nesting habitats.

Most of the sandpipers (*Scolopacidae*) breed in the high Arctic, where the tundra provides large amounts of readily accessible insects for the precocial young. This habitat has disadvantages in that it is fairly cold at all times, the breeding seasons are short, and it is a long distance from the wintering grounds of most sandpipers. Most sandpipers are monogamous, pairing during migration or on the breeding grounds. After this pairing occurs, the female lays the clutch of three or four relatively large eggs. At this point her energy reserves have been depleted, both from the long migration and from egg laying. This depletion is accentuated by the small size of shorebirds, which limits the amount of energy they can bring with them. (In contrast, many large waterfowl are known to accumulate energy reserves on the wintering grounds that provide energy for egg production.) Because the female needs to feed heavily, at this point the male begins almost continuous incubation of the clutch. Generally, he will incubate by himself for several days, until the female has replenished her reserves and can take her turn with incubation while her mate feeds. Because of these energetic constraints, and perhaps also because of the need to keep a pair together in case the clutch needs to be replaced due to predation, in most shorebirds the pair remains together throughout the breeding process.

This system results in two behaviours that serve as preadaptations for polyandry. First, males develop a strong attachment to a clutch of eggs, with a large amount of time and energy invested in the clutch during the egg laying and early incubation periods. On the other hand, females are selected for an ability to lay a set of eggs, then rapidly recover the necessary energy reserves to lay a replacement set, should that be necessary. Because of the male's commitment to incubation, the female does much of the foraging on her own during this period. Given these behavioural adaptations, all that is needed are the proper environmental conditions to serve as an EPP favouring polyandry. Not surprisingly, the EPP can be provided by unusually rich resource levels, which often occur in the Arctic due to early warm weather that increases the number of insects.

With the behavioural system outlined above and extremely rich resource conditions several options are available to the female. After pairing and laying the first clutch, she may be able to replenish her energy reserves rapidly enough to lay a second clutch. Because food is plentiful, both parents can incubate a clutch and find enough food to survive during brief forays away from the nest. This "double-clutching" doubles the reproductive potential of both sexes but retains a monogamous system.

Once females have adapted to laying clutches in fairly rapid succession, the presence of non-paired males may lead to a breakup of the monogamous pair bond. All mate then must occur is that a female ready to lay a second clutch be courted by an un-mated male with a territory. Under these circumstances, she can lay a clutch for this mate, then perhaps lay yet another for herself, or just have time to recover from the energetic investment. In this manner she can double or triple her reproductive output, while the male does not desert the eggs because of his investment in them. The result is what is known as *sequential* or *serial polyantry*, where a female may lay eggs for series of males, while each male may mate with only one female during a breeding season. Because the sex ratios for most species are usually approximately even, one might question whether or not extra, non-mated males actually occur. They may as a result of nest predation rates; if clutches are lost to predation faster than females can replace them, then females ready to lay may outnumber males with territories. As a result, males must compete for these limited females. Thus, even if an equal number of males and females occupy the breeding grounds, the number of females capable of breeding at a moment may outnumber the number of territorial males, thereby driving the polyandrous system. (The sex ratio of birds actually able to breed is often defined as the *operational sex ratio.*)

In a few species of shorebirds that nest in more southerly regions, a more confusing set of behaviours seems to result from these interactions. In these cases, a male may attract a female and get her to lay a set of eggs, but he will then try to mate with other females and get them to incubate the second clutches. He may then return to take care of the first clutch. As a result, some males may mate with more than one female, and some females may lay for more than one male a system known as *rapid multiple clutch polygamy.*

In a few species of shorebirds and the tropical jacanas, a slightly different system occurs where females defend a territory and several males sub-divide this territory and maintain continual pair-bonds with the female. This is known as *simulaneous polyandry.* It seems to have evolved from sequential polyandry in areas (such as southerly temperate breeding areas) where the breeding season is long enough or the predation rates high enough that there are advantages for a female to maintain some interaction with a male for which she has laid eggs. This may be achieved through territorial behaviour, and presumably increases the female's chances of laying any replacement clutches needed while also reducing the time needed for pairing. Shorebirds such as the Spotted Sandpiper (*Actitis maculariu*) that are sequentially polyandrous in the northern parts of their breeding range may show simultaneous polyandry in the more southern parts. In the tropical marsh-dwelling jacanas, this system is generally more stable, although migratory jacanas near their northern breeding limits may show sequential polyandry.

If the males in simultaneously polyandrous group receive their clutches from the female in succession, during which time each male monopolizes copulations with the female, the system is genetically very similar to sequential polyandry. There is evidence from the jacanas, however, that females copulate with several of the males within her territory on the same day and lay eggs in the nests of different males on subsequent days. In this case, a male may be raising a set of young with mixed paternity, although whether or not a particular male gains or looses by such a system depends on his overall participation in copulations. If all males copulate with the same frequency, each nest contains a similar mix of eggs in terms of paternity. Given that jacana nests seem to be subject to high nest predation rates, this mixing of paternity among nests may be a safer strategy for a male jacana, as the chances that at least a few of his offspring survive from one of the nests may be higher than the chances that a nest with all his offspring might succeed. The fact that these polyandrous system involve a nest for each male-female bond, even though the eggs may have mixed paternity, distinguishes them from other forms of polyandry to be discussed below.

The changes in the reproductive roles of males and females that occur in polyandrous systems such as those outlined above also affect patterns of sexual dimorphism. Not surprisingly, polyandrous females often are larger than males, presumably because they can be more efficient egg-laying machines when larger, but also because size may be important in female-female contests for males. In some cases, these female-female interactions or female roles in attracting males result in females that are brightly coloured, while males,

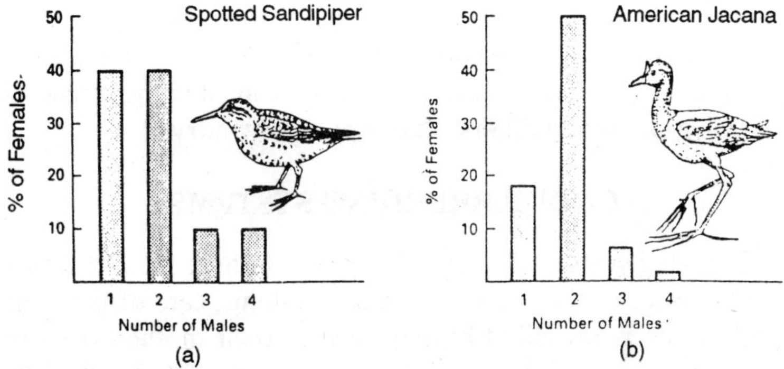

Fig. 2.13.   The distribution of number of male mates in two polyandrous species, the Spotted Sandpiper (*Tringa macularia*) and the American Jacana (*Jacana spinosa*).

relegated to incubation and other normally maternal duties, are dull coloured. While the egg-laying duties of the female seem to work against the development of bright colouration in most species, in the phalaropes (Phalaropodinae), painted snipe (Rostratulidae), and a few other species a phenomenon known as reverse sexual dimorphism occurs (although the term in some ways makes no sense). In these, the female is brightly coloured and the male dull, although all of the phalaropes are not known to be polyandrous. It is not surprising that female plumages may be somewhat more limited in polyandry than are those of males in polygyny if we again consider the relative productive costs of males and females. Even polyandrous females are limited to just as few males by the number of eggs they can lay; they can never reach the reproductive potential attained by some polygynous males, as noted above. This should then limit the expression of colourful female plumages.

The shorebirds are an interesting group for the study of mating systems because they show the range of variation from monogamy to polygyny on one hand, and to sequential or simultaneous polyandry on the other. Whereas polygamous mating systems seem to be connected either to habitats with predictably rich resource conditions or to species that move about in search of these conditions, more work will be required to understand why some species have evolved in the direction of polygyny while others developed polyandry.

GROUP BREEDING SYSTEMS

The mating systems described above include about 97% of the bird species of the world. In those systems, care of the eggs and young is provided by their father, their mother, or both (ignoring any sort of extra-pair copulations). Nearly all of the remaining 3% of bird species are characterized by reproductive systems in which a clutch of eggs is cared for by a group of birds that numbers more than two and which may consist of either multiple male or female parents, or birds that are not active breeders but that serve as nest helpers. These often rather complex breeding systems are often classified as communal or co-operative breeding systems, but they may include co-operative polygyny (where one male and several females share a nest and all have genetic investments in at least some of the eggs), co-operative polyandry (where one female and several males share a nest but all the males have a chance of fertilizing the female), or some form of co-operative monogamy. In a few species such as anis (*Crotophaga*), this co-operative monogamy involves groups of monogamous pairs that share the same nest (although they also often compete in trying to contribute the most eggs to it). More commonly, this co-operative monogamy includes a monogamous breeding pair and a group of non-breeding helpers in what are known as *helping* or *co-operative breeding systems*. In many cases, studies have not really revealed which birds are copulating among these group breeding birds, such that the actual categorization is often based on observation of simple behavioural interactions rather than real information on matings. We shall divide our discussion of these group breeders into two categories, one comprised of a few species that breed in groups that are rather temporary and not confined to territories, and the other representing the majority of group breeders, in which groups are often long-term, attached to territories, and highly co-operative.

Non-territorial Group Breeders

Non-territorial group breeding occurs in several forms among the ratites and tinamous. In some cases, groups consisting of one male and several females form, with all of the females contributing eggs to a nest over a period of several days. At this time, the systems is a form of polygyny, but once the clutch is complete, the females moves on to mate with another male while the male incubates the eggs and raises the young. This development represents harem defense polygyny for the male but sequential polyandry for the female, a system common in rheas and tinamous.

In contrast, an ostrich male mates with a dominant (termed *major*) female who initiates eggs laying and stays with the male throughout the nesting cycle. After the major female begins laying, other females also contribute to the nest. Because too many eggs are laid for the pair to incubate, many are rolled away by the major female. She apparently can identity her own eggs, thereby reducing potential losses to her own and her mate's fitness. The possible selective advantage for the breeding pair for the presence of these extra eggs may be in the satiation of nest predators, who can usually eat only a few of the giant ostrich eggs with each attack. The secondary females apparently lay the eggs despite their low chance of success because they constitute the only possibility of breeding for low dominance, probably young, females.

Breeding Groups on Territories

Breeding groups that maintain territories represent the other option in co-operative breeding systems. While this type includes a diversity of mating behaviour, all tend to share a few common characteristics that result in the maintenance of breeding groups.

Nearly all group breeding species occur in situations where their populations have saturated their environment, such that some limiting factor, often simply territorial space, is hard to find. Group-breeding species are more common in tropical or subtropical habitats, apparently because these are more likely to contain populations near carrying capacity. Long-lived species are also more likely to show group breeding, as are species with low reproductive rates, delayed maturity, and low dispersal. All of these are traits of populations that remain at or very near the saturation level of their environments.

While breeding territories seem to be the most common limiting factor that leads to group formation, more specific requirements may also act as a factor. Acorn Woodpeckers (*Melanerpes formicivorus*) seem to be limited by the larder trees that they make and maintain (Stacey and Koening 1984), while Green Wood-hoopoes (*Phveniculus purpureus*) are apparently limited by roosting cavities (Ligon and Ligon 1982). In some species it appears that the amount of actual territorial space may not be as important as the availability of marginal habitats in which nonbreeding birds can survive when breeding territories are full. The structure of the habitat may also be important, as birds that can readily hide within their habitat types may not have such strong selective pressures towards joining groups as those living in open environments (see below).

Many bird populations are large enough that more potential breeding individuals exist than breeding territories. Numerous removal experiments in different habitats suggest that large numbers of nonterritorial birds are wandering about looking for mates and territories. These are termed *floaters by some*, because of their drifting behavior, while others refer to them as the *subterranean* component of he population because they must sneak around avoiding territorial birds. These birds are usually younger, less dominant members of the population. After a year or two of floating, they normally develop enough status within the population to be able to acquire a territory and a male.

Species that show territory-based group breeding seem to be distinctive because the option of floating until sufficient dominance is achieved is not available to them. For this to be the case, there must be a lack of marginal habitats, those unsuitable for breeding territories, where these birds can live, and it must also be difficult for them to sneak around within other birds' territories. At its simplest, it appears that some young birds are faced with the option of dispersing to non-existent habitats, which implies a high risk of mortality, or staying out, which requires some sort of alliance with the existing territory holders if the bird is to conspicuous to sneak around within the territory.

This relationship between the lack of marginal habitat and group breeding is best exemplified by the classic study on group breeding, that done by Woolfenden and Fitzpatrick (1984) on the Florida subspecies of the Scrub Jay (*Aphelocoma coerulescens*). Florida Scrub Jays are confined to sandy habitats with scrub vegetation in Florida, where all available habitat is incorporated within territories. A young jay apparently can disperse with only a low chance of success. Instead, group breeding has evolved. Scrub Jays are common throughout the western United States, however, where they live in a variety of habitats. Young Scrub Jays in this region can live in marginal habitats until old enough to breed, and breeding groups have never been recorded. The population of Scrub Jays on Santa Cruz Island off the coast of California also never has breeding groups, apparently because breeding territories do not occur in all habitat types, so that juvenile birds have habitats where they can survive until they achieve enough dominance to acquire a territory.

Monogamous Breeders with Helpers. While the above

conditions are apparently necessary if it is to be adaptive for young birds to remain within breeding territories, they are not the only requirements for the development of some form of group breeding. Rather, the costs of keeping additional individuals on a pair's breeding territory must be less than the benefits to the breeding birds, while the young birds must get more out of joining a group than they would by living alone. Since most young birds join groups in which they do not breed, but instead serve as "helpers" to the breeding pair, they have effectively evolved a strategy of not breeding, at least for some time. Let us examine how a breeding pair and its helpers justify this apparent alliance, focusing only on those species where a monogamous breeding pair is assisted by nonbreeders (the classic case of "helpers at the nest" or cooperative breeding).

For group breeding to be adaptive to a monogamous pair with helpers, these helpers must actually contribute to the increased fitness of the breeding pair over their reproductive life-spans. If pairs with helpers produce fewer young than pairs without helpers, the genes favouring acceptance of helpers should rapidly disappear from the population. How this benefit occurs, though, may vary greatly from species to species. Because the birds that help are most often the offspring of the breeding birds, the parents may be increasing their own fitness by enhancing the survival prospects of their helper offspring long enough to allow them to establish dominance so that they can acquire territories of their own. Secondly, the helpers may actually assist in a variety of nest-related activities, including territory defense, nest building, nest defense, feeding the young, and repelling potential predators from the nest and young. The effect of this assistance may be to increase the chances of success for the parents, and a number of studies have shown that breeding groups fledge more young than breeding pairs. Depending on the activities in which helpers participate, their benefit may occur late in the breeding cycle, as studies in some species have suggested that the presence of helpers resulted in higher survivorship of fledglings, even when the fledging success of helper and non-helper nests was the same.

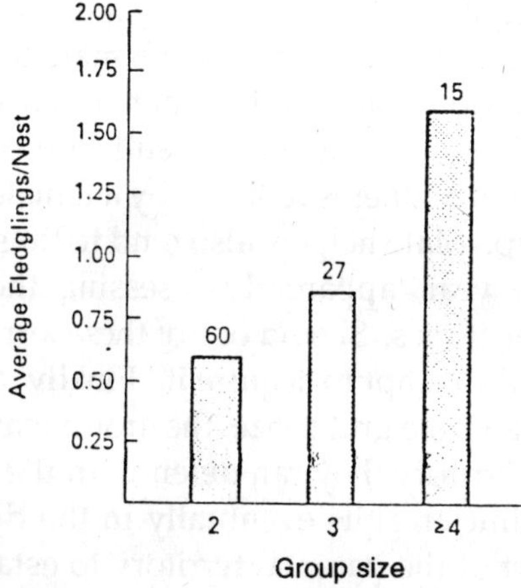

Fig. 2.14.   The relationship between nesting success and group size in the Green Woodhouse (*Phoeniculus purpureus*). Note that pairs with two or more helpers produce many more young than pairs.

What does a helper get by expending effort in helping even though it is not the genetic parent of the young? First of all, it remains alive; in most helping species the environmental constraints are such that a young bird faces a help-or-die choice. While helping, the bird gains experience in a variety of breeding activities which may increase its chances of nesting successfully if given the chance later in life. Because the young individuals aided by most helpers are their genetic siblings (brothers and sisters), which means that they share many genes in common, the helper may be gaining in fitness through what is known as *kin selection* (the increase in an individual's fitness through shared genes of relatives). A great debate raged over the role of kin selection in the evolution of helping behaviour. While its role cannot be discounted, in many cases it has been shown that helping would be and is adaptive even if the helping is done with unrelated breeding birds. While the helper is performing all of the activities described above, it is improving its status, which means that it is slowly getting itself into the position to obtain a territory of its own.

Territory acquisition by dominant helpers occurs in several ways. In some cases, the death of the male territory holder means that the dominant male helper (often his son) takes over. Because such mating could result in incest, when this occurs the helper's mother is replaced by a female from outside the family group. Male helpers also tend to do some roaming throughout the year, apparently assessing the situation in neighbouring territories. Should one of these territories become vacant, they will attempt to acquire it. Finally, as the number of helpers increases the group size, the group may also increase the size of the territory they can defend. In the Florida Scrub Jay, this sometimes results eventually in the dominant male taking over part of the parental territory to establish his own territory.

Although nearly all the evidence shows that helpers do help in some manner, individual strategies should be expected within this general mold. Work with the Scrub Jays suggests that first-year birds do relatively little helping, presumably because they have little chance of acquiring a territory anyway so it is adaptive to just lay low and take it easy. To avoid matings between parent and young, one of the sexes must disperse from the territory. As this is usually the female, helping systems often have more male than female helpers. While there are advantages to helping your parents, if they already have many helpers it might be advantageous for a young bird to help unrelated birds, as in this way the helper would be able to rise to a more dominant position within at least that territory.

The final reward for a bird that serves a stint as helper is to acquire a territory, breed, and then have helpers of its own. This suggests that, under such saturated conditions, it is adaptive to trade reduced nesting success early in life for a higher chance of breeding and attaining high reproductive success later in life. The bottom line in the evolution of all species is lifetime reproductive success (fitness), but with most species we tend to focus on annual production. species with helpers at the nest force us to look at lifetime strategies that are effective under such severe population constraints.

There is great variation with co-operative breeders in terms of the duties the helpers perform. In some species, helpers do everything the parents do but mate, while in others they are relegated only to defense of the nest or territory. An interesting situation that may serve as an evolutionary pathway to helping behaviour was found in the Green Jay (*Cyanocorax yncas*) in South Texas. Here, young birds are allowed to stay in the parental territory for over a year, where they assist in territorial defense but do not aid with nest-related behaviours. When a new set of young fledge, the year old young are then chased from the territory.

Other Group Breeding Systems. Of the 3% of bird species that regularly breed in some form of group, most exhibit monogamy with helpers. In a few species, a group of males and females share a nest or nests and apparently promiscuous matings occur. In several flocking species, breeding is done by a group consisting of one or more monogamous pairs plus the helpers, who may help at several nests. Anis (*Crotophaga*) are unusual in that several monogamous pairs share a single nest. All females may lay in the nest and all help raise the young, but the actual reproductive success of various pairs is sometimes uneven because the females throw eggs out of the nest. Because females do this only before they lay, and the older females often lay later than younger females, a skew in reproductive success often favours older pairs. In four or five species breeding groups are composed of one female and several males that share a single nest. This is known as *co-operative polyandry* because it involves both polyandry and helping, depending on who fathers the young and how many offspring there are. A final special case occurs in the Acorn Woodpecker (*Melanerpes formicivorus*), which breed in groups associated with their larder trees. These groups consist of a core of breeders plus helpers. This core of breeders may be a monogamous pair, one male and two females (sisters), or two males (brothers) and one female. This has been labeled *co-operative polygamy* or *polygynandry*. It functions in much the manner of typical monogamous co-operative breeding systems, although the larder trees seem to take the place of territories as the limiting environmental factor that favours group formation. It also differs from monogamous systems because a member of the breeding group that is lost to mortality is often replaced by a pair of brothers or sisters thereby resulting in these polygamous breeding groups. Although one might expect siblings to be highly co-operative because of the kinship factor, competition between females has been observed in the form of egg-throwing behaviour.

These non-monogamous group breeding systems seem to share the characteristic of spatial constraints found in monogamous species with helpers. The existence of unusual mating habits is harder to explain. In some cases, we simply may not understand the breeding relationships involved. For example, Harris' Hawks (*Parabuteo unicinctus*) breed in groups that were formerly considered to be co-operatively polyandrous. Recent work utilizing electrophoretic paternity testing, however, has shown that only one of the males in the group actually mates with the female, so the system is monogamy with helpers rather than polyandry. Further studies may reveal more examples like this. This Harris' Hawk is also unusual because recent work suggests that group foraging for rabbits may be a strong factor favouring group formation, rather than some limitation of habitat.

Where some form of group-related polygamy does occur, it may be related both to the severity of the constraints resulting in group breeding and the relatedness of group members. Among species with cooperative polyandry, the Galapagos Hawk (*Buteo galapagoensis;* may present the most extreme set of constraints in both these regards. This hawk breeds in groups of up to four males and one female that share a territory and nest; all males copulate with the female and co-operate in raising the young. These groups may produce more young on the average than monogamous pairs, but on a per male basis polyandrous lag far behind monogamous males in the production of young. Further evidence that group breeding has its limitations comes from the observation that once a group forms, no new males are added to it with the death of a group member. Rather the group declines in size until monogamy results.

While this suggests that polyandry is not the best of options for a male Galapagos Hawk, it has its rewards. Territorial birds survive at a much higher rate (90% annually) than birds off of territories (50 or less). A male that can join a group and acquire a territory early in life may have to live with lower annual reproductive success, but the increased survivorship must more than compensate for this during his lifetime. The alternative if all males are monogamous is for the average male to wait for a long time before gaining access to a breeding territory, if he can live long enough. On some of the islands on which this species lives, virtually all the island is incorporated into breeding territories, which makes life for non-breeders very difficult.

Why do male Galapagos Hawks share equal status in terms of reproduction, rather than having a dominant male with helpers? It appears that juvenile hawks are poor helpers for their parents, primarily because it may take a long time for a hawk to become an efficient forager. As a result, young are expelled from the natal territories at three to four months of age. Therefore, it is difficult for young males to form alliances with related individuals. If male groups are composed of unrelated individuals, the costs to a male of a system that would make him the subordinant bird may be too great for him to co-operate with the other males. In a system with ranked dominance, the sub-ordinant individual might act as a helper for many years until he was the sole surviving male. At this time, he could breed, but without the aid of helpers of his own. These costs would appear to be prohibitive. Instead, birds will co-operate in territory acquisition and other behaviour only if they have an equal chance at copulating with the female.

The Tasmanian Native Hen *(Porphyrrula mortierii)* is a moorhen that also breeds regularly in polyandrous groups. In this case, these groups are most often trios composed of two brothers and a female, where one of the brothers is usually the dominant breeder while the other mostly helps. This system is easier to understand because the helping brother receives the benefits of kin selection, and the chances are good that he will be able to breed by himself in a subsequent year. Without the benefits of kinship and with a long life-span, Galapagos Hawks apparently cannot afford to only serve as helpers.

Although it may seem that we have covered every possible option available to birds for mating, this is not the case. A variety of other strategies have been recorded. Several gull species regularly show female-female pairs. These apparently are able to produce viable young through copulations with paired males within the colony. They may occur because of a shortage of males within a large colonies.

Most species show at most two mating systems, usually monogamy plus a form of helping or polygamy. The Dunnock *(Prunella modularis)* is unusual because of the great variety of mating systems it exhibits *(Houston* and *Davies 1985)*. This small brown sparrowlike bird of northern Europe is distinctive because individuals of both sexes establish personal territories. They way in which these territories overlap greatly affects the type of mating system. When male and female territories are similar in size and location, monogamy results. When a male is able to defend a territory rich enough in quality that two or more females can establish their territories within his, polygyny results. In cases when habitat quality is low, females establish large territories that may include the territories of two males. In some cases, both these males mate with the female, who lays only one clutch of eggs. While this looks like co-operative polyandry, the males actually compete in a variety of ways that suggest an uncooperative polyandrous bonding.

Studies of mating and helping behaviour have provided a great deal of exciting information on avian reproductive strategies. With the recent use of electrophoretic techniques to determine parentage, we should be better able to define the systems and see the individual rewards available to different group members. In addition, many of these species require long-term studies to elucidate the actual breeding system involved. Many studies are underway and should provide much interesting information in the future.

BROOD PARASITISM

We have shown that reproduction is both a lot of work, given the energy required to manufacture and incubate eggs and to raise young, and a gamble, since at any moment all this effort may be lost to predation. The presence of an eggs stage when parent birds are strongly attached to relatively inanimate objects provides the opportunity for a reproductive strategy generally unavailable in other animals, one known as *brood parasitism*. In this system, one species, the parasite, lays its eggs in the nests of another species, the host, which provides incubation and parental care to the foster young. In birds with altricial young that require much care, the cost to the host of raising a parasite can be high (as are the benefits to the parasitic offspring). In birds with precocial young that care for themselves in many ways, both the costs and benefits of brood parasitism are reduced. The system occurs only in birds because in animals that do not provide care for their young (such as lizards), such parasitism is of little benefit, while in mammals the close ties from birth between a mother and her offspring make it nearly impossible for another female to sneak her young into the litter.

A brood parasite gains the possible advantages of having its young raised by other birds, which reduces the energy required for reproduction, and it is able to spread its clutch among a number of nests, which increases the chance that at least some young will avoid nest predation and survive. Given the latter factor, it is not surprising that brood parasitism is most common in tropical regions. Of course, raising parasitic offspring is not advantageous for the host birds, so evolutionary responses have often developed to reduce the frequency of such parasitism. Additionally, a parasite cannot put eggs in the nests of all species due to differences in egg size, in the begging behaviour of the nestlings, and in the diet of the hosts. It appears that the diet of the host must include insects, as brood parasities that in the nests of frugivorous birds are rarely successful.

About 1% of birds species are brood parasities that collectively parasitize a large number of other species. The largest group of parasities is among the cuckoos (*Cuculidae*), in which about half of 130 species are brood parasities. Nearly all of the parasitic cuckoos are found in the Old World. Five cowbirds (*Icterinae*) are parasitic, along with the honeyguides (*Indicatoridae*), two genera of finches from the family *Polceidae* and a duck *species*. The duck, the finches, and one of the cowbirds (see below) do not seem as costly to their hosts as the cuckoos and most of the cowbirds. For example, the Black-headed Duck (*Heteronetta atricapilla*) can raise itself once it has reached a day or two of age, and thus has little affect on its host's reproductive success. There is some evidence that nests of the estrildid finches that are parasitized by viduine finches may raise more total young because of the stimulus provided by the viduines (Morel 1973). Cuckoos and cowbirds are generally more detrimental to the nesting success of the host; these parasities show both generalized and very specialized host selection behaviours. The cowbird *Molothrus rufoaxillaris* of tropical America parasitizes only the cowbird *Molothrus badius*, while the temperate *Molothrus* alter is known to parasitize over 100 species.

Parasitism, particularly from cowbirds and cuckoos, generally reduces the nest success of the hosts. In some cases, the existence of a single parasitic egg results in a total loss of the host's young (see below), while in others it may cause only partial loss.   Because parasitism is usually infrequent in populations [a sweeping survey by *Lack* (1963) found that about 3% of nests in England were parasitized], population dynamics are usually not affected. When parasites come into contact with small populations that have not previously encountered parasitism, however, host populations can suffer. Noteworthy examples are the Kirtland's Warbler (*Dendroica kirtlandii*) of Michigan, which did not have to deal with brood parasitism until agricultural land-use practices resulted in an expansion of the range of the Brown-headed Cowbird (*Molothrus ater*), and the Yellow-shouldered Blackbird (*Agelaius xanthomus*) of Puerto Rico, which has recently had to deal with parasitism from the the Shiny Cowbird (*Molothrus bonariensis*). Both species are now considered threatened, in part because of the effects of cowbird parasitism, although other factors have made them susceptible to this problem.

Of course, the low rates of parasitism found in natural situations may be the result of natural selection favouring behaviour among the hosts to reduce parasitism due to its adverse effects. Many potential host species are know to drive parasitic species away from their nests or nesting territories. Some species are able to determine when a parasitic egg has been deposited in their nest and respond to it. Such species are called *discriminators*, and they usually reject the egg, either by throwing it from the nest, deserting the nest, or building a new nest on top of the parasitized clutch. In contrast, some species (*acceptors*) seem to be unable to identify the foreign egg. The selective action of discriminators against alien eggs within their clutch results in selection for egg mimicry; a discriminator cannot reject a parasite's egg if it cannot tell which egg is the intruder. Of course, a generalized parasite like the Brown-headed Cowbird cannot mimic all 100 of its potential hosts; it must succeed by having high survival rates in the few species for which its egg is a mimic and by utilizing acceptor species. The specialized parasitism of some species may reflect their concentration only on host species that their eggs mimic.   An interesting, though poorly understood, mechanism to get around the problem of laying mimetic eggs without being limited to a single host occurs in some of the cuckoos. In these species, individual females lay eggs that mimic the eggs of a particular species, and selectively put them in the appropriate host nests. Yet, within a parasitic species several races (termed *gentes*, singular *gens*) may occur that parasitize different hosts with different eggs types. The genetics involved in this unusual pattern are not yet understood.

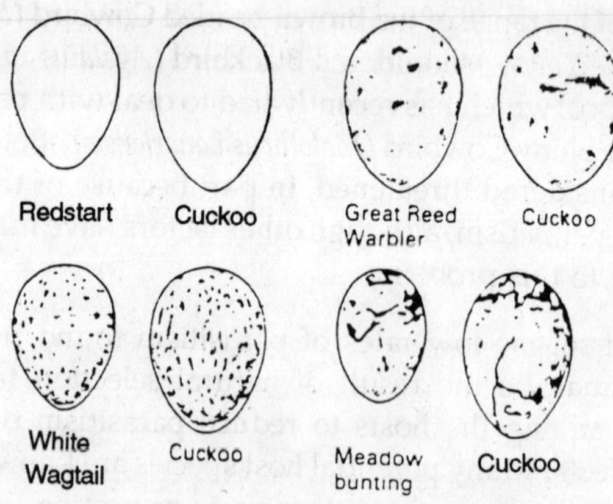

Fig. 2.15.   Some examples of mimetic eggs laid by cuckoos.

A final line of parental defense against parasitism could occur after hatching, if the young do not provide the proper stimuli for feeding or other parental behaviour. Because most passerines have solid-coloured mouthparts when young a good parasite should also have such colouring. This is found in the parasitic cuckoos, even though non-parasitic species have different looking offspring. The viduine finches that parasitize estrildids mimic the mouth colours and markings and begging behaviours of their estrildid hosts.

Brood parasites have evolved a variety of responses to the protective adaptations of their potential hosts, as well as other adaptations to ensure their success. Parasitic cuckoos are large birds that apparently mimic hawks, a possible way for them to avoid being chased away from potential nests. Cuckoos have been known to destroy a clutch of eggs or kill a set of young, apparently in order to get the parents to lay a new clutch that the cuckoo can parasitize.

Most parasites lay eggs that develop rapidly, which gives them an early hatch and advantages over their nestmates. There is some evidence that cuckoos lay an egg with a partly developed embryo to further aid development. Although growth rates of the nestlings are similar to those of their nestmates, many of the parasities, particularly the cuckoos, fledge at smaller sizes than their nestmates. Cuckoos fledge at 50%-60% of adult body weight, while most passerines fledge at near their parental weight. This gives the parasite further advantages in monopolizing parental care. Finally, some young parasites will kill nestmates. Young cuckoos actually push nestmates out using their back, while honeyguide young have a special point on their bill that enables them to kill their nestmates. Parasitic species with such aggressive young usually lay just one egg per nest, for obvious reasons.

Fig. 2.16.   A wagtail *(Motacilla)* feeding a young cuckoo *(Cuculus)*; note the difference in size.

With the reduced costs of parental care, many parasitic species have the potential for higher reproductive success than species that provide care. Estimates for total eggs laid in a season range upto 25 for both cuckoos and cowbirds. These may be laid in "clutches" of 2 to 5, perhaps due to their evolutionary predecessors having laid clutches of this size when nesting for themselves. Given that no parental care is provided by either sex, it is not surprising that parasitic species are usually polygynous. Some may show territorial behaviour, although others show primitive forms of lekking. Some rather extreme forms of sexual dimorphism occur among the parasitic viduine finches. Despite all of these adaptations that favour high reproductive output in brood parasites, they are generally no more successful than non-parasitic forms. In general, the world is not overrun with brood parasites, presumably because of the many defense mechanisms developed by host species.

Although many of the adaptations of parasite and host have been studied in detail, the steps in the evolution of this strategy have not been uncovered. Many species use previously used nests, and some are known to take over already constructed nests. This behaviour, accompanied by the laying of an egg or two, may have served as one of the first preadaptations in the evolution of brood parasitism. Because females will sometimes "dump" an egg in a nearby nest of a conspecific, that has been suggested as a pathway, buy the shift from facultative parasitism to obligate parasitism is difficult to understand. In some duck species in which individuals regularly lay eggs in the nests of other species, large "dump" nests sometimes occur and are subsequently deserted by the original nester.

Parasite-host interactions are often intricate, usually because of the potential costs to the host. Among the viduine finches, parasitic specilization on a particular host species may involve not only morphological adaptations such as the appropriate markings on the nestlings, but also some behavioural flexibility in such characteristics as learning of song. In some viduines, males learn the song of their host, which apparently serves as an attractant to females also reared by that host. There is some evidence that this vocal mimicry may be leading to speciation, perhaps without the allopatric distributions usually required for avian speciation. The viduine-estrildid interaction is also unusual in that the parasitism is not always detrimental to the host, an individual parasite might lay several eggs in a single nest, and the young of both species are adapted to a diet of seeds.

TABLE 2.2

A general summary of the interactions occurring between oropendolas, cowbirds, and other animals in Panama

Host colony with wasps or bees	*Host colony without wasps or bees*
A. Wasps bees provide hosts with protection against ectoparasitic flies (Philornis) and vertebrate predators	Hosts are open to attack by flies and vertebrate predators
thus	**thus**
B. Hosts discriminate against parasitic cowbirds	Hosts accept parasitic cowbirds
because	**because**
C. Cowbird chicks outcompete host chicks and lower hosts' fitness	Single cowbird chicks increase host fitness by removing parasitic fly larvae from host chick
but	**but**
D. Wasps and bees often desert site and hosts are then open to attach by flies and vertebrate predators.	When food is limited, two or more cowbird chicks decrease host fitness despite eating fly larvae.
E. Host colony functions only when wasps/bees are active, and thus get but one chance per year to reproduce.	Host colony is independent of wasps/bees, and has two or more chances to reproduce per year.
F. Too many nests around a wasp or bee colony may cause the branch to break with a total loss to all.	No such danger.

Note. Host chick fledging rate varies in both situations from year to year, but on the average both are equally successful.

Another case in which a brood parasite actually may be an advantage to the host was reported by Neil Smith (1968) working on several species of oropendolas and caciques (*Icterinae*) that are parasitized by the Giant Cowbird (*Scaphidura oryzivora*) in Panama. When these large orioles build their nesting colonies in the vicinity of bee or wasp colonies, they appear to be discriminators that will reject cowbird eggs that do not accurately mimic their own eggs. Parasitized oropendola nests produce fewer young under these conditions than unparasitized nests, presumably because of food limitations. Under these circumstances, natural selection favours discriminating behaviour among oropendolas and mimetic eggs among cowbirds. When oropendolas nest away from such bee and wasp colonies, however, they no longer behave as discriminators that reject cowbird eggs. Rather, they allow cowbird eggs to remain in the nest. It turns out that parasitic botflies can be detrimental to oropendola young. When the nests are near bees and wasps, these insects keep botflies away from the nests. Without bees and wasps, fotflies parasitize oropendola young, eating their flesh and causing death. Because cowbird nestlings are more precocious and active than oropendola young, cowbird nestlings will snap at and eat botfly adults, eggs, and larvae. As a result, oropendola nests with a cowbird under these conditions produce more young, on the average, than nests with only oropendola chicks and botfly larvae. Thus, there can sometimes be an advantage to rearing a brood parasite! Other potential interactions between the nesting birds and predators of bee / wasp nests make this story even more complicated and interesting

The recent expansion of the ranges of several parasitic species has made studies on parasite-host interactions of potential importance in avian management. We have already noted that populations already reduced in size by other factors may be susceptible to increased nest parasitism. If long periods of time are required for a species to adapt to parasitism, such parasities may become an extinction-threatening problem. Understanding the variety of interactions and behaviour involved is therefore of both scientific and applied interest.

FINAL COMMENTS ON REPRODUCTIVE BEHAVIOUR

The previous material has provided a brief overview of the great variation in general reproductive behaviour that occurs within the world of birds. Given that there are only two sexes, an incredible number of options seems to exist, particularly when one takes into account how strategies may vary with the age of a bird, its population levels, or temporally through the breeding season.

Despite the variations already discussed, a few unusual reproductive behaviours have been left out. One of the most unusual incubation patterns is that of the brush-turkeys or megapodes (Megapodiidae), a group of gallinaceous birds from the Australia-New Guinea region. Many of these accomplish incubation by using the heat of decomposition in mounds of decaying vegetation or the heat stored by intensely radiated beach sand. While in some species the eggs are simply deposited and left alone, in others the mound of vegetation is tended throughout incubation to control the conditions in which the egg develops.

The classic study of mound-tending was done on the Mallee Fowl (*Leipoa ocellata*) of Australia. In this species, the male tends the mound for up to 11 months of the year. He excavates a pit of 1 m x 3 m, then fills it heaping full with vegetation, the mound often reaching 60 cm above ground level. This is then covered with a layer of soil. When the heat of decomposition approaches 29°C, females will lay their eggs in the vegetation. Although the heat produced by decomposition tends to remain around 33°C, climatic conditions may alter the changes, temperature, thereby affecting egg development. To compensate for these the male regulates eggs temperature by exposing the eggs when cooling is needed or covering them to protect them from colder conditions. He may also expose the eggs to direct sunlight when that is optimal. Experiments comparing male-manipulated nests with untended vegetation or sand showed how well the male maintained nearly constant conditions..

Another unusual behaviour that is worth noting occurs in some of the hornbills. These cavity-nesting species are unusual because the egg-laying female is sealed into the cavity early in

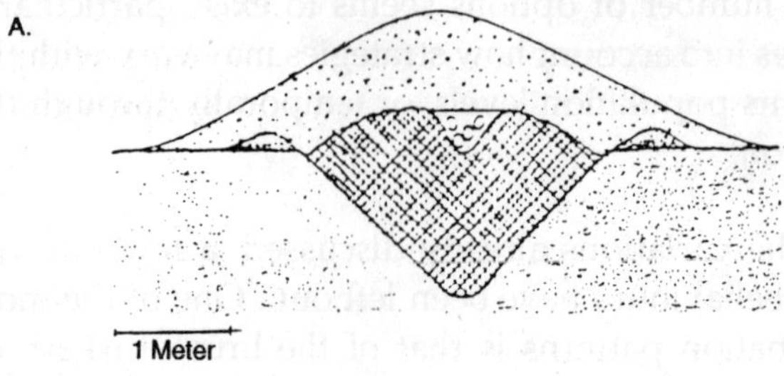

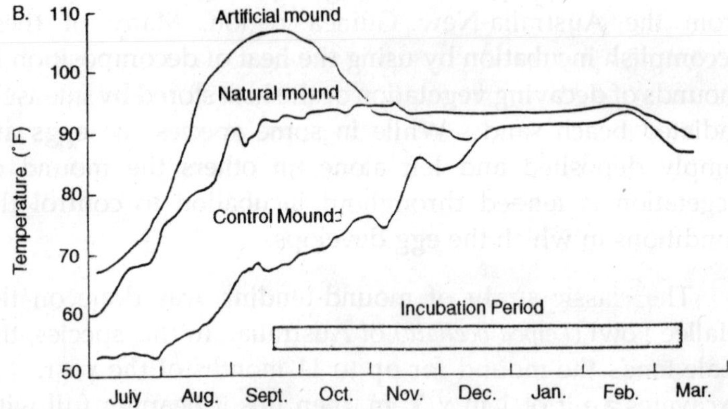

Fig. 2.17.   (Top) Structure of the nest of the Malle Fowl (*Leipoa ocellata*). The pit contains decaying vegetation, while the covering is sandy soil. (Bottom) Temperature control within a Mallee Fowl nest (natural mound) compared to an unregulated mound of sand (control) and an unregulated mound of sand and vegetation (artificial).

the breeding cycle. From this time until she leaves, when the nestlings are fairly well grown, she depends on the male to bring her food. This seal is made of mud and presumably serves as a predator deterrent. Food for the female during incubation and brooding, as well as food for the young during early nestling development, is provided by the male. When the young are partly grown, the seal is broken and the female leaves the nest, then it is resealed until time for the young to fledge.

Recent studies on reproductive behaviour in birds have found that nearly all species possess complex and interesting behaviour. Although the focus of research has often been on non-monogamous systems because of their seeming complexity, many authors have shown that monogamy is anything but the conservative, perhaps monotonous and mundane system that it appears to be on the surface. Rather, subtle variations occur among individuals with small changes in age, population levels, environmental conditions, and experience. With all the work in progress, continuing significant advances will be made in our understanding of avian reproductive strategies in the future.

3

Breeding and Nesting

In reproducing their kind, birds frequently show great ingenuity, whether it's the blackbird nesting in the garden shed, the stork on the chimney pot or the falcon on a sky-scraper in New York City. Many have adapted well to a world now dominated by man, but millions of years before homo sapiens first walked the earth birds had already learnt to build nests and rear young in virtually every natural niche on earth. Despite the rigours of having to protect fragile egg in frequently hostile environments, they nest in every continent, from arctic ice to waterless deserts where no man could endure. All lay eggs, but not all make nests. Some are born blind, naked and helpless, but other hatch 'ready to go.' Most are good parents, but others show surprising lack of care and a few parasitic species have learnt to hand the entire business of baby care over to foster parents.

The Largest Tree Nest in the World

Bald Eagle's - up to 2.9 m (9.5 ft) wide and 6 m (20.0 ft) deep: If we take the popular meaning of nest — a structure made by a bird from nsatural materials such as twigs and grasses, and containing a cup or cavity in which eggs are laid or incubated— then the largest are undoubtedly made by some of the birds of prey which use the same site year after year. The very largest of these are generally made by the bald eagle (*Haliaeetus leucocephalus)* and the biggest known as 2.9 m (9.5 ft) wide — more than enough for the tallest man in the world to lay across with his arms outstretched — and an incredible 6 m (20 ft) deep. This monostrous fortress at St. Petersburg, Florida was almost certainly added to by several pairs, the usual maximum depth for any — even the largest eagle eyrie — being about 3 m (10 ft). The next weighed more than 3,000 kg (2.9 tons). Another bald eagle nest at Vermilion, Ohio reached 2.59 m (8.5 ft) across and 3.66 m (12 ft) deep after 35 years. Eventually it crashed to the ground during a storm, killing the eaglets inside. It had an estimated weight of 1,814 kg (1.78 tons).

The bald eagle characteristically builds a nest of sticks and twigs high in living trees and usually lays two eggs. Its former range was virtually throughout North America from the tree-line in the north to northern Mexico and Baja, California, in the south. The main reasons, for its decline included the widespread use of DDT, which became highly concentrated in the tissues of its preferred fish food. This hampered the bird's breeding success. Also significant were encroachment on habitats through human population growth and some trophy shooting. It is still listed as endangered in some states but there has been a notable recovery since the banning of DDT and the species' position as the USA national emblem has helped its status. Nonetheless its position is still serious. As recently as 1983 US authorities uncovered a smuggling business worth 1 million dollars a year of bald eagle carcasses and plumages used to make replicas of Indian artefacts such as fans, whistles, rattles and head-dresses. Eighteen other species were also involves.

Most huge bird-of-prey nests occur in remote or inaccessible areas where the birds have nested undisturbed for centuries. Those of the Australian white-bellied sea eagle *Haliaeetus leucogaster)* may be 2-3 m (7-10 ft) across and 4 m (13 ft) deep and traditional nests of the osprey along the Red Sea coast are at least 1 m (3.25 ft) across and 2 m (6.5 ft) high. The largest nests in Britain are those of the golden eagle *(Aquila thrysaetos)*: in 1954 British naturalist Seton Gordon saw one in Scotland which was 4.6 m (15 ft) deep and had been in use for 45 years.

The World's Largest Nest-Mounds

The Megapodes' — to 11 m (36 ft) across and 5 m (16.4 ft) deep:
If we take the word nest in its widest possible sense, to mean any place or structure in which eggs are incubated, then the largest are undoubtedly those of the megapodes (*Megapodiidae*), a small family of three ecological groups — the scrub fowls, brush turkeys and mallee fowl, those of the common scrub fowl (*Megapodius freycinet*) sometimes being 11 m (36 ft) across and 5 m (16.4 ft) high. These may have been worked for many years and may be shared by several females, even of different species. It is not known whether these amazingly skilful birds ever acquired the art of brooding of discarded it, but mounds were first reported from the Philippines in 1521. yet it was not until the 19th century that such reports were authenticated. One mound discovered on an island by John Macgillivray of HMS *Rattlesnake* had a circumference of 45 m (150 ft).

Scrub fowls, big-footed relatives or the pheasants (*Phasiantidae*), mate for life and a pair inhabits a large territory. It is mostly the male, which works the mound. This begins with the digging of a hole up to 5 m (16 ft) in diameter and 1-1.5 m (3-4.5 ft) deep. The mound may contain 300,000 kg (295 tons) of matter and the birds can move very large objects. A 1 kg (2.2 lb) scrub fowl was once seen to shift a rock of 6.9 kg (15.2 lb). During the winter this hole is filled with vegetation swept up from the ground over a radius of about 45 m (150 ft). When this is moistened by rain the whole is covered by a layer of sandy soil 50 cm (20 in) or more thick. Sealed from the air, by solar heat. When fermentation heat is too much in the spring the birds need to open the mound frequently to let the heat escape and maintain the ideal incubation temperature of 34°C (93.2° F). In midsummer the sun's heat is most important, but this may be too much and then the thickness of soil has to be increased to prevent 'overcooking'. But by autumn the sun's heat is less and the fermentation heat is exhausted so the birds open and flatten the mound daily to let the sun's rays warm the eggs more easily. In this way the birds achieve precise temperature control over several months.

None of the 12 species megapode of Australia, New Guinea, Indonesia and Polynesia uses body heat to incubate eggs. The required heat comes from the decomposition of vegetable matter in the mounds, directly from the sun or even from volcanic upwellings. Where the mound temperature is naturally fairly constant some species leave the eggs to hatch on their own, but most megapodes employ great skill in maintaining a remarkably constant incubation temperature so that the eggs are not spoiled through chilling or over-heating. The precise way in which they do this is uncertain, but one suggestion is that they gauge the incubation temperature with ultra-sensitive tongues.

In some areas scrub fowls are regarded as rain-prophets as they start to scratch their mounds together at the approach of rain in spring. But they will also begin work if rain falls in the autumn and will attend to mounds through practically the whole year, raking them over for essential temperature regulation and aeration.

Direct solar heat may be more important for the well-studied mallee fowl (*Leipoa ocellata*) of southern Australia as it breeds in dry places where, in the absence of rain, rotting of mound material is very slow. But for those species which live in dense, sunless jungles heat from decomposition is crucial. Yet some moisture is important in every mound to prevent the egg shells cracking.

The number of eggs laid varies from 5 to 35, the great variation being not only between females but also in the same individual from year to year. The interval between the laying of eggs depends on the nutritional state of the female but is often several days. They may be laid over several months but each begins the nine-week incubation immediately so that some chicks may leave the mound before the last eggs are laid.

The eggs are laid in tiers and in 1842 John Gilbert discovered that they are stood upright on their smaller apex and that the shell is unusually thin and fragile to help the buried chick escape. In the same year, Sir George Grey, the Governor of South Australia, discovered that each mallee fowl egg laid is completely enveloped in soft sand.

Megapode chicks lack both egg teeth and special hatching muscles, but they are very advanced. In the egg the head remains tucked between the legs and strong movements of the wings and feet crak the shell in several places allowing the chick to break free.

Chicks emerge feet first, the reverse of normal procedure, so that they can scramble up to the surface more easily, and only after the shell has broken away does the head lift. The large feet give the megapode its name.

On hatching, the chicks are also particularly damp and slimy and presumably this helps them in their almost perpendicular surfacing. Once up they can run swiftly within a few hours and fly within 24 hours. They never see their parents and live completely independent and solitary lives.

The Largest Social Nest Sociable Weaver's —up to 300 chambers: The weaver family is well-known for its 'hanging sock' nests with tubular entrances as long as 70 cm (27.5 in) and as much as 10 cm (4 in) wide. One of the most remarkable is that of the sociable weaver (Philetairus socius), of south-west Africa which first constructs a roof of coarse straws in a tree or on a telephone pole and then makes as many as 300 nest chambers below. The ensuing mass can dominate an entire large tree and attract other species. The tiny pygmy falcon (Polihierax semitorquatus) is one guest but, although it feeds on small birds, it does not molest its hosts.

Probably the most remarkable of the social-breeding parrots is the South American monk parakeet (*Myiopsitta monachus*) which builds huge colonial twig nests in the tops of trees. Within the main structure each pair has its own nest chamber. In *Birds of La Plata (1920), Hudson* noted that such a clump can 'weigh a quarter of a ton and contain enough material to fill a large cart.'

The largest roofed nest hammerhead stork's — up to 2 m (6 ft) wide and 2 m (6.5 ft) deep : More popularly known as the hammerkop, the African hammer-head stork (*Scopus umbretta*) defies accurate classification and is not closely related to the storks. It prefers to build its huge nest of twigs, grass and mud about 12-15 m (40-50 ft) up in a tree. An although the nests are usually 1-2 m (3.25-6.5 ft) across, often of similar depth and take some two months to make, this industrious bird may make several in one season. No one knows why it goes to such great lengths to raise its young.

First a platform is constructed of sticks cemented with mud. Then the wall are raised and roofed over with a metre of thatch which may be decorated with feathers, bones, snakeskins and other debris. The whole thing may contain over 10,000 sticks and easily support the weight of a man. The internal chamber, which is about 30-50 cm (12-20 in) in diameter and height, is very carefully shaped and large enough to house both parents and young. The entrance tunnel is 13-18 cm (5-7 in) wide and 40-60 cm (16-24 in) long and heavily plastered with mud to give a smooth surface. If faces away from the tree trunk so that predators such as genets and snakes find it very hard to get in. Where trees are scarce the nest may be built on a cliff face or the ground, but man-made structures such as dams are also used.

The three to six young, fed by both parents, spend about seven weeks in the nest before fledging. When both birds feed the young they may leave them for long periods, and presumably the thick nest walls then provide good protection. Even after fledging the young remain near the nest for a month or so roosting in it at night. Many other species are also attracted to the massive nests, whether occupied or not. During construction some nests have been take over by Verreaux's eagle (*Aqudla verreauxii*), and grey kestrels (*Falce ardostaceus*) and barn owls *(Tyto alba)* often evict the rightful owners from the finished nests. Many smaller birds such as weavers, mynahs and pigeons attach their own nests to the main structure. Other hole-nesting species such as the Egyptian goose *(Alopochen aegyptiacus)* and African pygmy goose *(Nettapus auritus)* soon take over the vacated nests, so overall the hammerkop provides important important nest-sites for many birds which might otherwise have to fore-sake the habitat.

Small Builders of Big Nests

The most extraordinary large nests for small birds are made by oven-birds *(Furnariidae)* (10-26 cm/4-10 in long) in South America. That of the rufous-breasted spinetail *(Synallaxis erythrothorax)* is made from thorns, with tunnels to the nest chamber. It is roughly oblong or retort-shaped and as much as 75 cm (30 in) long by 50 cm (20 in) wide. It has a large platform on the side which leads via a tunnel through a tangle of thorns to a thatched nest-chamber lined with downy leaves. That of the rufous-throated thornbird *(Phacellodomus rufifrons)* has further chambers added in subsequent seasons and gives the impression of a large colonial nest. It is unlikely that spare chambers are used other than by non-breeding members of a previous brood. Also voluminous is the thorny nest of the 21 cm (8.3 in) firewood gatherer *(Anumbius annumbi)* of Argentina. It measures about 70 cm x 30 cm (28 in x 12 in) and is made of big sticks with a crooked passage at the top leading down to the nest chamber. But the largest in this varied family is probably the huge, thorny nest of the white-throated cachalote *(Pseudoseisura gutturalis)*, being up to 1.5 m (5 ft) in diameter. The enclosed structure has a cavity big enough for an eagle or vulture and strong enough for a man to stand on without damaging it. The brown cachalote *(Pseudoseisura lophotes)* of Paraguay, Uruguay and Argentina makes a nest the size of a barrel. Famous naturalist *W.H. Hudson* noted that he could stand on the nests of some of these species and make no impression. Although these huge nests are conspicuous feature of open South American countryside little is known about the breeding because of the impenetrability of the structures.

The longest nest burrow the Rhinoceros Auklet's 8 m (26 ft) : Why the 36 cm (14 in) Rhinoceros auklet (*Cerorhinca monocerata*) of the North Pacific should excavate such a long burrow is a mystery. This misnamed bird is actually a puffin and its range is from the coast of Kamchatka and the Kuril Islands and from the west coast of Alaska south to Washington. In winter it is found in Japan and lower California. Its single egg hatches after about 40 days and the chick fledges after 40-50 days, having been fed on fish.

The Strangest Nest-sites

There are many fascinating nest-sites across the world, but the strangest must include that of the violaceous trogon (*Trogon violaceous*) of Central and South America. This bird takes over a wasp's nest, eats the adult wasps, then digs out the comb to make a nest cavity.

The water thick-knee or dikkop (*Burhinus vermiculatus*) of Africa is equally unusual in frequently laying its eggs on the dried droppings of large mammals such as the hippopotamus.

The Highest Tree Nest

The Marbled Murrelet's- 45 m (148 ft) : Kinglets (Regulus) and Lewis's woodpecker (Melanerpes lewis) nest at 30 m (100 ft) up in trees but the highest nest found so far is that of the marbled murrelet (*Brachyramphus marmoratus*) at 45 m (148 ft). A very strange situation indeed for a 24-25 cm (9.5-10 in) auklet. The single egg is laid in a small cup, made mainly of guano, on a pad of moss in large coniferous trees in remote areas upto 10 km (6 miles) inland. Only five nests of this common seabird have ever been reported. Whereas other auks breed in colonies, often huge, only the marbled and Kittlitz's murrulet (*Brachyramphus brevirostris*) nest in solitary pairs. No one had ever seen the nest of this bird until 7 August 1974 when one was found in a Douglas fir in Big Basin

State Park, Santa Cruz County, California. Kittlitz's murrulet is also exceptional for an auk in that it lays its egg on bare ground above the tree-line and far from the sea.

North American sequoias are the tallest trees in the world and may exceed 90 m (300 ft), but their tops seem to be avoided by birds and there are no records of any bird nesting at such a height.

The Smallest Cup Nest in the World

The bee hummingbird's- thimble-sized: There is no doubt that the tiniest cup-shaped nests are made by hummingbirds. These are mostly fixed to twigs by cobwebs and built by the female alone. The narrowest is the thimble-sized cup of the world's smallest bird — the *bee* hummingbird (*Calypte helenae*) of Cuba and the Isle of Pines. But this is slightly deeper than that of the vervain hummingbird (*Mellisuga minima*) of Jamaica and Hispaniola, whose nest of lichen, silk and cotton is only about the size of half a walnut shell.

The female hummingbird alone incubates the eggs and cares for the young. Except in three or four species, male hummingbirds do not even know the whereabouts of the nests. Some, such as the hermit hummingbirds, build hanging nests, attached by cobwebs to the undersides of large leaves, such as palms and ferns, so that the leaf forms the inner wall and they are sheltered from tropical downpours. Others, such as lancebills, metal-tails and hillstars, use cobwebs and glue to suspend their nests from the ceilings of caves or rocky overhangs. The deepest hummingbird nest was the only one known of the blue-fronted" lancebill (*Doryfera johannae*), which was a pendent structure attached by cobwebs to a rocky overhang near the bottom of a 75 m (250 ft) shaft.

Many birds make no nest at all and lay their eggs on bare rock or earth, or even in a tree cavity. Some scrape together a few woodchips or merely make a depression in vegetation or shingle. The 25 cm 10 in) *fairy tern* (*Sterna nereis*) of Australasia lays its single egg on the rare branch or leaf of a tree, or on coral boulders and cliffs. In the absence of a nest to contain it, the young has strong claws which enable it to cling tightly and even rang upside down.

Palm swifts also make tiny nests, one of the smallest being that of the 10.5 cm (4 in) pygmy palm swift *(Tachornis furcata)* of Colombia and Venezuela. The Asian palm swift *(Cypsiurus batasiensis)* and African palm swift *(Cypsiurus parvus)* build any open nests on the insides of palm leaves (they are restricted to where the fan palm grows). In Africa the minute eggs are glued to the nest with saliva. The nest is made from feathers and fibres and has a small lower rim for the bird to perch on while it incubates vertically.

The Smallest Nest in Britain

The Goldcrest's: The smallest nest in Britain is, not surprisingly, made by the smallest herd in Britain — the goldscrest *(Regulus regulus).*   It is about 8-9 cm (3.5 in) in diameter and mostly placed underneath and generally rear the end of a branch of first or other tree. The materials used are moss and lichens interwoven with willow down, cocoons, spiders' webs, wool, grasses and a few hairs, spherically shaped and with a deep, tight cup. The well-camou-faged structure is usually 1-5 m (-17 ft) from the ground and frequently lined with carefully placed feathers.

However, two of Britain's other very small birds — the common wren *(Troglodutes troglodutes)* and the long-tailed tilt *(Aegithalos crudatus)* — make relatively large structures, both domes. In proportion to the size of the bird, the wren probably makes the largest British nest, and the male frequently constructs several 'cock nests' in an attempt to attract a male. The beautiful long-tailed tit's nest, made with mosses, lichens and cobwebs, has the highest number of feathers of any lining: up to 2,000 have been counted.

The most Valuable Nests in the World

The cave swiftlets': Cave swiftlets (Collocalia) are best known for providing the Chinese with their esteemed birds' nest soup. While all swifts and swiftlets use saliva to glue their nests together, species of *Collocalia* use saliva as the principle ingredient, though some incorporate feathers, leaves and other materials in varying proportions. Those most prized for soup-making are made of pure saliva, and include the little nests of the grey-rumped swiftlet *(Collocalia francica)*, which do not require extensive cleaning. The crop of such nests may be both valuable and extensive. In one year alone over 3.5 million nests were exported from Borneo to China. The nest sell for about £3 each and it takes two to make a single bowl of soup selling for up to £10 a bowl. The nests have no nutritional value and are tasteless, but are highly regarded as an aphrodisiac.

The colonies in vast Asian caves may contain several hundred thousand individuals whose nests of dried saliva are often stuck to the roofs and walls of caverns in total darkness up to 400 m (1,300 ft) or more from the entrance. Collection of them can be very hazardous, involving ropeways and ladders up to 100 m (330 ft) high. But the swiftlets have no trouble in negotiating the bird-filled blackness because they have developed the faculty of echolocation to a higher degree than any other bird species.

The well-studied edible-nest swiftlet *(Collocalia inexpectata)* has two peaks of maximal breeding activity: October-December and February-April, and the nests take about 30-40 days to build. Even without the cropping by humans, breeding success is poor as they can fall prey to cave crickets which eat the eggs and sometimes the young. Many other nestlings fall and are eaten by snakes, while those in houses may be eaten by rats and shrews.

The vast colonies also provide another valuable resource—the droppings (guano) which accumulate on the cave floor and are mined for use as fertilizer. But neither this nor taking the nests endangers the swiftlets' population. Rights to the caves are jealously guarded and nest removal has been carefully controlled for many centuries, even long before the first white man set foot in the area.

The Largest Egg of Any Living Bird
The Ostrich's - up to 17.8 cm (7 in) x 14 c m (5.5 in): The largest egg of any living bird and the biggest single cell in the animal world today is that of the ostrich (*Struthio camelus*), the world's largest living bird. There is considerable variation among the races but the North African birds generally lay the largest eggs, up to 17.8 cm x 14 cm (7 in x 5.5 in).

As with all species, egg-size is linked to individual female size, heavier individuals producing larger eggs, and is genetically controlled and inherited. Thus exceptional eggs do occur throughout Africa. One from Lake Jipe at the south-west corner of Tsavo (West) National Park in Kenya was 17.1 cm (6.7 in) long, 13.6 cm (5.35 in) across, 49.1 cm (19.3 in) round the long circumference and 43.7 cm (17.2 in) round the short circumference. It weighed 1,974 gm (4.35 lb) fresh weight. This was the largest of 600 eggs weighed and measured in East Africa and considerably larger than any of 100 measured on an ostrich farm at Oudtshoorn in South Africa. Most North African ostrich eggs are about 15.5 cm (6.1 in) long by 13.5 cm (5.3 in) across and weigh about 1.65 kg (3.63 lb).

Ostrich eggs are very strong, with a shell some 1.6 mm (0.063 in) thick, resisting the attentions of most predators, even playful lions, but Egyptian vultures (*Neophron percnopterus*) have learnt to crak them with stones. One egg is equal in volume to some two dozen hens' eggs and takes about 40 minutes to boil.

Despite its great size, the ostrich egg represents only about 1% of the female's body weight. Other species lay eggs which represent up to 25% of body weight. Smaller specimens are laid by females breeding for the first time, but the oldest birds sometimes revert to laying smaller eggs. Scarcity of food may also restrict the size of eggs and small eggs are less successful at producing strong fledglings.

Large as they are, ostrich eggs are much smaller than preserved eggs of the extinct *Aepyornis maximus*.

The Largest British Egg

The Mute Swan's - up to 12.2 cm (4.8 in) long: The largest British egg is laid by Britain's largest free-flying bird, the mute swan (*Cygnus olor*). The average size is 11.43 cm (4.5 in) long by 7.31 cm (2.88 in) across, but some specimens of up to 12.2 cm (4.8 in) by 8.8 cm (3.47 in) have been recorded.

Fig. 3.1.   The mute swan lays Britain's largest egg, yet its incubation is not the longest.

Laying begins in the latter part of April. Incubation takes 34-40 days, sometimes 43, and is almost entirely by the female (pen). There is just one brood of 5-7 but 8-17 have been recorded, the larger numbers usually involving more than one female.

The Largest Egg Proportionate to the Layer

The little spotted Kiwi's- 25% of the female's weight : As a proportion of the female's weight, the largest egg is laid by the little spotted kiwi (*Apteryx owenii*)of New Zealand (now known on only three offshore islands). The average 35 cm (113.8 in) female has a body weight of about 1.2 kg (2.6 lb), about 20% heavier than the male, and her eggs weigh 275-370 g (9.7-13 oz), with an average of 310 g (10.9 oz) — about 25% of her body weight. Average dimensions are 108.5 x 71 mm (4.27 x 2.79 in). Eggs of the North Island brown kiwi (*Apteryx australis*) are larger at 330-520 g (11.6-18.3 oz), with an average weight of 440 g (15.5 oz) (125.5 x 78.5 mm 4.94 x 2.76 in), but represent just 18-22% of the female's body weight.

Eggshell thickness varies considerably between species and this significantly alters the internal volume. The shell of the larger emu (*Dromaius novaehollandiae*) egg (635 g/22.3 oz, 525 ml/1.05 pt) is about four times and the shell of the ostrich (*Struthio camelus*) egg (1,240 g'43.7 oz, 985 ml/1.97 pt) about eight times the weight of the kiwi eggshell. Total weights, therefore, exaggerate the size differences between the kiwi egg and the eggs of larger ratites, as the weight of the ostrich egg is nearly three times the weight of the kiwi egg, but its internal volume is only about 2 $\frac{1}{2}$ times as great. The average 53 g (1.86 oz) domestic chicken egg and a 435 g (15.3 oz) kiwi egg are both laid by birds weighing about 2.4 kg (5.29 lb). Not surprisingly, kiwi eggs take a long time to incubate 65-85 days, almost entirely by the male.

The little spotted is the smallest of the three species of flightless kiwi. The largest is the Stewart Island race of the brown kiwi (*A. a. lawryi*), which has females of at least 3.5 kg '7.7 lb) and is the only kiwi active by daylight. Kiwis are much smaller than the other ratites because they have no natural predators in New Zealand and therefore size is unimportant in defence. The high-protein diet also makes size unimportant. As well as worms, this includes woodlice, millipedes, centipedes, slugs, snails, spiders and a wide range of insects, as well as seeds and berries — all highly nutritious.

Captive females have laid up to five eggs in a continuous series with an interval of about 33 days between each. In the wild, clutches are usually 1-2 eggs (rarely 3) and the egg is heavily yolked (61%) like those of other species whose chicks hatch active, open-eyed and fully-feathered.

Even though hen ostrich lays 12-15 of the largest eggs in the world, weighing in total 20-25 kg (44-55 lb), such clutches represent only 20-25% of body weight. Hummingbirds lay two eggs, each of which weighs about 13% of the female's body weight, but the proportion is even greater in the smaller species.

In total contrast, the smallest of all clutchs as a proportion of body weight is that of the emperor penguin (*Aptenodytes forsteri*) whose single-egg clutch represents a mere 1.4% of the female.

The Smallest Egg in the World

The bee hummingbird's - 6.35 mm (0.25 in) long : In terms of average weight or volume, the question of which hummingbird lays the world's smallest egg is unresolved as comparatively few eggs have been comprehensively and accurately measured, and among these there has been considerable variation. But in terms of length alone the egg of the world's smallest bird — the bee hummingbird (*Mellisuga helenae*) of Cuba and the Isle of Pines — is the shortest, varying between 6.35 mm (0.25 in) and 11.4 mm (0.45 in) long. Those of the vervain hummingbird (*Mellisuga minima*) of Jamaica and Gonave Island, Hispaniola are generally no smaller than 10 mm (0.39 in) long and 0.365 g (0.0129 oz) in weight.

However, abnormally small hummingbird eggs have also been reported for several species. These 'sports' or 'runts' are fairly common throughout the bird world and are generally thought to have been laid prematurely. One exceptionally small example of the Costa's hummingbird *(Calypte costae)* of south-west USA and north-west Mexico measured only 7.36 mm (0.29 in) by 5.33 mm(0.21 in) and contained no yolk. In the same nest at Escondido was a second egg which was slightly incubated and measured 12.7 mm (0.5 in) by 8.4 mm (0.33 in) Amazingly, the finder managed to blow the tiny egg and it is now preserved in the Western Foundation of Vertebrate Zoology at Los Angeles, California. The bird which laid the two eggs was just 8.6 cm (3.4 in) long.

Hummingbirds generally lay two (sometimes one) elongated, white eggs which may be very small to us but represent about 13% of the female's bodyweight. Incubation takes 14-23 days and the nestling period is 18-38 days.

The Smallest Egg in Britain

The Goldcrest's-12.2 mm (0.48 in) long: The smallest egg in Britain is laid by Europe's smallest breeding bird — the goldcrest *(Regulus regulus).* Average size is in the region of 12.2-14.2 mm (0.48-0.57 in) long by 9.4 - 10.2 mm (0.37-0.40 in) wide. There are often two broods of 7-11 eggs, sometimes 12. Laying begins in the latter part of April and incubation is by the hen.

The eggs of the long-tailed tit *(Aegithalos caudatus)* are almost as small, averaging 14.2 x 11 mm (0.56 x 0.43 in).

The Largest Clutch

Grey Partridge's — average 15-19 eggs : The largest single clutches are laid by those species with a high mortality and low average life expectancy. This is particularly marked among the ground-nesting partridges, pheasants and grouse, some species of which frequently lay over 20 eggs. The bobwhite quail *(Colinus virgininianus)* is often said to lay the most eggs, and indeed it produces between 7 and 28, but its average is just 8-15. This is considerably less than the wide-spread grey partridge *(Perdix perdix)*, which usually averages at least 15-16, and in Ostrobothia, Finland the average is 19. Within many species clutches become larger with increasing latitude and longer summer days as there is a shorter breeding season with less time for successive broods but more daylight in which to feed a large family. The largest grey partridge clutch to a single hen reported in Britain was 25 in Sussex in 1974. Twenty-four chicks hatched and all were still alive at six weeks. Although it is through that clutch-size in most species is linked to the physical condition and age of the hen, it now seems that the amount of food the parents can collect for the potential brood is more important. There is also evidence that clutch-size is hereditary and that some individual females are either more experienced or more capable than others in caring for large broods.

Hole-nesting birds often lay more eggs than closely related species in more open sites, presumably because the enclosed eggs and young are safer. For example, the *blue tit (Parus caeruleus)*, which has adapted so well to the widespread use of nest boxes, generally has the largest clutch of any nidicolous (having young birds that remain in the nest after hatching) species. An average of 11-12 eggs is normal for good habitats in Europe, but individuals may lay up to 19. In tits and many other species there is also some seasonal variation in clutch-size, the larger clutches generally being laid early to benefit from peak food abundance.

Many supposedly large clutches are due to 'dump-nesting' where two or more females have laid in the same nest. This is particularly common among gamebirds and other *Galliformes* as well as ducks, and the generally unmanageable nest is usually abandoned. This sometimes happens when females have no nest of their own or occasionally practise parasitism. Incubation is very occasionally attempted, though not when the egg pile reaches intimidating proportions. For example, one North American redhead *(Aythya americana)* nest contained

87 eggs! Some redheads are entirely parasitic, making no attempt to build a nest and laying all their eggs in the nests of other duck species. Others are partially parasitic and lay eggs in the nests of other ducks, before settling down to raise their own family. Yet others never practise parasitism at all.

Some species, such as anis *(Crotophaga)* nest communally and share the incubation. This is quite distinct from dump-nesting. Up to 29 eggs have been found in one ani nest and the socially dominant pair does most of the breeding. The anis of the Americas are related to the parasitic *cuckoos (Cuculidae)*.

Ostriches *(Struthionidase)* have a fascinating system of communal nesting. The cock defends a large territory and acquires a mate who lays her eggs in a scrape. A few days later other hens arrive and also lay in the nest. Surprisingly, the first, or major, hen gets off her nest to let them in, but only she and her mate will incubate — up to 40 eggs! But even an ostrich cannot cover so many eggs and some get pushed to the edge. This was once thought to be a way of regulating the temperature of the eggs, but investigation has shown that the outer eggs get overcooked in the sun.

The major hen is able to recognize her own eggs (perhaps by the pore pattern in the plain white shell) and she rolls those of the minor hens to the edge. The 'rejected' eggs form a front-line defence against predators such as jackals and vultures, which can eat only a few eggs or chicks. This reduces the chances of the major hen's eggs being eaten and presumably strengthens the species as the offspring of the dominant major pair are likely to be more successful than those of the minor birds. But the minor hens gain in that although they have failed to get mates it is likely that at least some of their eggs will hatch.

By removing their eggs as they are laid, some birds may be fooled into laying way beyond the normal clutch-size. In this way a hole-nesting *common flicker (Colaptes auratus)*, a

North American member of the woodpecker family, was once induced to lay 71 eggs in 73 days. also in the woodpecker family, the *wryneck (Jynx torquilla)* was made to lay 48 eggs instead of the usual 5-14. A house sparrow *(Passer domesticus)* l aid 51, mallards *(Anas platyrhunchos)* 80-146 and a *bobwhite quail* 128. Such stimulated laying is put to good use among domesticated birds such as the common chicken, which will lay for much of the year. And in protected environments, domesticated birds may be exceptionally successful with unusually large clutches. A *muscovy duck* once hatched 25 ducklings from a clutch of 31.

Many birds with relatively low mortality and high life expectancy lay single eggs. some of the longest-lived birds, such as the three *great albatrosses (Diomedeidae)* and various large raptors, may breed only once in two years. The single egg laid in the summer produces a chick which is fed through the winter to fledge a year later — the lowest reproductive rate of any bird.

The *redlegged partridge (Alectoris rufa)*, on the other hand, has a most unusual method of attaining high reproductivity. It is unique among British birds at least in that one female may lay two clutches in separate nests — one for her mate to incubate and one for herself so that each will raise a brood at the same time. The average clutch size is 10-15 but up to 28 have been recorded. The high rate of egg production is offset by increased losses, for, unlike the grey partridge, the redleg never covers its eggs when left and many are taken by predators. Surprisingly, clutches left for weeks and eventually incubated by the male are hatched as successfully as those incubated immediately.

The Most Clutches, Eggs and Fledglings in One Year

The number and size of clutches, and of chicks surviving from each brood, are all governed by seasonal abundance of food. Some species, particularly passerines, have a prolonged breeding season and the maximum number of broods raised in the wild is generally six. But there are reports of captive zebra finches *(Poephila guttata)* raising as many as 21 consecutive broods. In the wild this finch is an opportunist breeder, laying in dry parts of Australia at irregular intervals after rainstorms and continuing to breed until the weather deteriorates. But adults of all species are restricted by their own survival needs such as moulting and building up fat reserves for winter, and, for most, late broods do not produce as many surviving young as earlier ones.

Because of their semi-wild state and adaptation to the protecting human environment, feral pigeons *(columba livia)* often enjoy a protracted breeding season and may well hold the record for the number of clutches laid in one year by a free-flying bird, in Britain at least. However, many of their eggs do not produce fledged young. Within Britain, older, more experienced blackbirds *(Turdus merula)* probably fledge as many young as most other birds — often 17 from four broods of 4-5. Birds such as the grey and redlegged partridges *(Perdix perdix* and *Alectoris rufa)*, which often lay over 20 eggs, generally have only one brood per season and suffer a high mortality rate. Yet the redlegged partridge may occasionally fledge more than the blackbird because it has a unique system whereby the male and female raise broods simultaneously.

Another prolific breeder is the moorhen *(Gallimula chloropus)*, which commonly raises 2-3 broods of 5-11 and is most expert in their care. Predation is high at the beginning of the season but there are many repeat layings as well as 'dump nesting.' Both parents share the incubation and feeding of young. The chicks leave the nest quickly — within 2-3 days — can swim and dive immediately and may even be fed by young of an earlier brood. There may be several special brood nests, depending on the number of young. Some have ramps and are built soon after the chicks hatch.

The moorhen is unusual among monogamous birds in that the female plays a dominant, even aggressive, role. The largest females get the pick of the males, but, rather surprisingly, they prefer small males as these need less food to keep healthy and can put on comparatively more fat. Such males need to spend less time feeding and are better placed to concentrate on incubating so that their mates can start more clutches and replace losses.

Though laid singly in the nests of different hosts, the eggs of parasitic species may produce a larger number in a year. Females of several species of African *cuckoo*, as well as the *European cuckoo (Cuculus canorus)*, commonly lay as many as 20-24 eggs in a season.

The fastest and slowest egg laying

A few seconds to 2 hours: The actual laying, or voiding, of an egg from the end of the oviduct (vagina) is made possible by wave-like contractions of the vaginal wall and in most birds takes 1-3 minutes. However, some parasitic species such as cuckoos *(Cuculidae)*, which must lay quickly to minimize chance of detection can lay in just a few seconds. On the other hand, turkeys *(Meleagridinae)* and geese *(Anatidae)* may take 1-2 hours.

The interval between the laying of each egg in a clutch also varies. The first egg is often laid as soon as the nest is finished, but some species start before the nest is complete and others delay for a day or two after nest completion. A regular time of day is adhered to for many species. *Songbirds* seem to prefer dawn, but *pigeons* and *doves (Columbidae)* like the early afternoon and a few species the evening. Although one egg every 24 hours is common, a three-day interval is not uncommon among birds of prey *(Falconiformes)* and owls *(Strigiformes)*, for whom varying prey supply is linked to survival. When food is short the older and larger birds get the lion's share and are more likely to survive. But even slower is the masked booby *(Sula dactylatra)* with a seven-day interval between the two eggs. Where food is in short supply from impoverished waters, the booby brood is often reduced to one by sibling murder.

The Most Variably Coloured/Patterned Egg

The common guillemot's: No two birds' eggs are 'identical in appearance, though shape and size are remarkably similar within most species. Even in colour and pattern there is little significant variation, though in a few species great variety is a distinct advantage. Of these the common guillemot *(Uria aalge)* is by far the most notable , with a huge variety of egg shades and patterns. The ground colour ranges from white to creamy yellowish, ochreous, blue or deep blue-green, with the most extraordinary variety of markings, inter-lacing lines, sports, blotches or uniform masses of colour ranging from bright red or brown to deep black and greenish black and, occasionally, there are no markings at all. The *Brunnich's guillemot (Uru lomvia)* also has a wide variety of shell markings, but the closely related razorbill (Alea torda) has much less variable eggs.

The guillemot's great egg variation helps to prevent confusion over egg ownership in the species' large, dense colonies on the cliff ledges. It has been clearly shown that guillemots identify their own eggs on the basis of individual combinations of ground colour and marking patterns. Great-tailed grackles *(Quiscalus mexicanus)* may also recognize their eggs in this way among the crowded coconut-palm colonies of Central America.

The single guillemot egg is pear-shaped to help prevent it rolling off the cliff ledge, though many still do. Interestingly, the rolling radius changes from 17 cm (7 in) for fresh eggs to 11 cm (4 in) for fully incubated eggs, but this is a fortuitous outcome of a change in the egg's centre of gravity as the embryo develops, rather than a special adaptation. The large egg (12% of female's body weight) is rested across the feet when incubated. Its shell is thicker at the narrow end, where it is in contact with the rocky nest-ledge, and thinner at the broad end, where the chick emerges. Parents find the right chicks in the crowded colonies through voice identification.

Eggshell is secreted in the uterus and consists largely of proteins and minerals, largely calcium carbonate in the form of calcite. Colour is derived from pigments secreted by cells in the wall of the oviduct, particularly of the uterus. The colours are deposited at different depths in the shell according to the position of the egg in the oviduct at the time of secretion. Much of the ground colour is provided by pigments in the spongy layer of the shell and must therefore be secreted at the upper end of the oviduct, while the blotches, speckles and scrawls on the surface are secreted lower down shortly before laying. The colours are derived from blood haemaglobin and bile pigments. Some shades are achieved by over-lapping of colours. Originally all eggs were probably white as the ancestral reptilian method is to bury them in sand or loose soil. Pale eggs are retained by primitive families such as cormorants *(Phalacrocora-cidae)*, pelicans *(Pelecanidae)* and albatrosses *(Diomedeidae)* and by more advanced species which nest in holes. Eggs laid away from daylight have no need for cryptic colouring. Hole-nesters, such as titmice *(Paridae)*, with speckled eggs, may have taken up the habit after a period of nesting out in the open.

Parasitic species also lay a wide variety of eggs to mimic those of their hosts. The brown-headed cowbird *(Molothrus ater)* of North America and Mexico is recorded as parasitizing 206 species, some regularly. But the screaming cowbird *(Molothrus rufoaxillaris)* of South America parasitizes just one host — the bay-winged cowbird *(Molothrus badius)*.

Although cuckoo species such as the *European cuckoo (Cuculus canorus)* parasitize many different hosts, individual cuckoos generally lay eggs to match just one host and therefore do not produce a wide range of egg patterns and colours. It is believed that a female cuckoo inherits her egg colour exclusively from her mother and that the egg colouration is not affected by the father's genetic input. She chooses the right nests by seeking out the host species that reared her and is probably attracted by the habitat in which she was raised.

The Longest Incubation Periods

Wandering Albatross and Brown Kiwi - up to 85 days: Being warm-blooded, bird embryos must be kept at a constant, relatively high temperature when they leave their mother's body. Thus, except for the megapodes (*Megapodiidae*), which use incubation mounds and parasitic species such as cowbirds (*molothrus*) and cuckoos (*Cuculidae*), which let other birds incubate their eggs for them, all birds keep their eggs at the proper temperature by covering them so that their body heat is transferred more or less directly to the developing embryos. The incubation period is strictly defined as the time between the laying of the last egg in a clutch to the hatching of that egg, and among species reliably studied so far the longest periods have been up to 85 days for the brown kiwi (*Apteryx australis*) and wandering albatross (*Diomedea exulans*). But that is an extreme and the incubation ranges of these two species are generally 74-84 and 75-82 days respectively. The royal albatross (*Diomedea epomophora*) has a similar range of 75-81 days, with a tendency towards the higher figure. Generally, the heavier the egg of a species, the longer the incubation : each doubling of egg weight increases incubation time by 16% on average.

Most birds develop a special brood patch for incubation. This comprises 1-3 adjacent abdominal areas which are normally relatively free of feathers all year round, but any down which occurs there, and some marginal contour feathers, are shed as the skin thickens slightly and the density of blood vessels increases.

Sometimes birds will sit on the incomplete clutch but not apply full body heat. The required temperature of 34-39°C (93-102°F), depending on species, is achieved only after a warm-up period, the length of which may be related to tightness of sit and time required for the incubation patch to develop.

Some 75% of species aid for a constant temperature of 35°C (95°F).

Most birds cover eggs for 60-80% of the incubation period, regulating the egg temperature according to atmospheric conditions. Periods spent sitting vary greatly from under an hour for many passerines such as the European robin (*Erithacus rubecula*) to several hours in most seabirds, such as gulls and terns (*Laridae*). Offshore-feeding seabirds such as shearwaters and storm-petrels (*Procellariidae*) sit for 2-12 days at a stretch and the great albatrosses (*Diomedeidae*) for $2\,^1/_2$ - 3 weeks. But the male emperor penguin's (*Aptenodytes forsteri*) 64- day continuous sitting beats them all. In some species the time spent sitting decreases as the season advances, probably due to rising air temperatures.

Maintenance of correct humidity is also important during incubation, most eggs losing about 15% of their weight before hatching. This is lost through the porous surface, which also allows important ventilation. Turning the eggs, too, is important: some species do this up to 11 times an hour to prevent 'sticking' of membranes and avoid the adverse effects of any temperature gradients in the nest.

In 54% of species sexes share the incubation. For 25% it is the female alone, for just 6% the male alone, and it varies between male, female and both in about 15%. Single birds sit more continuously, but where both sit hatching is not necessarily quicker. Some parents, such as bushtits (*Psaltriparus*) of North and Central America, are assisted in the incubation by adults without nests of their own. Others, such as anis (*Crotophaga*) and acorn wood-peckers (*Melanerpes formicivorus*), have communal nests with several females laying and males and females sharing the incubation. The waxbills (*Estrilda*) are the only species in which the male and female incubate side by side on the nest.

The percentage of egg volume taken up by the yolk food supply at the start of incubation depends upon how advanced the chicks need to be at hatching. For most passerines, which hatch blind, naked and helpless, a 12-15-day incubation is normal, with the yolk taking up only about 20% of the egg.

So-called precocial species (hatching young which are soon able to leave the nest) such as shore-birds, ducks and gamebirds, which hatch in 3-7 weeks, require a larger yolk of about 35% total egg volume. Such chicks emerge from the shell with a full coat of protective down and are soon able to fend for themselves.

An embryo consumes the yolk until just before hatching, when the remains of the yolk sac pass into the body of the chick through the umbilical opening and, in some instances, may continue to nourish the bird for several days after hatching, giving the parents or chick time to establish the new feeding routine.

The Longest Incubation in Britain

The fulmar's - up to 56 days: The longest incubation recorded for any bird on the British list is 56 days for the fulmar (*Fulmaris glacialis glacialis*), though a period of 52-53 days is more likely in a range said to start at 42 days and perhaps exceed 56. The manx shearwater (*Puffinus puffinus*) has a range of 47-55 days, though 52-54 is more likely. Next come the gannet (*Morus bassanus*) with 39-45 days, and white-tailed sea eagle (*Haliaeetus albicilla*) with 35-45.

Although resembling a gull, the fulmar belongs to the relatively primitive family of tube-noses or petrels (*Procellariidae*). It lays just one egg and this is not generally replaced if lost to a predator, bad weather or accident. Despite the long incubation the chick is hatched blind, but covered in long, thick down. Typical of the family, the fledging period is very long — 3-5 months — varying with the erratic food supply, the chick eventually becoming heavier than the parents. It may take eight years to reach maturity and actual breeding condition, but visits the colonies for a couple of summers before this in order to establish territory and perhaps a pair bond. Many live for 30 years or more so there is not need for a high replacement rate to maintain the population. The fulmar is the only British seabird which stays at its nest-site throughout the winter, when it will continue to defend its cliff territory.

The Atlantic population of fulmars has undergone a dramatic but largely unexplained increase this century. This started in 1872 from the species' single stronghold on the remote island group of St Kilda, off the Outer Hebrides, and they are now found around the entire coastline of Britain, though in smaller numbers in the south-east. The population is about 600,000 pairs, but with young and non-breeders at cliff-sites in winter may be over 1 $^1/_2$ million birds. They now even nest at inland quarries, on ledges of large coastal buildings and in sand-dunes, having taken all suitable cliff-faces. The increase has been linked with more food in the form of offal from a growing fishing fleet, but this has not been consistent throughout the bird's range.

The name fulmar probably derives from the word foul, the bird having developed the art of spitting foul-smelling stomach oil at molesters.

The Shortest Incubation Periods

Small passerines - 10 days : Ten days seems to be the shortest possible incubation period for a fairly small small number of passerines which lay eggs weighing under 1 g (0.035 oz). These include the hawfinch (*Cocothraustes cocothraustes*), blackcap (*Sylvia atricapilla*), lesser whitethroat (*Sylvia curraca*), redpoll (*Acanthis flammea*), great-spotted woodpecker (*Picoides major*) and the black-billed cuckoo (*Coccyzus erythropthalmus*). But these birds may also incubate for 11 days or even longer, according to how long the egg was in the oviduct and whether or not it was fully formed when laid.

Many more small birds have 12-14 day incubation periods, their young generally hatching form eggs with relatively small yolks. Birds which hatch in a more advanced state are sustained by proportionately larger yolks over longer incubation periods.

The Fastest and Slowest Hatching

30 minutes to 6 days: The actual process of 'escaping' from the egg — hatching — can take anything from 30 minutes or so in small passerines to six days for the larger albatrosses (*Diomedeidae*). But for most small to medium-sized birds it takes from a few hours to most of a day on average. In most clutches the eggs do not hatch together, but in sequence, reflecting the intervals in laying.

Preparation for hatching begins quite early in the incubation. As the embryo's moisture gradually evaporates through the porous eggshell an air pocket gathers at the blunt end of the egg, between the inner and outer shell membranes. The developing chick uses this for breathing over several days prior to hatching, though it still depends on a part of the embryo called the allantois for oxygen exchange.

The actual 'tools' or hatching — the temporary egg tooth on the tip of the upper mandible, and the special hatching muscle — also form early in the incubation, and the eggshell becomes conveniently weaker as it provides minerals for the developing chick's skeleton.

First the embryo realigns its body to lie along the length of the egg, the best position to force its way out from the blunt end. Convulsive thrusting forces the bill against the shell, and with the aid of the horny egg tooth the shell is soon cracked or 'piped'. After the chick has broken out the hatching muscle withers away and the egg tooth either drops off or is reabsorbed into the bill, according to species. From pipping to full emergence is the hatching period. During this, the chicks of some species are helped by the parents, which may pick at the shell or poke into the cracks. It is said that ostriches crack their eggs with their breastbones and even drag the chicks out with their bills. Some parents will begin bringing food in response to the chick's peeping even from inside the unbroken shell.

Quickest to Fly

Cabot's Tragopan and Megapodes - within 24 hours of hatching : It would not be right to say these species are the fastest at *learning* to fly because they seem to be born able to fly without any training or exercising of flight muscles. Most birds are born with down and are said not to be fledged until they have acquired their first true contour feathers, including the primaries which will propel them in flight. This generally takes at least 10 days and ground-nesting species, whose young leave the nest soon after hatching, are usually much quicker at getting into the air than birds which are born naked and helpless and remain in the nest after hatching. This is taken to the extreme in the megapodes *(Megapodiidae)*, whose young hatch in special incubation mounds and must fend for themselves immediately. They never see their parents and instead of the usual down they are born with extremely advanced contour-like feathers which enable them to fly and escape from predators almost at once.

Equally advanced on hatching is Cabot's tragopan *(Tragopan caboti)* of China a member of the pheasant family. Its chicks hatch with a thick coat of coarse, shaggy down, but with primaries so far advanced they are immediately able to flutter up to perches. They also climb well and within a day or two can fly as well as any young passerine. Yet, although incubation is generally long among pheasants, it is not particularly long for this advanced bird, being just 28 days. This species is also unusually arboreal and lives in dense forests with very thick undergrowth which is often saturated. In this habitat a well-developed plumage is a distinct advantage.

Cabot's tragopan is listed as endangered, but reserves have recently been established within its historic range. Only one nest has ever been described from nature and that was 9m (30 ft) up in an old squirrel drey.

The satyr tragopan *(Tragopan satyr)* is also highly precocial at hatching and within two or three days can fly up to an elevated perch to roost under its mother's wings. It too nests in trees, and is widely hunted and trapped with nooses in Nepal, legal protection being little help.

Temminck's tragopan (*Tragopan temmincki* — the Chinese crimson horned pheasant) of the eastern Himalayas can also fly up to elevated perches within two or three days of hatching. Yet the chicks grow slowly, are sensitive to disease and chilling and do not attain full adult male plumage until the second year, the adult female plumage coming first. This tragopan has a wider range and is more secure, but it too is extensively trapped and shot for its feathers and flesh, and much of its forest habitat is being cut for timber or razed for agriculture.

The shortest fledging time for altricial young — those helpless when hatched — is about eight days for some of the warblers (*Sylvinae*) and finches (*Fringillidae*).

Britain's Quickest to Fly

Corn bunting and bearded TIT — from 9 days old : The shortest fledging period in Britain is that of the corn bunting (*Emberiza calandra*) at 9-11 days. This is advantageous for a species which breeds late in the season — late May to July — yet usually squeezes in two broods. Its incubation is short too, at 12-13 days. The cirl bunting (*Emberiza cirlus*) and yellow bunting (*yellow-hammer*) (*Emberiza citrinella*) also breed late and have short fledging periods of 11-13 days.

Bearded tit (*Panurus biarmicus*) chicks also sometimes fledge in 9 days, but may take 12, the shorter time perhaps being linked with the early part of a long breeding season — April to July. They frequently manage more than two broods and the 13-day incubation is by both parents.

Among Britain's ground-nesting species, the ptarmigan (*Lagopus mutus*) is the quickest to fledge, at 10 days. Laying takes place May-June, according to variable weather, and the 25-day incubation gives time for the precocial chick to develop as it must leave the nest soon after hatching. It is fed by both parents as quick growth is particularly important during the short summer at high altitude.

Slowest to Fly Great

Albatrosses — 9-12 months : The slowest birds to fledge are the great albatrosses (Diomedeidae), especially the royal (Diomedea epomophora) and the wandering *(Diomedea exulans),* which generally take 9-12 months.

At first the bulky young birds sit very quietly in the nest. They have two successive coats of down and are brooded by a parent for the first three to five weeks , after which they are visited and fed at intervals on regurgitated food. As the time approaches to leave they fidget and flap. Most fly straight from the nest, especially if inland, but they are likely to make abortive flights and land clumsily in unsuitable places. Sometimes they are drawn to village lights, and late-developing young may   swim out to sea if they can reach the water. But strong chicks fly well-immediately and never see their parents or birthplace again. When fledging is as long as this the parents can breed only once in two years.

Britain's slowest bird to fledge is the mute swan *(Cygnus olor),* which leaves the nest after a day or two but does not fly for some four months. The cygnets are tended by both parents.

The Fastest to Breeding Maturity

Quails — from 5 weeks: The five species of quail in the genus *Coturnix,* including the common quail *(Coturnix coturnix),* reach maturity before any other bird — one reason why they have been bred in large numbers for the table. They are able to reproduce from as little as five weeks, through in most quails adult plumage is not attained before 10-12 weeks. Some other bird species are also capable of reproduction while still in immature sub-adult before they are sexually mature.

Most small songbirds breed for the first time-when just under one year old, in their second summer, and many others near the end of their second year. But some individuals within a species mature as year earlier than others.

Although physiologically capable of breeding, some species delay breeding for years as they need to develop their food-finding skills before they are capable of feeding themselves and a family. Hunting species in particular rely very much on experience, so young predatory birds continue to depend on their parents while practising their skills. Some manage this in two or three weeks but the African crowned eagle *(Stephanoaetus coronatus)* continues to receive food from its parents for up to a year even though it can fly.

The Slowest to Breeding Maturity

Albatrosses — from 6-10 years old : Despite the fact that the larger albatrosses *(Diomedeidae)* — notably the royal *(Diomedea epomophora)* and the wandering *(Diomedea exulans)* — have very long incubation and fledging periods, they still delay for the record period of 6-10 years before breeding for the first time. There are even reports of a first breeding at 12 years old, through the birds are physiologically capable of reproducing before this. And even after all this time first efforts are frequently unsuccessful.

After fledging, the transequatorial migrants set out on long journeys which must include a period of starvation as they cross the tropics. The early years are spent at sea, and they sometimes gather in nurseries where there is a good food supply. As they get older they return to land for increasing periods wandering in search of new nest-sites, though most return to their birthplaces. Further years are spent displaying and excavating nests before they even attempt to breed.

THE SEARCH FOR A MEAL

Finding enough to eat dominates every day in every wild bird's life and what they choose or find to eat determines their subsequent behaviour and, eventually, shape. Also, the way in which birds have been able to exploit so many different habitats on earth is a reflection of their great variety, versatility and adaptability through food specialization. In finding a meal, their bills in particular have become highly adapted tools, though some species, such as crows *(Corvidae)*, have been successful in retaining a fairly standard bill for a more general, or omnivorous, diet. The value of food specialization is that it enables closely related species to live side by side with minimal competition. But the greatest draw-back for the specialist is that if the food supply disappears rapidly, sometimes through man's impact on the environment, there is an increased risk of extinction.

Nonetheless, in their feeding behaviour birds have never failed to surprise us and frequently react well to sudden, dramatic changes. Who could have foreseen the way in which so many species quickly became tame enough to feed at garden bird tables, or how species such as tits *(Paridae)* would even learn to take the cream from our milk? But even without man's help birds are surprisingly resourceful. A kestrel *(Falco tinnunculus)*, which usually feeds on small birds and rodents, was seen eating apples during a severe winter.

The Most Specialized Diets

Over 30 species of bird have learnt to use tools to obtain food, but most of these do not have very specialized diets, and if necessary they can switch to other food supplies. But the greatest specialists, usually those species with highly-adapted and often peculiarly-shaped bills, are often condemned to rely on very restricted or localized foods.

A textbook case of such food specialization is that of the North American. Everglade kite *(Rostrhamus sociabilis plumbeus)*, which eats only snails of the genus *Pomacea*. The Everglade kite has three races in Cuba, the Isle of Pines, Mexico, Guatemala and South America, where it remains generally abundant in its preferred habitat. But the fourth race — *R.s. plumbeus of southern Florida* — is now down to just 300 individuals through feeding exclusively on the large freshwater 'apple snails' (*Pomacea paludosa*). A bird of marshes and lake-margin rush beds, the 'snail kite' suffered a drastic decline-along with the snails when most of the Florida wetlands were drained earlier this century. In 1965 there were only about 10 individuals left, on the shores of Lake Okeechobee, but thankfully this bird is hardy and resilient, and with protection and good management has established stable breeding populations in the Ever-glades National Park and two other refuges.

The bill and feet of *Rostrhamus* are specially adapted for extraction of snails from their shells. The long toes and claws are ideal for graspring the round, slippery shells and the long, slender, strongly down-curved upper mandible serves to penetrate the columellar muscle by which snails hold on to their shells. In Guyana snail kits *(R.s. major)* have been seen to wait until a captured snail raises its operculum (the plate covering the shell opening where the body emerges) of its own accord, before stabbing their bills into the soft muscle and then waiting for the pierced muscle to relax before shaking the shell free. But in Florida they specialize in inserting the bill under the closed operculum to sever the columellar muscle.

Another specialist raptor is the lammergeier or bearded vulture (*Gypaetus barbatus*), which feeds on nutritious bone marrow. The carcass bones are too big to swallow whole so the lammergeier has learnt to drop them from great heights to smash them on rocks. It uses the updrafts of air which prevail along the high mountain cliffs of Europe, Asia and Africa to lift it up repeatedly, thus saving precious energy.

The lammergeier's long, slim tongue is shaped like a narrow trowel to scoop the marrow from the core of the bone. Dangerously sharp fragments of bone, with marrow attached, can be swallowed safely because the lammergeier has powerful, acidic digestive juices which soon break them down. Tortoises are also sometimes eaten.

Some species have developed total or partial immunity to poisons in butterflies and caterpillars. For example, most species have learnt to avoid the North American monarch butterfly, which derives poison from the leaves of the milk-weed plant. The caterpillars store the poison in their bodies and it is retained when they pupate and then become butterflies. The black-headed oriole *(Icterus graduacauda)* has learnt to eat only those parts of the butterfly which contain small amounts of poison. If it ingests too much it is sick. But the black-headed grosbeak *(Pheucticus melanocephalus)* can eat the monarch butterflies indiscriminately. How it does this remains a mystery.

Sapsuckers are three species of North American woodpecker in the genus *Sphyrapicus*, which drill horizontal or vertical lines of slightly uptilted round or square holes into a wide variety of coniferous and deciduous trees and shrubs and return later to lap-up the sugar-rich sap which has dribbled down. They use the brush-like tips of their extensible tongues. They also eat the green inner bark, called cambium, as well as a wide variety of insects. Other bird species are attracted to the holes, both for the sap and the insects which gather there.

The Most Bizarre Feeding

Blood-drinking by the sharp-beaked ground Finch : The vampire-like activities of the small sharp-beaked ground finish *(Geospiza difficilis)* were not discovered until a television camera team visited the remote Galapagos Islands in the 1980s. To their great surprise, on Wolf Island they found the bird drinking blood from holes which it had pecked in the wings of nesting masked boobies *(Sula dactylatra)*, which do not seem particularly bothered by this. For their part, the finches not only gain extra food but also valuable moisture in a very arid environment. Normally seed eaters, the finches have also learnt to exploit the boobies by rolling their eggs away so that they crack apart on the rocks and provide further food.

Britain's Most Specialized Feeder

The Crossbill : The Scottish crossbill's (Loxia scotica) crossed bill may look awkward, but it is one of the most efficient natural tools in the world. In fact *scotica* is just one of a number of crossbill species and sub-species which have crossed bills of varying size and strength, which correspond to the heaviness of the preferred conifer cones to be cracked apart for their seed. For example, whereas the massive-billed parrot crossbill *(loxia pytyo-psittacus)* of north-east Europe and West Siberia specializes in the toughest of green Scotch pine cones, the relatively delicatebilled white-winged crossbill *(Loxia leucoptera)* of Europe, Asia, North America and Hispaniola prefers larch cones.

The precise way in which the crossbill's bill works has been debated for centuries, but it is clear that the bird does not always need to use all the bill's extraordinary functions in cone-seed extraction. In fact, it can turn to other foods, such as suet put in bird feeders, insects and seeds of fruit (after slicing the pulp in two with a scissor-like action). The full technique employed by this remarkable bill is used only in opening green, dignity closed, unripe cones. After plucking the growing cone, the crossbill takes it to a horizontal perch where it holds it down firmly with one of its exceptionally large and powerful feet. The tip of the downward-curving, longer upper mandible is wedged between two of the cone scales so that the curve of the lower mandible rests on the outside of the cone. Then the bird twists its head to force the scales apart. Simultaneously, the two mandibles are moved sideways by special jaw muscles.

When the scales have been parted, the exceptionally large, protrusible tongue reaches in to detach the firmly anchored, unripe seed with the aid of a special cartilaginous cutting edge at the tongue tip. Such power is exerted by the bird when tackling cones in this way that the loud cracking noises can be heard for some distance.

Although the crossbill chick's upper mandible begins to lengthen by the fourteenth day, no crossing takes place for about four weeks so that the parents can continue to feed it. It is only the horny sheath of the bill, not the underlying bones, which is bent and curved.

Fig. 3.2.   Crossbills have evolved extremely specialized bills in order to prise the seed from the cones of coniferous trees.

Other Remarkable Feeders

Hawfinch and Oystercatcher : For its size, the hawfinch (*Cocco-thraustes coccothraustes*) has the most powerful bill and jaw muscles of any bird. The 18cm (7 in) finch concentrates on the soft seeds of elm and hornbeam in Britain, but it can easily deal with the very hard stones of cherries and olives. Its massive bill has four rounded knobs at the base — two on the inner side of each mandible — and by placing a stone between them the force for cracking it is shared equally by the muscles on each side of the head. To crack an olive stone it has to exert a force of nearly 50kg (110 lb).

Fig. 3.3.   For its size the hawfinch has the most powerful bill and the muscles of any bird.

The oystercatcher *(Haematopus australegus)* is a shellfish specialist with a bill that is laterally flattened towards the tip. This chisel-like tool is inserted into the shells of bivalves such as mussels and oysters and then severs the powerful adductor muscle that hold the shell together. But sometimes a shell closes very tightly on the bill before the muscle is cut. Then, if the shell-fish is still anchored firmly to the seashore, the bird drowns when the tide comes in.

The Largest Items Swallowed Whole

Irrespective of size, birds can be incredibly ambitious feeders and swallowers of surprising objects. The ostrich *(Struthio camelus)*, for example, is well-known for swallowing inedible objects in captivity, from padlocks to beer bottles. In the wild they must swallow stones to aid digestion of vegetable matter and these are usually fairly small pebbles, but one swallowed as ruby and led to the discovery of a mine.

Much larger, extinct birds, such as moas, were also vegetarians which had to swallow stones to aid digestion. Some remains are so well preserved the contents of the gizzards remain untouched in the skeletons. They reveal that the grasses, leaves, berries and twigs were pulped in the gizzard by up to 3 kg (7 lb) of stones!

One species of imperial pigeon *(Ducula)* stretches the base of its bill like a snake swallowing prey, to gorge on whole nutmeg fruits of 5 cm (2 in) diameter — larger than its own head. But whereas most opportunist fruit eaters must regurgitate large stones, specialist birds such as the fruit pigeons can pass nutmeg seeds 2.5 cm (1 in) in diameter through their intestines.

One of the world's greediest' birds is the common cormorant *(Phalacrocorax carbo)*, which sometimes takes salt and freshwater fish of astonishing size, including conger eel 0.76 m (2.5 ft) long. One bird died from exhaustion after trying to swallow an immense mullet.

The Most Wide-Ranging Diet

Crows *(Corvidae)* and gulls *(Laridae)* eat a very wide range of plant material, live vertebrates and invertebrates as well as carrion and waste products provided by man. Their medium size and general shape of bill are probably more suited to an omnivorous and opportunist diet than those of any other species. But no one can accuse the North American ruffed grouse *(Bonasa umbellus)* of being fussy: it is known to have eaten part of at least 518 species of animal and 414 species plant!

The Greatest Appetites

The amount of food any bird eats in a day is related to its body weight, metabolism, calorific value of the food and how much energy the bird can store. Although large birds obviously eat more in absolute quantity than small birds do, in general, the smaller the bird the more it eats in proportion to its body weight. Smaller birds generally have a more active lifestyle and higher metabolism which uses up energy more quickly. For example, although an average pelican *(Pelicanidae)* might eat 1.8 kg (4 lb) of fish in a day and an average eagle *(Accipitridae)* up to 1 kg (2.2 lb) of meat, this constitutes only about a half and a quarter of body weight respectively.

In contrast, hummingbirds *(Trochilidae)*, which are the smallest of all birds and probably have the highest metabolic rate of any living creature, must consume twice their weight or more in nectar and insects each day, even though this may be less than 15 g (0.5 oz) of food.

But there are exceptions to the general rules, with sedentary birds in warm climates needing far less energy than very active and migratory species who spend at least part of the year in very cold regions. Most fruit eaters live in warm countries where fewer calories are needed, but the calorific value of fruit is generally low so that relatively large quantities must be eaten. For example, waxwings *(Bombycilla garrulus)* in northern Europe eat three times their body weight in cotoneaster berries each day. Other gluttonous fruit eaters are sometimes made drunk by the juice of overripe fruits which have fermented.

Birds will also eat very large numbers of single items in one day. A dabbling duck's stomach often contains 50,000-100,000 seeds or plant parts, and 185 blue mussels have been taken from the digestive tract of a common eider *(Somateria mollissima)*. The stomaches of insectivorous birds may contain over 5,000 small insects such as mosquitoes and ants or several hundred caterpillars — impressive quantities considering that birds can pass food through their digestive tracts in 30-90 minutes and such numbers represent only a fraction of an average day's consumption.

Insectivorous birds take even larger quantities of food when raising a family. Pied flycatchers (*Ficedula hypoleeuca*) will visit the nest over 30 times an hour and make a total of 6,000 provisioning journeys to rear the brood. But other hole-nesters with larger broods must make even more trips. Automatic counters set up at nest-boxes show that tits (*Paridae*) always make many thousands of visits in rearing a brood, the actual number varying with the size of the main local food items. One pair of great tits (*Parus major*) fed their young 10,685 times over 14 days but another pair 'only' 4,655 times before the chicks flew on the 19th day. Even more active was a wren (*Troglodytes troglodytes*) which fed her young 1,217 times in a 16-hour day while still having to feed herself.

The Longest Fast

The Emperor Penguin's — average 115 days : Except during hibernation, no living bird fasts for anything like as long as the male emperor penguin (*Aptenodytes forsteri*) of the Antarctic. He begins the fast before incubation starts, and while the female is away at sea feeding he must incubate the single egg for some 64 days despite the very low temperatures of mid-winter. The colonies are often far from the sea (the only source of food) on the pack ice and travelling to and from them extends the fasting period. Thus, the average male's fast is 115 days and the longest so far recorded an incredible 134 days. The birds do not all fast at the same time and, overall, they fast during six months of the year.

To endure such a fast in the most hostile climate on earth the average 30 kg (66 lb) make will put on a lot of fat, but many other special adaptations are also necessary.

The only other birds which can fast for such long periods are the few species which hibernate and, of these, the most remarkable is the North American poorwill *(Phalaenoptilus nuttallii)*, a relative of the nightjars *(Caprimulgidae)*, which can sleep in a torpid condition in rock crevices for some three months.

4

Incubation Requirements

When the egg is laid development has already begun. It is then suspended until the egg is warmed again sufficiently. The time between this renewed warming and emergence of the chick (the incubation period) varies between species, and is roughly in proportion to egg size. In small song birds the incubation period can be as short as 10-14 days, in the domestic fowl it is 20-21 days and it extends to 42 days in a large sea bird such as the gannet or up to 50 days in the Mallee fowl and the brush turkey while in the royal albatross it lasts about 80 days.

In all species the rate of development and the level of hatchability vary with the conditions of incubation, one of the more important of these being temperature. For obvious economic reasons the requirements of the developing embryo - the optimum conditions of incubation — have been assessed in the greatest detail in eggs of the domestic fowl and in artificial incubators. Work in this field (the consideration of incubator conditions associated with good hatchability) has been summarized by *Landauer* (1967) and by *Lundry* (1969). Under good incubation conditions and after being held under the correct storage conditions eggs from good stock can be expected to give a hatchability of well over 90% of fertile eggs set. In this chapter the conditions within artificial incubators which are consistent with a high level of hatchability will be considered first for the fowl, and then the effects of these conditions on other domesticated species. Incubation procedures of a number of species of wild birds will then be discussed.

Requirements of the Embryo of the Domestic Fowl

Heat

In a forced draught incubator an optimum incubation temperature has been found to be 37.8°C. There is, however, evidence that hatchability is unaffected if the incubator temperature is dropped by 1 $^1/_2$°C after 16 days of incubation; and at this time it is known that embryo temperature has begun to rise above that of the incubator. Research of a number of Russian workers, summarized by *Lundy* (1969), claims that fluctuations in incubator temperature provide an added increase in hatchability.

Humidity

Incubator humidity determines the rate of water loss from the egg. Maximum hatchability can be obtained with a relative humidity between 40% and 60%. However, a reciprocal relationship has been found between temperature and humidity, such that if the humidity level is raised the incubation temperature can be lowered and vice versa.

The gaseous environment

Correct ventilation of the incubator is as important as its temperature and humidity. With regard to the oxygen content of the air there may be an optimum concentration for maximum hatchability at 21%. With regard to carbon dioxide the picture is still unclear. It is known, however, that a level of carbon dioxide above 1% results in a decrease in both hatchability and growth rate, and also that the sensitivity of the embryo to carbon dioxide decreases with age. During late embryonic development a rise in the Pa_{co2} of the blood has been found by *Visschedijk* (1968c) to be necessary for normal development.

Egg position and change of position

Eggs must be turned periodically, and egg-turning is particularly important between the fourth and seventh days of incubation. Lack of turning can result in premature adhesions between the extra-embryonic membranes and distortions in subsequent development. The amount of turning also appears to be important, a minimum being about three times a day while more than 24 times a day is unnecessary; adequate turning reduces mortality especially between days 5 to 16 and 19 to 21 (Kaltofen, 1961). When turned, eggs should not be rotated always in the same direction as this can result in rupture of the yolk sac, disruption of the chorion, allantois and shell membranes, twisting of the chalazae and rupture of the blood vessels. In incubators with mechanical turning, the egg is kept upright and the tray is moved alternately from side to side through an angle of about 65°.

It has been found important also to incubate eggs either horizontally, or with the large and upwards. With this limitation turning of the eggs in a number of different planes of rotation has been shown to have a beneficial effect on hatchability. When eggs are incubated with the small end uppermost there is an increase in embryonic malpositioning (head in the small end of the egg) which leads to a reduction in hatchability.

Incubation Requirements of Other Domestic Species

For reasons which will become clearer when incubation under natural conditions is considered below, hatching success, which is high when eggs of the fowl are incubated artificially, may fall off when the same conditions are provided for other species. Where other domesticated species are concerned the difficulties appear to arise from levels of humidity more than any other factor. Although the differences here are not fully understood, practices recommended for the artificial incubation of duck, goose and turkey eggs usually include a level of humidity higher than that required by the fowl. It is, for instance, frequently recommended that eggs of the duck and goose should be sprinkled with water at intervals during incubation. *Insko* (1961) recommends a higher level of humidity for turkey as compared with chicken eggs for the first part of incubation (61-63% as compared with 55-61%), and 65% for the first 24 days, rising to 70-75% for the last few days has also been recommended (Ministry of Agriculture, Fisheries and Food, 1960). With regard to temperature Romanoff (1935) found hatchability to be best in the turkey between 36° and 38°C, and *Martin & Insko (1935)* obtained their highest hatchability (in still air incubators with the bulb of the thermometer level with the top of the eggs) with average temperatures of 38.1°, 38.6°, 39.2° and 39.4°C for the first, second, third and fourth weeks of incubation respectively.

The level of humidity recommended for ducks is higher than that for the turkey. For the domestic duck 70% is suggested for the first 24 days, then 60% until pipping, and then a return to the original level. For the runner *duck Romanoff* (1943) found the optimum temperature to be 37.4°C, with a drop of 0.3°C at the time of hatching, and the optimum relative humidity to be 70% or higher until day 24 and 60% or lower after that. An even higher level of humidity has been found the most effective for the game farm mallard duck; *Prince et al.* (1969) obtained their best hatch with a temperature of 37.5°C and a humidity level of 70-80%. *Kaltofen* (1971) also raised the incubator humidity to 80% at the time of pipping, for the Pekin duck. In geese the recommended temperature is 37.2°C with a relative humidity of 70%.

Romanoff (1934) considered whether the eggs of game birds (in this case the ring-necked pheasant and the bobwhite quail) could be incubated together successfully. Experimental results showed this to be inadvisable as, in a still air incubator with the temperature recorded at the level of the upper side of a hen's egg, the optimum temperatures for the two species were different. Hatchability in the pheasant was highest at an incubator temperature of 38.9°C for the first week, 38.3°C for the second and 37.8°C for the third, while it was 38.3°C throughout incubation for the quail, with a permissible small rise towards the time of hatching. The optimum humidity for the pheasant fell from 75% at the beginning of incubation to 65% at the end, while in the quail it rose from 65% at the beginning 75% at the end. For the quail the permissible range was found to be narrower than for the pheasant, in respect to both temperature and humidity.

Incubation in Wild Birds

Introduction

Incubation requirements of wild birds are very similar to those of the fowl. But whereas work on incubation in the fowl arises from attempts to incubate eggs artificially, what we know about incubation in wild species arises largely from research on the breeding biology of specific species, in which the mode of adaptation within their ecological niches is usually considered. An attempt will be made to bring information on wild species into the focus provided by research on the fowl. From this is appears that incubators eliminate two types of selection pressure which affect wild species. Thus, incubation in the wild is adapted partially as a result of climatic pressures (as from extreme cold or heat), and also as a result of predation, which can threaten the parent, and the eggs. Bearing these additional hazards in mind, we shall be concerned briefly, with the problem of how the physical requirements of incubation are met under natural conditions in a few selected species.

Incubation sites

Nest or incubation-sites are almost infinitely variable between species, descriptions of these are readily available in the ornithological literature and a review of incubation practices is outside the scope of the present volume. Therefore incubation sites will be described for a few species where there is much information or which raise specific issues to be dealt with below.

Guillemots have no nest. They breed in densely packed colonies on the ledges of steep cliff faces, and the single egg is incubated on the bare rock. It is shaped in such a way as not to roll except in a half circle (the ledges may be very narrow). The egg is tended continuously, by one parent or the other. The parents incubate in rows, shoulder to shoulder. In this way they present a barrier against predation which is noticeably weakened if, for some reason, one incubating pair is lost.

Gulls are also colonial nesters, in this case on the ground and each pair defends a territory around its nest. *Tinbergen* (1953) describes the herring gull nest as a well-rounded and well-lined nest cup, partly scraped out of the ground and partly built up of straws and moss which are arranged in it, or round the edge. Again, incubation is undertaken by both sexes, they incubate continuously (about 95% of the time; and the eggs are normally left only when the birds fly up in response to intruders. Under normal conditions three eggs are laid.

The *yellow-wattled lapwing* nests in the dry season, in tropical conditions. It is a solitary species confined to Pakistan, India and Ceylon. The breeding behaviour has been described by *Jayakar & Spurway* (1965a, b). The pairs watched by them nested in the neighbourhood of water. The nest is a small concavity surrounded by a small circular bank of gravel and twigs. Four eggs are laid in each nest. One or other of the parents remains on the nest during the night. During the daytime the eggs are left for long periods apparently unattended, although one or both adults invariably appear if the nest is approached by intruders.

The breeding biology of the *greater crested grebe* has been described by *Simmons* (1955). In this species the nest-site appears to be dictated by the threat of predation. Grebes are exclusively aquatic birds and very clumsy on dry land. They are expert divers and the nest is sited in or near the water, and may be partially floating on it. When approaching, the parent swims right up to it. It is built of sodden rotting weed. During the laying period, or when approached by intruders, the incubating bird swiftly covers the eggs with this material before sliding into the water. The nests are frequently awash, and when the water splashes over the rim they are built up with more weed. The egg surface consists of a white chalky substance which covers the blue shell.

Like many other small song birds, the *wren* builds nests with a quality noticeably lacking in those described above, that is, insulation. Work on nest-building in the wren has been reviewed by *Armstrong* (1955). The structure, which may be built into a cavity, or into surrounding twigs on a stump, is roofed and has an entrance hole in the side. It is built by the male, often using materials which blend into the background sufficiently to make it hard to locate, and is lined with 'feathers and other downy material' by the female. The eggs are incubated exclusively by the female, who occupies the nest intermittently in the daytime. The insulating properties of nests have been considered by *Drent* (1972), together with the energy cost which the process of incubation imposes on the parent and the possible part played by the nest in reducing this.

In *megapodes,* such as the *Mallee* fowl and the brush turkey the use of insulation is carried to its extreme. In both species the male constructs a mound of mixed earth and leaves which then heats up. In the brust turkey it is built at a rainy period, and is about 4m wide and 1m high. Before laying each egg the female excavates a hole in the mound, deposits the egg in an upright position and covers it over with the nest material. A clutch can consist of about 24 eggs laid at intervals over several weeks.

Warming of the eggs
Data collected by *Drent* (1972, 1973) show that incubation temperatures do not vary between species by much more than 4°C. He points out that this similarity is an expression of the similarity in body temperature in adults of different avian species.

In almost all species the eggs are brought up to the incubation temperature by contact with the parent; in most species one or both parents develops a brood patch, or patches. Brood patches are vascular, defeathered, and later, oedematous areas which appear on the bird's ventral surface at the time of egg laying. Hormonal changes underlying this development are discussed by *Hinde* (1967) and *by Jones* (1971). There are many descriptions of ways in which incubating parents bring their brood patches into the right relationship with the egg or eggs, see for instance *Simmons* (1955) for the grebe, *Drent* (1970) for the herring gull and *Beer* (1961) for the blackheaded gull. Certain species, for example cormorants, do not develop a brood patch, and in some, such as the gannet and the booby incubation is carried out by enclosing the single egg between the webbed feet. According to *Howell and Hartholomew* (1962) the mean internal temperature of the egg of the red-footed booby, similarly incubated, is 36°C and the foot temperature 35.8°C. In megapodes the heat is provided by the process of fermentation, supplemented, in the Mallee fowl by the heat of the sun. The mound of the brush turkey (Baltin, 1969) is tended throughout the incubation period by the male parent, which periodically excavates 'control' shafts : these are holes scratched in the mound, about 20-30 cm deep, in which the male sticks its head for a few seconds, before refilling them. The behaviour of the Mallee fowl is similar, except that in the arid scrub lands where it lives, leaves are scarce and the mound is constructed largely of sand. For this reason fermentation is less and *Frith* (1962) reports that three types of work are required from the male to keep the temperature constant. Early in the season, when fermentation is rapid the mound temperature rises rapidly. The male opens the mound every morning, to within a few inches of the eggs, fills in the hole and leaves. This results in a drop in temperature. Later fermentation is slower, fewer holes are excavated, and the eggs are covered with a foot or two of soil. Later still, the rate of fermentation decreases again, but the sun becomes hotter. The bird then works to keep the mound cool by piling soil over it, sometimes to a height of about $1\text{-}1/_4$ m. Sometimes the mounds are completely dug out in the early morning, the sand spread out to cool, then piled up again. According to *Baltin* (1969) the mean temperature in a brush turkey mound is between 33.3°C and 33.9°C; beginning at about 36.7°C it falls through the five months of incubation to about 28.8°C.

In both the last two families (Megapodiiadae and Sulidae) the egg temperature is more likely to resemble that of an egg in a forced draught incubator, in the sense that heat is applied rather evenly over its surface.   In most species the reverse is the case; a normal nest resembles more a still air incubator, where the upper surface of the egg is maintained at the higher temperature.   For example, Baldwin and Kendeigh (1932) recorded temperatures at different places in the wren's nest. they found that the highest temperature is in the area where the egg makes contact with the brood patch. The outside of the clutch is held at an intermediate temperature, and the nest bottom is considerably cooler. This stimulation is largely reversed when the incubating parent leaves the nest..

Egg-shifting behaviour resulting in reorganization of the clutch, may also counteract temperature gradients within the rest. There is more evidence that incubation is not an all-or-nothing affair; the parents' incubation behaviour varies with egg temperature. Franks (1967) has observed the responses of incubating ringed turtle doves to artificial eggs where the temperature could be changed. When the egg temperature was high gular flutter occurred, and elevation of the feathers and shivering was observed when egg temperatures were lowered. Gular flutter is a mechanical aid to evaporative cooling. *Baerends et al.* (1960) and *Drent et al.* (1970) have demonstrated that egg temperature is one of the feed-back stimuli regulating incubation behaviour. They have shown that the herring gull's response to changes in egg temperature is aimed at maintaining the parent's body temperature. In this way egg temperature is regulated while allowing for almost continuous coverage of the eggs.

The problem of cooling

The problem faced by incubating birds in hot climates is a very specific one which has been touched on already at certain stages in the megapodes. It is discussed by *Drent* (1972); in his view the response of the parent to extreme heat and exposure to the sun is to remain at the nest; the problem of keeping the eggs cool is thus shifted to the problem of keeping the parent cool. Drent points out that different ways of doing this have

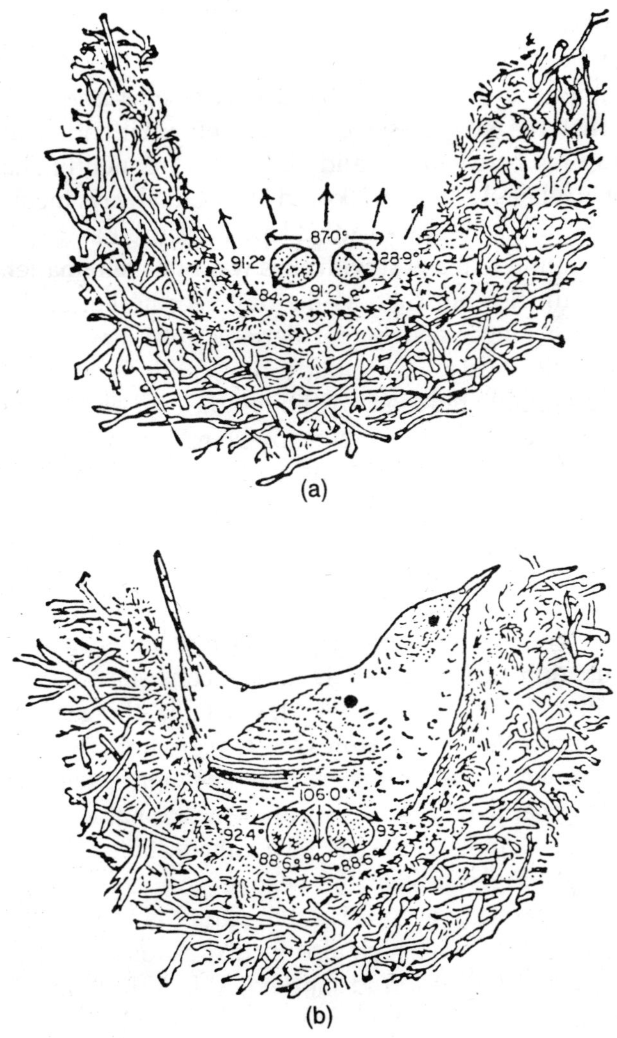

Fig. 4.1. Temperature gradients in the nest of an eastern house wren during a recess *(above)*, and during a session *(below)*.

evolved in different species and include the use of poorly feathered body regions which can act as dissipators of heat (including the legs and feet, panting or gular flutter, the periodic wetting of the abdominal feathers and the orientation of the sitting bird away from the sun. It appears that at extremes of air temperature the eggs cannot be protected by shading alone, but may have to be cooled by contact with the brood patches. For example, in the double banded courser *Maclean* (1967) found that the single egg could be left unattended at air temperatures between 20ˋ and 30°C, was frequently shaded and not incubated between 30o and 36°C, but always incubated above 36°C. In the yellow-wattled lapwing *Jayakar* and *Spurway* (1965a, b) found that a nest with eggs was often left unattended after the sun rise but that, as the day became hotter, it was attended for longer and longer periods until it was covered almost continuously. On cloudy days periods on the nest were substantially reduced. Although the bird on duty never 'shaded' the eggs from the sun (by standing over them) its behaviour suggested that it was cooling rather than heating them.

The continuity of attentiveness

Under artificial conditions attempts are made to maintain incubator temperatures within quite narrow limits throughout the incubation period and this aim is consistent with high hatchability. However, in the wild, where hatching success is also high incubation temperatures are less constant. Fluctuations in nest and egg temperatures are found even in species where one or the other parent is on duty at the nest for at least 95% of the time. For example, *Drent* (1972) has demonstrated a tendency for the egg temperature of the herring gull to vary slightly with air temperature even when the parent is incubating. In the pigeon guillemot (which also incubates for about 95% of the time) *Drent* (1965) found marked, although, fairly short-lived, variations in egg temperature not only when the parents changes 'shifts' on the eggs, but also within shifts, which are interrupted for short periods for defecation, preening or drinking. *Beer* (1961) reports that the blackheaded gull rises and resettles on the eggs frequently during incubation bouts (on the average once every 11.3 min), and similar behaviour may be observed in the guillemot, a species where the parents take turns in incubations, and the egg is never left.

In a large number of species incubation takes place for only part of the time. During off-duty periods in the wren the nest and egg temperatures fall quite considerably. In such species also there must be temperature fluctuations during attentive periods when at rather frequent intervals the sitting bird rises, shuffles about and resettles *Halftorn*, 1966, for three *Parus* species,). In the brust turkey *Baltin* (1969) found the temperature to drop sharply when the male excavates 'control' shafts in the incubation mound. After dropping, it returns rather slowly to its normal level.

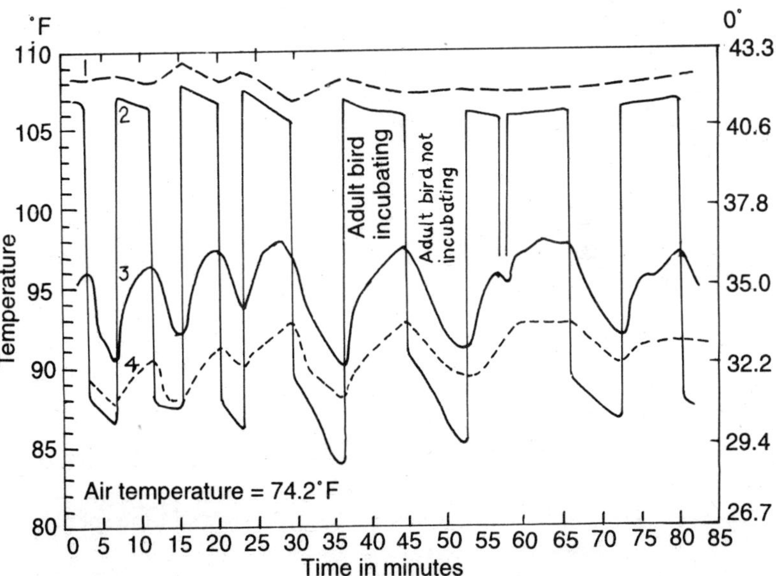

Fig. 4.2.   Natural fluctuations in temperature of the eastern house wren's egg under normal conditions in the nest. (1) Body temperature of adult bird; (2) temperature at top of nest just above the eggs; (3) egg temperature; (4) temperature at bottom of nest beneath the eggs.

In species which incubate intermittently the length of attentive periods has been shown to vary with air temperature; the time spent off the eggs increases with increases in air temperature, while the length of the individual sitting bouts decreases.

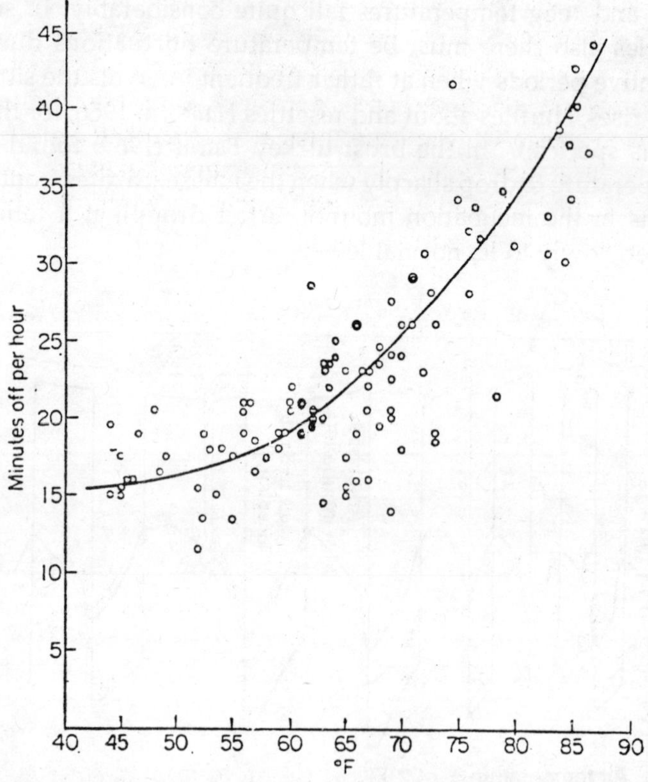

Fig. 4.3.   The correlation between air temperature and time spent off the eggs.

In titmice Haftorn (1966) found the onset of incubation to be intermittent — a factor which makes the duration of the incubation period difficult to assess precisely. Haftorn measured egg and nest temperatures from the beginning of the egg-laying period. During this time in the coal tit he found that the female roosts in the nest box. On entering she settles down on the eggs. After an interval, she stand up again and sleeps in this position. The consequent fall in egg temperature occurs progressively later from night to night, and during the last night or two before the end of egg laying she incubates the egg continuously.

Clearly the maintenance of a constant temperature is not a necessary condition for hatching, although it must be a necessary condition for a constant, or a minimal, incubation period. One problem not yet settled, is whether rather frequent small temperature changes are of any significance in incubation. Russian work summarized by *Lundy* (1969), has claimed that fluctuations in incubator temperatures provide an increase in hatchability in the fowl; it could be of some interest to investigate these fluctuations further. We do not know whether they arise in the wild simply as a compromise between the different requirements of the eggs and of the parents, different species adapting in slightly different ways to achieve this compromise, or whether such fluctuations stimulate or assist embryonic development.

Humidity
It has already been noted that humidity requirements appear to vary between domestic species; in the nests of wild species, however, little or no attention has been paid to direct measurements of humidity during incubation. Nests where humidity has been measured have been those of the domestic fowl, incubating under natural conditions. For example, *Burke* (1925) sampled a steady flow of air from a perforated egg-shaped receptacle in the nest; he found that the amount of moisture given off by the hen is fairly constant throughout the incubation period. Similar measurements have been made by *Koch* and *Steinke* (1944) and by *Chattock* (1925) who showed that humidity within the nest is higher than that outside, although both vary in the same way.

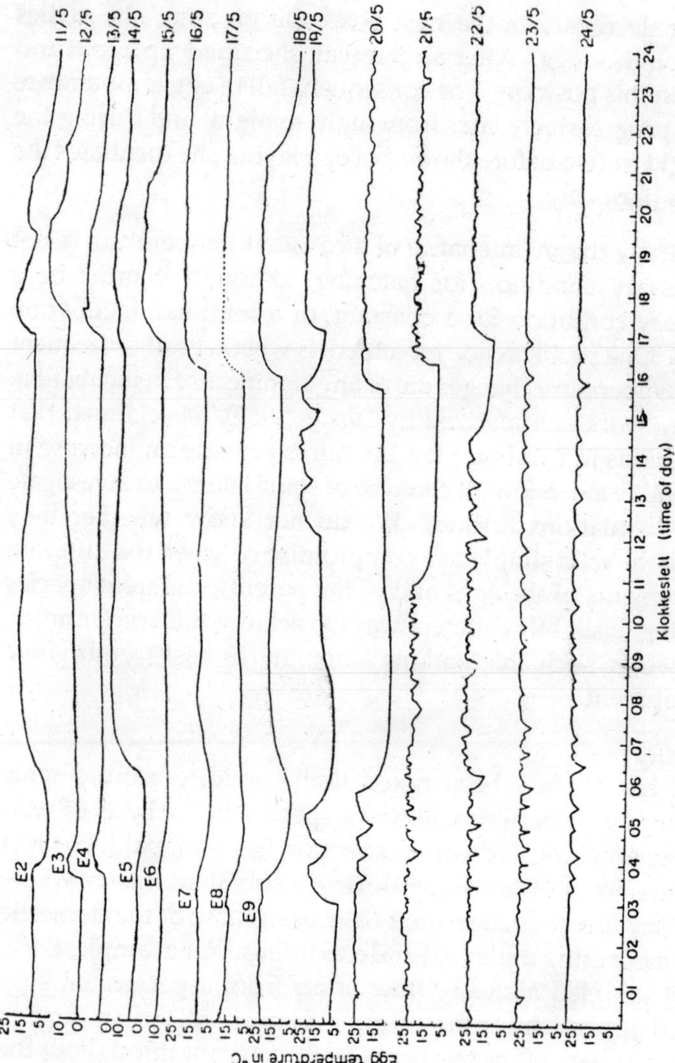

Fig. 4.4.   Egg temperature during 14 successive 24-hour periods in the egg-laying period and initial incubating period in a coal tit nest. Continuous line indicates egg temperature and dotted line outdoor temperature. E2, E3, etc. caption show egg-laying times. Dates are indicated on the right side. X = the female was flushed off the nest.

Also working with natural incubation in the fowl *Lamson* and *Edmond* (1914) considered weight loss from the egg as an indication of humidity and this method has been developed by *Drent* (1973). Drent considers that weight loss of the egg per unit surface area per unit time is a measure which can reflect humidity conditions although temperature and shell characteristics also have to be taken into account. The measure should be a reliable indication of water loss from the egg, since gaseous exchange involves only a negligible change in weight. The daily weight loss during natural incubation of eggs of a wide range of weights has been determined and the specification of the normal range in water loss for an egg of given weight calculated.

The gaseous environment

Oxygen and carbon dioxide concentrations in the mound of the brush turkey have been measured by Baltin (1969). They were found to be variable, but to rise and fall in a definite pattern during each day. The concentration of carbon dioxide was often found to be above 10% ($Pco_2 \cong 76$ mmHg) while the oxygen concentration fell in extreme cases to 7% or 8% (Po_2 60 mmHg). Baltin considered that 'control' shafts excavated by the cock are of significance mainly in regulating gas concentrations within the mound. The CO_2 levels reported are similar to those found in the air space of the egg of the fowl during the last few days of incubation (Visschedijk, 1968c). Gaseous exchange and the oxygenation.

There is early work on the gaseous environment in nests of the domestic fowl, when the female is incubating naturally. For example, *Lamson* and *Edmond* (1914) placed a perforated wooden 'egg' in the clutch of a sitting hen and sampled air from it. They found that the CO_2 concentration within the nest increases from the beginning to the end of the incubation period, reaching a final level of 3-5%. *Burke* (1925) obtained similar results.

Measurements made in nests of other species appear to be

rare. It has, however, been suggested that the frequent rising and shuffling carried on by incubating parents (see below) might serve to 'air the nest' to some extent. It titmice (hole-nesting species which build particularly well-insulated nests, this behaviour has been observed by *Haftorn* (1966). While incubating, the female great, marsh or coal tit frequently rises, stands over the eggs with head down in the nest cavity, and with the tail up. She then works over the nest lining with trembling or shaking movements of the bill before shifting the eggs and settling on them with characteristic 'resettling' movements. Baldwin and Kendeigh (1932) reported that the incubating wren stirs about in the nest every few minutes during the day and the night. The possibility of a relationship between nest insulation and parental activity of this kind could be examined by sampling the air before and after bouts of activity.

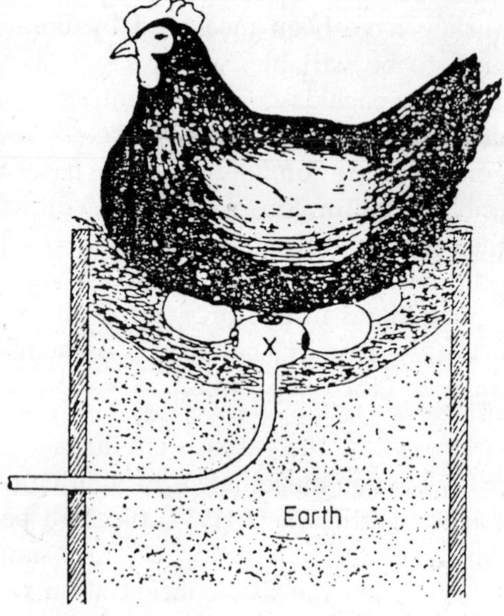

Fig. 4..5.   Method of drawing a sample of air from under a sitting hen.
X = a perforated wooden egg.

Egg shifting or turning

In most nests the eggs are frequently turned, or shifted about; eggs may be rolled, moved to another part of the nest, or simply jostled. Under natural conditions turning is likely to be accompanied by a drop in temperature, by vibratory and acoustic stimulation, and (during the hours of daylight) by an increase in photostimulation.

Egg shifting has been observed in the domestic fowl, when incubating. *Eycleshymer* (1907) made nests with felt sides and glass bottoms; by marking the eggs he found that they were turned more than 5 times a day. *Chattock* (1925) registered the position of each egg in a number of nests of the fowl. He found that no hen failed to shift her eggs very thoroughly to all parts of the nest, and intervals between bouts of turning varied from about 10 to 55 minutes.

Among wild species herring gull and blackheaded gull nests have been observed from below through glass panels. Drent found that the parent rose and resettled on the eggs every 20 minutes or so and that the eggs were moved once in about 7 resettlings. Beer found a resettling rate of about four per hour, mostly without moving the eggs. In nests of titmice observed by *Haftorn* (1966) eggs were shifted every few minutes.

Tinbergen (1953) pooled the data from several herring gull nests and found that changes in egg position were irregular, or largely random. Since then *Lind* (1961) has considered factors underlying egg position. In the black-tailed godwit he found this to be determined partly by the frictional resistance of the nest, and partly by the egg's centre of gravity, which changes in the course of incubation. For the first few days the centre of gravity is situated almost centrally, but after 4-6 days it changes, depending on the positions of the air space, the yolk and the embryo. Towards the end of the incubation period it becomes fixed, as the air space becomes larger and more asymmetrical. The egg can therefore be turned into any position, but towards the end of incubation it tends, after being moved, to roll back into the nest cup with the heaviest part towards the bottom. In this way the parent's rather random egg-shifting behaviour results in the eggs being turned sufficiently at early stages when this is necessary for the embryo's correct development, while later they lie predominantly with the lighter side upwards and pip at the top. The embryo is then in the 'back up' position — a position which facilitates normal breathing.

According to *Poulsen* (1953) egg turning occurs in nearly all species. Exceptions include the palm swift which builds on the underside of palm leaves, glues its eggs to the nest and broods them vertically and also megapodes where the eggs are buried upright in the incubation mound, and remain in this position until hatching. In the swift it would seem that the egg contents could be shifted by leaf movements. In megapodes *Baltin* (1969) has observed that the air space is movable. Possibly this, or other structural peculiarities, obviates the necessity for turning in these species.

Hatching success

Hatchability in incubators cannot be compared directly with that found in the wild. However, when losses due to bad weather conditions, food shortages (which can result in parents deserting their nests), predation and infertility are taken out, hatchability in the wild is often high. For example in bobwhite quail living in semi-wild conditions *Stoddard* (1931) found a high percentage, varying from 85% to 93% over a number of years. In the herring gull *Drent* (1970) reports a hatching success of about 85% of fertile eggs not taken by predators. Gulls, however, fly up from their nests in response to disturbance, a factor which is likely to reduce hatchability. In the relatively undisturbed colony of gannets on the Bass Rock, *Nelson* (1966) found that about 82% of eggs hatched, about 7.3% were infertile and about 11% were 'lost' mainly on account of aberrant parental behaviour, or through an accident (eggs left unattended were usually taken by gulls).

In song birds, where the nests may be less vulnerable, hatching success appears to be slightly higher. A survey of nesting titmice from 1947-1953 has been summarized by *Lack* (1955): excluding nests robbed by a variety of predators about 90% of eggs of the great, blue and coal tit hatched. *Seel* (1968) found 92% of house sparrow eggs and 96% of tree sparrow eggs to be fertile. Omitting losses due to desertion or destruction, but including infertile eggs, hatchability was 88% and 93% respectively.

Lack (1966) discusses hatching success in a number of species, the most successful of these being the blackbird where *Snow* (1958) reported that about 90% of eggs hatched in the Botanic Garden in Oxford (small territories) and between 92% and 95% in Wytham woods (larger territories). In the Botanic Garden snow related hatching success with parental age; where both parents were more than one year old 95% of eggs hatched. Failure to hatch was nearly always due to infertility.

5

The Egg

The parts of a newly laid hen's egg are the shell, shell membrane, albumen and yolk. In an egg that has been undisturbed for a short time the yolk floats in the albumen with a whitish disc, the blastoderm about 3.5 mm. in diameter, on its upper surface. If the yolk is rotated, its center of gravity is such that it will return to its former position in a few minutes, with the blastoderm on top. The blastoderm is the living part of the egg, from which the embryo and all its membranes are derived. It is already in a fairly advanced stage of development when the egg is laid. The yolk and blastoderm are enclosed within a delicate transparent membrane (*vitelline membrane*) which holds the fluid yolk mass together. We may now consider some details of the structure and composition of the parts of the egg.

The shell is composed of three layers: (1) the inner or mammillary layer, (2) the intermediate spongy layer, and (3) the surface cuticle. The mammillary layer consists of minute spherulitic crystals of calcite about 0.01.015 mm, in diameter, welded together, with conical faces impinging on the shell membrane. The minute air spaces between the conical inner ends of the mammillae communicate with the meshes of the spongy layer, which makes up two thirds of the thickness of the shell, and which is bounded externally by the extremely delicate shell cuticle. The spongy layer consists of a mat of very small crystals of calcite. It is indented with pores on its outer surface, which are particularly numerous over the region of the air sac. The pores communicate with the meshwork of the spongy layer and thence with the air spaces of the mammillary layer. The colour of the hen's egg is due to pigments localized in the outer part of the spongy layer; they are derivatives of hemoglobin and are secreted by the uterine glands as the shell is being finished. The shell cuticle is a very thin layer of protein containing scattered globules of fat, but is otherwise structureless. It is continuous over the entire surface, including the pores. The entire shell is permeable to gases, and thus allows embryonic respiration and evaporation of water.

The shell membrane consists of two layers, a thick outer layer (40-60 μ) next to the shell and a thinner one 13-17 μ) nest to the albumen. Both are composed of a matted network of fibers which are chemically intermediate in properties between keratin and chitin. The fibers of the inner layer are of the same size (2.3 μ) and cross one another in all directions. Those of the outer layer are variable in size (8-12 μ), the mesh is looser, and there is a star-like condensation of fibers at certain points which are embedded in the calcite of the shell. This union serves to strengthen the shell, which would otherwise be brittle and fragile. At the blunt end of the egg the two layers are separated and form a chamber containing air that enters after the egg is laid (Figure 5.2).

The physical chatracteristics of the albumen are too well known to require description. A dense layer immediately nest to the vitelline membrane is prolonged in the form of two spirally coiled opalescent cords towards the blunt and narrow ends of the egg respectively; these are the chalazae, so called from the Greek for "tubercles" (Romanoff) or "hailstones" (Bartelmez). The two chalazae are twisted in opposite directions. In a hard-boiled egg it is possible to strip off the albumen in concentric spiral layers from left to right from the broad to the small end of the egg.

The yolk and blastoderm are enclosed within the delicate vitelline membrane; the yolk is a highly nutritious food destined

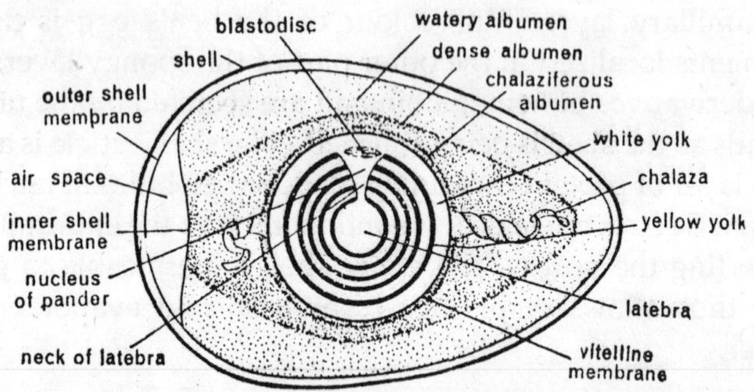

Fig. 5.1. Diagram of the hen's egg.

to be gradually digested and absorbed by the living cells of the blastoderm and used for the growth of the embryo. It is not of uniform composition throughout, but consists of two main ingredients known as the yellow and the white yolk. The yellow yolk makes up the greater part of the yolk sphere; the main part of the white yolk is a flash-shaped mass, the bulb of which, known as the latebra, is situated near the center of the whole yolk, the neck rising towards the surface and expanding in the form of a disc (nucleus of Pander) situated immediately beneath the blastoderm at its margin this disc is continuous with a thin peripheral layer of white yolk that surrounds the entire mass. In addition there are several thin layers of lighter-coloured yolk concentric to the inner bulb-shaped mass.[1] If an egg be opened, a delicate hair inserted in the blastoderm to

---

[1] Whether these concentric rings of light-coloured yolk are true white yolk or merely yellow yolk with less carotinoid content is not settled. The assertion that the thin layers that define the concentric stratification of the yellow yolk are of the nature of white yolk is traceable to Meckel V. Hemsbach, Leuckart, and Allen Thomson. His was not able to satisfy himself that the characteristic elements of the white yolk occur within these thin concentric lamellae (*Untersuchungen ueber die erste Anlage des Wirbeliltierlebes*). Recent work indicates that the presence or absence of visible lamellae within the yellow yolk can be controlled by adjusting the diet of the bird, whereas the true white yolk is not visibly affected by such dietary changes.

mark its position, and then boiled hard, a section through the hair and center of the yolk show the above relations quite clearly. The white yolk does not coagulate so readily as the yellow yolk, and it may be distinguished by this property as well as by its lighter colour.

Both kinds of yolk are made up of innumerable spheres which are, however, quite different in each (Figure 5.3). Those of the yellow yolk are on the whole larger than those of the white yolk (about 0.025-0.100 mm. in diameter) with extremely fine granular contents. Those of the white yolk are smaller and more variable in size, ranging from the finest granules up to about -.07 mm. The larger spheres of the white yolk contain several highly refractive granules of relatively considerable size as compared with those of the yellow spheres and such granules may have secondary inclusions. As we shall see later, the smaller granules of the white yolk extend into the germinal disc (forerunner of the blastoderm) and grade into minute yolk granules contained within the living protoplasm.

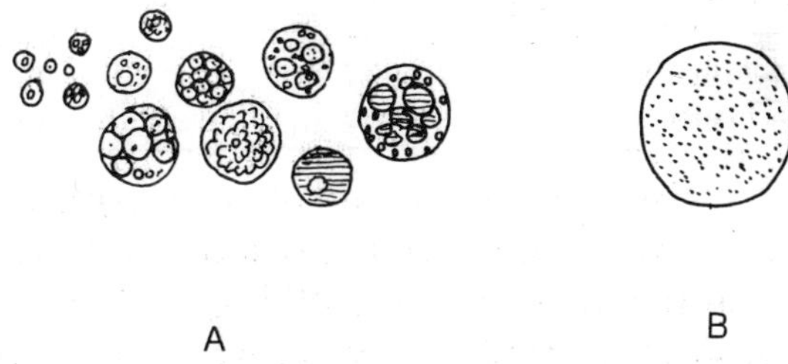

Fig. 5.2. Yolk spheres of the hen's egg; highly magnified. (After Foster and Balfour)

A. Varieties of white yolk spheres.
B. Yellow yolk sphere.

The earlier investigators from the time of Schwann regarded the white yolk spheres as actual cells (Schwann, Reichert, Coste, His). His especially laid great stress on this interpretation; he believed that they were derived from the cells of the ovarian follicle which migrated into the ovum in the course of oogenesis, that they multiplied like other cells,and took part in the formation of certain embryonic tissues. Subsequently he abandoned this position as untenable. The white yolk spheres are now universally regarded as food matters of a particular sort rather than cells.

Chemical Composition of the Hen's Egg

The yolk and albumen are complex mixtures of many different substances, organic and inorganic, containing all of the compounds necessary for the growth of the embryo. The figures given below are for average amounts; the individual egg is, of course, subject to great variation:

GENERAL COMPOSITION OF THE YOLK

	Amount in Grams	Percent of Total
Total	18.7	100.0
Water	9.1	48.7
Solids	9.6	51.3
Organic Matter	9.4	50.2
Proteins	3.1	16.6
Lipids	6.1	32.6
Carbohydrates	0.2	1.0
Inorganic Matter	0.2	1.1

The lipids make up the greater part of the solid constituents of the yolk and serve as a major source of food for the embryo. The consist of the true fats (glycerides) (623 per cent), phospholipids (32.8 per cent), sterols (such as cholesterol) (4.9 percent), and traces of cerebrosides. Two proteins have been identified in the yolk: a phosphoprotein, called ovovitellin, which makes up about 77 per cent of the total protein, and ovolivetin (23 per cent), which has a high content of sulfur. The phosphoproteins are the main nutritive proteins utilized by the young of both birds and mammals (Hewitt). The carbohydrates of the yolk consist mainly of free glucose (70 per cent) and of polysaccharides (chiefly mannose and galactose) combined with the phospholipids, phosphoproteins, and cerebrosides. Other organic components include pigments (carotenes, xanthophylls, riboflavin, etc.), vitamins, and enzymes. Of the minerals, phosphorus is the most abundant and occurs mostly in organic compounds, principally lecithin. Calcium, magnesium, chlorine, potassium, and sodium are present in lesser quantities, as are very small amounts of sulfur and iron.

GENERAL COMPOSITION OF THE ALBUMEN

	Amount in Grams	Percent of Total
	32.9	100.0
Water	28.9	57.9
Solids	4.0	12.1
Organic Matter	3.8	11.5
Proteins	3.5	10.6
Lipids	Trace	-
Carbohydrates	0.3	0.9
Inorganic Matter	0.2	0.6

The proteins are the chief solid constituent of the albumen and consists of ovalbumin (75 per cent), ovomucoid (13 per cent), ovomucin (7 per cent), ovoconalbumin (3 per cent), and ovoglobulin (2 per cent). The second and third named are glycoproteins (i.e., combined with a carbohydrate); the others are simple proteins. The exact nature of the carbohydrates is obscure, but they are polysaccharides and are believed to contain mannose in various combinations with glucosamine or galactose. The only pigment reported for the albumen is riboflavin ("ovoflavin"), which gives it a faint yellowish tinge. Sulfur, potassium, sodium, and chlorine are the most abundant minerals, in the order named; there are lesser amounts of phosphorus, calcium, and magnesium, and traces of iron.

The albumen contributes not only to the nutrition of the embryo, but serves as a protective envelope, shielding the embryo both mechanically (from physical inquiry) and chemically (from bacterial contamination). Chemical protection is afforded by the presence of avidin, a biotin-binding protein which is abundant both in the albumen and in the hen's oviduct from which it is secreted. The albumen also has a lytic action on some kinds of bacteria, which may be due to a discrete substance (*"lysozyme"*). Thus, even if a contaminant should succeed in passing the shell-barrier, it would probably be lysed or its growth would be inhibited by the avidin.

The shell consists of a collagenous matrix impregnated with lime salts: about 95 per cent calcium and magnesium carbonates and phosphates (of which amount the calcium salts constitute 98 per cent); about 5 per cent collagen and water; traces of lipids, iron, and sulfur.

The shell membrane consists of keratin, or a closely allied substance, along with a small amount of water and traces of minerals. The vitelline membrane is believed also to contain a keratin-like material along with mucin or collagen.

Formation of the Egg

The organs of reproduction of the hen are the ovary and oviduct of the left side of the body. Although the right ovary and oviduct are formed in the embryo at the same time as those of the left side, they degenerate more or less completely in the course of development so that only functionless rudiments remain. This would appear to be correlated with the large size of the egg and the delicate nature of the shell, as there is not room for two eggs side by side in the lower part of the body cavity.

The ovary lies at the anterior end of the kidney attached by a fold of the peritoneum (mesovarium) to the dorsal wall of the body cavity. In a laying hen, ova of all sizes are found from microscopic up to the fully formed ovum ready to escape from the follicle. Such an ovary is shown in Figure 5.3; the gradation in size of the ova will be noticed up to the one fully formed and ready to burst from its capsule. The figure shows a reptured follicle, from where the ovum has escaped.It will be seen that the part of the definitive hen's egg produced in the ovary is the so-called yolk. The blood supply of the very vascular ovary is derived from the dorsal aorta, and the ovarian veins open into the postcaval vein.

The oviduct is a large coiled tube which begins in a wide mouth with fringed borders, the *ostium tubae abdominale* (funnel or infundibulum) opening into the body cavity near the ovary. It is attached by a special mesentery to be dorsal wall of the body cavity, and opens into the cloaca. The following divisions are usually distinguished: (1) the funnel or infundibulum, (2) the albumen-secreting portion or magnum, (3) the isthmus, (4) the uterus or shell gland, (5) the vagina (Figure 5.3). The albumen-secreting portion includes all of the coiled tube; the isthmus is a short section next to the dilated uterus, and the vagina is the short terminal portion opening into the cloaca (Figures 5.3 and 5.4).

The formation of an egg takes place as follows: the yolk, or ovum proper, escapes by rupture of the follicle along a preformed band, the stigma (Figure 5.3), is picked up by the infundibulum, which swallows it, so to speak, and is passed down by peristaltic contractions of the oviduct. The escape of the ovum from the follicle is known as the process of ovulation. During its passage down the oviduct it becomes surrounded by layers of albumen secreted by the oviducal glands. The shell membrane is secreted in the isthmus and the shell in the uterus. The ovum is fertilized in the uppermost part of the oviduct and the cleavage and early stages of formation of the germ layers take place before the egg is laid. The time occupied by the ovum in traversing the various sections of the oviduct is as follows :

The ovum enters the oviduct within 15 minutes after its liberation from the follicle, passes through the funnel in 18 minutes, spends approximately 3 hours in traversing the magnum or albumen-secreting portion, one hour in the isthmus, and 20 to 24 hours in the uterus. Laying normally occurs as soon as the egg is completely formed. The tendency of some hens to lay eggs in "clutches" or groups, instead of laying at

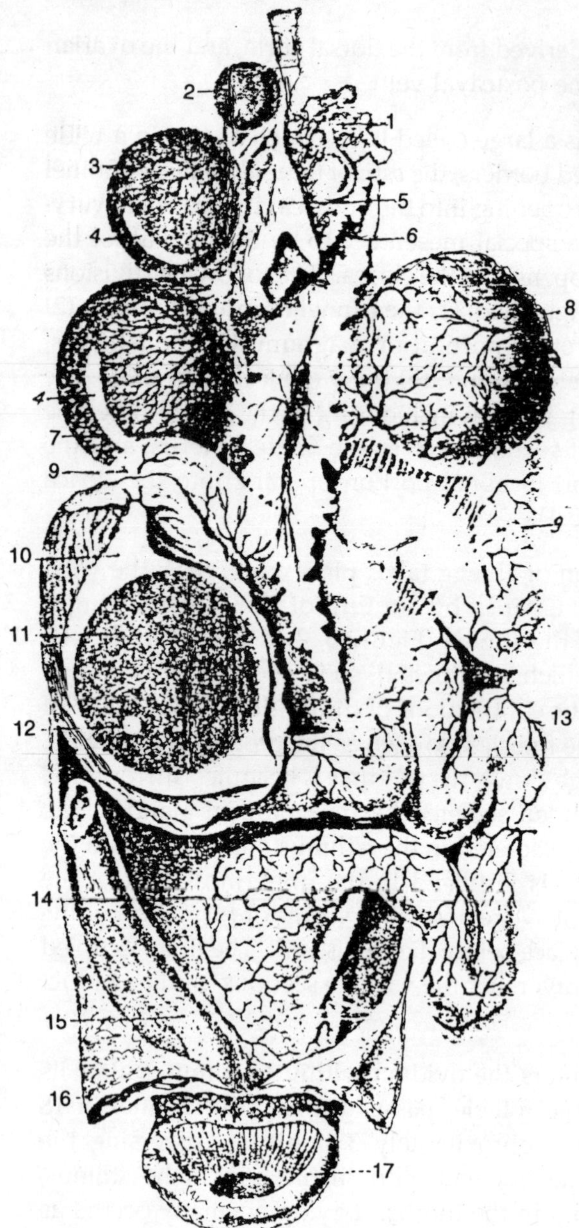

Fig. 5.4. (Caption on page 189)

the usual daily interval, is due to irregularities in the rate of ovulation and not to the retention of fully formed eggs in the uterus.

Some of the details of these remarkable processes deserve attention: the definite periodicity of ovulation, which occurs at approximately daily intervals, has long been known to be governed in some way by the amount of illumination to which the hen is subjected. Furthermore, if a hen is anesthetized and the oviduct is extirpated so as to remove such possible causes of ovulation as manipulation of the ovarian follicles by muscular activity of the infundibulum, then the liberation of mature ova still continues at the expected intervals. These facts suggest that ovulation is under hormonal control, and indeed. Fraps and his co-workers have successfully induced premature and multiple ovulations by injecting hens with pituitary hormones from the horse luteinizing and follicle-stimulating hormones; gonadotropic hormones from pregnant mares' serum). Another hormone which proved to be highly effective was crystalline progesterone. The release of an ovum could be induced by as much as 17 hours previous to the time of its expected normal ovulation. The most sensitive follicle was the one which was most mature, and the effectiveness of the hormones was

Fig. 5.4.   Reproductive organs of the hen. The figure is diagrammatic in one respect, namely, that two ova are shown in the oviduct at different levels; normally but one ovum is found in the oviduct at a time.

1. ovary, region of young follicles. 2 and 3 Successively larger follicles. 4, Stigmata, or nonvascular areas along which the rupture of the follicle takes place. 5, Empty follicle. 6, Cephalic lip of ostium. 7, Funnel of oviduct (ostium tubae abdominale). 8, Ovum in the upper part of the oviduct. 9, Region of the oviduct in which the albumen is secreted (magnum). 10, Albumen surrounding on ovum. 11, Ovum. 12, Germinal disc. 13, Lower segment of albumen-secreting portion. 14, Lower part of the oviduct ("uterus," shell gland). 15, Rectum. 16, Reflected wall of the abdomen. 17, External opening of cloaca.

progressively less with younger follicles. Thus, it appears probable that normal ovulation is caused by the sudden release of an appropriate hormone into the blood stream. The source of the rhythm governing its release and the identity of the hormone which normally operates have not yet been clarified in birds. However, by analogy with the situation in the rat, we may infer that a neural timing factor (localized in the anterior hypothalamus?) acts through the pituitary gland, thus causing the periodic release of increased quantities of gonadotropic hormone sufficient to bring about ovulation (Everett and Sawyer). Light and the sexual hormones are no doubt also involved in the ovulatory cycle, because of their effects on the activity of the pituitary.

The actual expulsion of the ovum from the follicle is accomplished by a contraction of muscular fibers in the follicular wall. The tension thus produced increases the turgidity of the follicle, compresses blood vessels near the stigma so that they "fade out," and causes the stigma to rupture at one of its ends where the stress is greatest.

The upper part of the oviduct, particularly the infundibulum, becomes very active at the time of ovulation. By virtue of its muscular activity, the funnel makes clasping movements, enclosing any object which it encounters, immature as well as mature follicles entirely at random. The funnel does not appear to exert any pressure on the follicles while grasping them, so that it is doubtful whether such muscular activity is a causative factor in ovulation. The funnel of the oviduct can pick up ova that have escaped into the body cavity, but in some cases ova that escape into the body cavity undergo resorption there.

As soon as the ovum passes through the funnel, it enters the magnum of the oviduct which is lined with secretory goblet cells intermixed with ciliated columnar cells. The albumen is elaborated by the goblet cells as strands of mucinous gel which are then wrapped around the ovum in successive spiral layers as it passes down the oviduct.

Only about 50 percent of the white of the egg is formed by the albumen-secreting portion of the oviduct; this is in the form of a dense layer formed of matted fibers of mucin; the shell membrane is deposited directly on this while the egg is in the isthmus; and the more fluid portion of the albumen constituting 50 per cent or more of its entire bulk enters through the shell membrane while the egg is in the uterus. The uterine secretion is not an albumenous substance, but rather a mineral solution consisting principally of water, potassium and bicarbonate ions, and smaller amounts of sodium and chloride ions. The concentration of potassium ions is unusually high (about ten times that of normal blood serum) and probably related to the heavy demand for potassium by embryos and young animals. The uterine secretion diffuses through the shell membrane, dilutes the thick albumen somewhat, and in turn receives dissolved albumens, resulting at equilibrium in the outer layer of so-called thin albumen.

The chalazae do not appear until the egg has been in the uterus for some time. Contrary to the views of early workers, they are not formed by a special secretory mechanism in the cephalic end of the oviduct but apparently arise by mechanical means. The most plausible explanation is that of Conrad and Phillips: the albumen is laid down throughout the magnum as a homogeneous gel, but as the ovum continues to rotate in its spiral descent through the oviduct, there is a gradual breakdown of the colloidal gel structure of the innermost layer of albumen until it becomes a sol. This conversion into a sol, which is completed at about the time that the egg enters the uterus, permits the ovum to rotate freely within its albumenous envelope. The yolk tends to maintain a constant position (due to the lesser density of the hemisphere which contains the germ spot); but the albumen continues to revolve as the egg turns within the uterus, so that the mucinous fibers of the inner liquefied layer of albumen becomes twisted at each end of the yolk and form the chalazae. Added weight is given to this explanation by removing egggs just as they enter the uterus and incubating them within a rubber "artificial uterus" with a proper amount of artificial uterine fluid; under such conditions, no chalazae are formed unless the rubber "uterus" is rotated. The fact that the line joining the attachments of the chalazae is at right angles to the main axis of the ovum (that passing through the germinal disc) indicates that there must be some antecedent condition that determines the position of the ovum in the oviduct and, therefore, the orientation of the ovum to its albumenous wrapping. This is probably the position of the ovum in the follicle, i.e., the relation of the germinal disc to the stigma, for the follicular orientation is apparently more or less preserved in the oviduct. The question is of considerable importance because as we shall see, the axis of the embryo is later bisected by a plane passing through the chalazae, and is therefore certainly determined at the time that the albumen-gel is laid down, and Bartelmez even traces it back to the earliest stages of the oocyte.

Abnormal Eggs

These are of two main kinds: those with more than one yolk, and enclosed eggs (*ovum in ovo*). Double-yolked eggs are obviously due to the simultaneous, or almost simultaneous, liberation of two yolks, are their incorporation in a single set of egg membranes. The two yolks are are usually separate in such cases and are derived, presumably, from separate follicles. But two yolks within a single vitelline membrane have been observed; such are in all probability products of a single follicle. Cases of three yolks with a single shell are extremely rate. The class of enclosed eggs includes those in which there are two shells, one with the other. In some cases the contents of the enclosed and the enclosing eggs are substantially normal, though of course the enclosing shell is abnormally large; in others the enclosed egg may be abnormal as to size (small yolk), or contents (no yolk). In all cases described, the enclosing egg possesses a yolk (Parker). Abnormal eggs of these three classes are of either ovarian or oviducal origin; double-yolked eggs and eggs with abnormal yolks are due to abnormal ovarian conditions; enclosed eggs to abnormal oviducal conditions, or to both ovarian and oviducal abnormalities. Assuming the normal peristalsis of the oviduct to be reversed when a fully formed egg is present, the egg would be carried up the oviduct a greater or less distance and might there meet a second yolk. If the peristalsis became normal again, both would be carried to the uterus and enclosed in a common shell.

Oogenesis

Oogenesis, or the development of ova, may be divided into three very distinct stages. The first stage, or period of multiplication, is embryonic and ends about the time of hatching (in the chick); it is characterized by the small size of the ova and their rapid multiplication by division. The multiplying primitive ova are known as *oogonia*. At the end of this period multiplication ceases and the period of growth begins. The ova, known as *primary oocytes*, become enclosed in follicles; the size of the ovum constantly increases and the yolk is formed. The third period, known as the period of maturation, is characterized by two successive exceedingly unequal divisions of the egg cell, producing two minute cells, the polar bodies, that take up part in the formation of the embryo, but die and degenerate. The process of maturation begins in the fully ripe follicle and is completed after ovulation in the oviduct, while the ovum is being fertilized.

In the young chick all the cell cords and cell nests become converted into primordial follicles. During the egg-laying period there is a continuous process of growth and ripening of the primordial follicles, which takes place successively; the immense majority at any given period remain latent, but all stages of growth of egg follicles may be found in a laying hen.

A primordial follicle consists of the ovum surrounded by a single layer of the cuboidal epithelial cells (granulosa or follicle cells); the fibers of the adjacent storma have a concentric arrangement around the follicle forming the theca folliculi. The ovum itself is a rounded cell with a large nucleus placed excentrically so as to define a primary axis of the ovum. In the protoplasm on one side of the nucleus is a concentrated mass of protoplasm, the yolk nucleus, from which rays extend, and minute fatty granules.

Holl derives the follicular cells in birds from the storma, but on insufficient grounds. According to D'Hollander, they are derived, like the primitive ova, from the germinal epithelium, in which he agrees with the majority of his predecessors. He states that the period of multiplication of the oogenia ends about the time of hatching that the period of growth of the oocyte begins at about the fourteenth day of incubation (seven days before hatching), and before the formation of the primordial follicle, which begins on the fourth day after hatching. Thus the periods of multiplication and growth overlap.

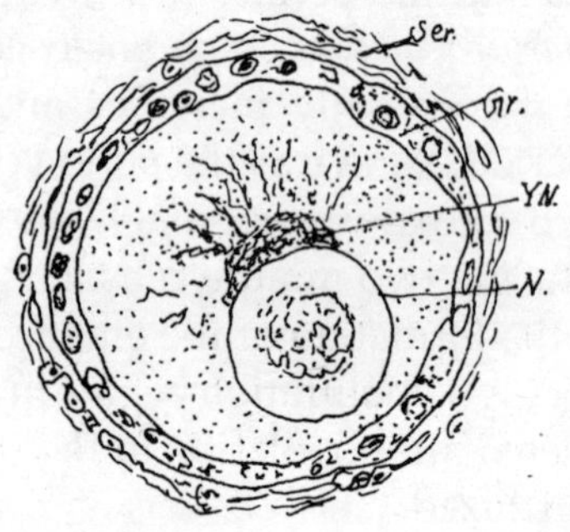

Fig. 5.5.   Primordial follicle from the ovary of the hen. (After Holl)
Gr., Granulasa, N., Nucleus, Str., Stroma, Y.N. Yolk nucleus.

Although the nucleus is strongly excentric in position in the youngest oocytes, it occupies a more nearly central position in those slightly older. When the ovum is about 0.66 mm. in diameter, its nucleus moves to the surface along the shortest radius and comes to lie almost in contract with the vitelline membrane. It becomes elliptical, and later the outer surface is flattened against the vitelline membrane, the inner surface remaining convex. The point on the surface to which the germinal vesicle migrates is situated away from the surface of the ovary, and thus in the position of the pedicle of the follicle, when the latter projects from the surface of the ovary. This determines the position of the future germ disc. The nucleus increases in size with the growth of the ovum; in the youngest oocytes its diameter is about 9 μ; in the ripe ovum it is flattened and may measure 455 μ in diameter by 72 μ in thickness.

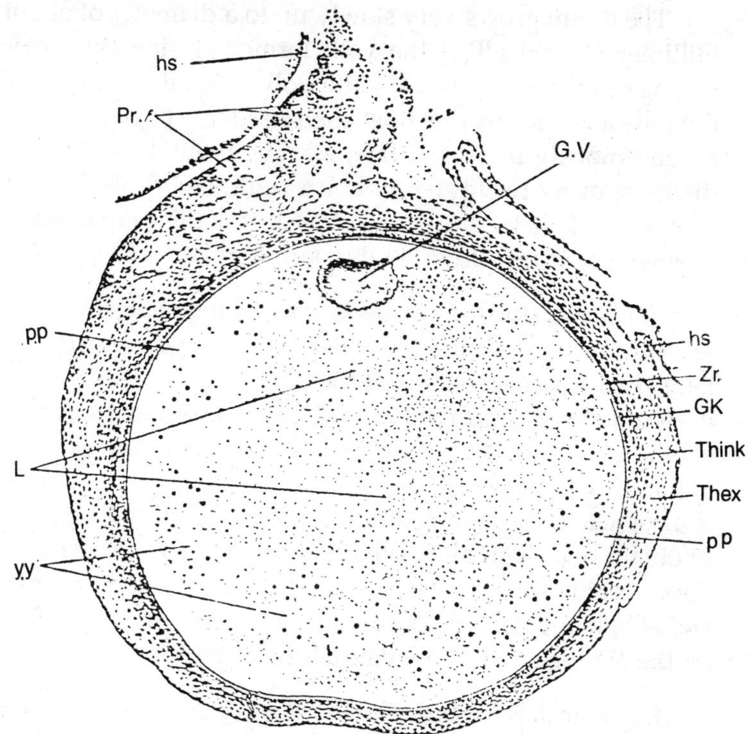

Fig. 5.6.   Section of an avarian ovum of the pigeon. The actual dimensions of the ovum are 1.44 x 1.25 mm.

f.s., Stalk of follice. G.V. Germinal vesicle, Gr., Granulosa. L., Latebra. p.P., Peripheral protaclasm pr. f., Primordial fllicles. Th. ex., Theca externa. Th. int., Theca interna. Y. Y, Yellow yolk. Z.r., Zona radiata.

While the nucleus is still near the center of the egg a very dense deposit of extremely fine granules is formed around it, and gradually extends out towards the periphery of the cell, but does not involve the peripheral layer of protoplasm. This cluster of granules ("yolk nucleus" is called a *"mitochondrial cloud"* by *Brambell*. The mitochondria swell up and become converted directly into spherules or white yolk. The central aggregation of yolk granules thus formed represents the primordium of the latebra or central mass of the white yolk.

The ovum grows very slowly up to a diameter of about 6 millimeters, and all of the yolk formed during this period belongs to the category of white yolk. Certain of these ova, but only a few at any one time, then suddenly begin to grow at an enormously increased rate, adding about 4 mm. to their diameter every twenty -four hours until the full size of about 40 mm. in diameter is attained. It is during this period that the yellow yolk is laid down in the periphery.

Riddle has studied this period by the ingenious method of feeding the stain Sudan III, which has an especial affinity for fat, to laying hens at definite time intervals. The stain attaches itself to fatty acids of the food which are taken up unchanged by the egg. The consequence is that during any period of Sudan III feeding a red stained layer of yolk is formed: so that it is possible by regulating the dose and interrupting the feedings to obtain ova with alternate bands of stained and unstained yolk. In this way he was able to show that a layer of yellow and of lighter-coloured yolk about 2 mm. in combined thickness on the average is laid down each twenty-four hours.

Later studies have shown that the concentric stratification in colour of the yellow yolk is largely governed by the diet of the hen. If a uniform and well-balanced mash is available at all times, then strata do not appear in the yellow yolk. But if hens are kept on a restricted feeding period (6 hours) or are fed a mash which is poor in the carotinoid pigments (chiefly xanthophylls) which colour the yolk, then the yolk becomes stratified in colour. In view of these facts, it does not appear probable that the concentric lamina of light-coloured yolk are a true "white yolk" as previously described, but rather represent yellow yolk which is lacking in carotinoid pigment.

The germinal vesicle lies in a thickening of the peripheral layer of protoplasm known as the germinal disc or blastodisc, which is continuous, like the remainder of the peripheral protoplasm, in early stages with the protoplasmic reticulum that forms the walls of the yolk vacuoles. The blastodisc increases in extent and thickness, and the peripheral protoplasm disappears over most of the yolk. An inflow of the peripheral protoplasm into the disc appears very probable by analogy with the bony fishes where this process can be studied with great ease.

The method of formation of the neck of the latebra and the so-called *nucleus of Pander*, or peripheral expansion of the neck, follows more or less directly from the preceding account: As the ovum increases in diameter during the period of rapid growth, yellow yolk is deposited all around its circumference except in the region occupied by the blastodisc. Since the blastodisc remains at the ever-changing circumference of the growing ovum, a column which contains no yellow yolk is formed along the radius extending from the central mass of white yolk to the definitive site of the *blastodisc*. The column is known as the *neck of the latebra*. When the ovum is fully grown, the exact boundaries between the protoplasmic blastodisc and the yolk are not determinable. The disc itself is charged with small yolk granules which grade off very gradually into the white yolk lying around and beneath the disc.

The ovum is not a perfect sphere, but is ellipsoidal, with its longest axis coinciding with the longest axis of the egg. In addition to this asymmetry of the axes, more yellow yolk is deposited in that half of the ovum which lies opposite the blastodisc than in the half which contains it. Thus, the central mass of the latebra comes to lie somewhat above the true center of the ovum, and the resultant shift in the center of gravity probably accounts for the fact that the yolk tends to float with the blastodisc on top.

The mode of nutrition of the ovum and the formation of

the vitelline membrane remain to be considered. The nutrition is conveyed from the highly vascular theca folliculi by way of the follicular cells, or membrana granulosa, to the ovum. The nutriment enters by diffusion; it has been reported that Golgi apparatus is extruded from the follicular cells directly into the ovum and persists in its new environment but there is no evidence of immigration of solid food particles, let alone entire cells, into the growing ovum. Prior to the last stage of yolk formation, a definite membrane is formed between the ovum and the follicular cells, the zona radiata or primordium of the vitelline membrane.

The discussion as to whether the zona radiata is a product of the ovum itself or of the follicular cells seems to be largely academic and will not be summarized here. There seems to be sufficient evidence of a primary true vitelline membrane secreted by the ovum itself, though this may not represent the entire zona radiata of older ova. Yolk is formed probably by a combination of secretion from the follicular epithelium and a diffuse of nutrients from the blood steam through the follicular epithelium and zona radiata, with final synthesis into yolk taking trace with the ovum. The follicular epithelium may be thought of as a selective filter which controls the proportions of ions and food materials which diffuse into the ovum.

6

Early Development

During the growth period the germinal vesicle has increased to an enormous size (455 x 72 μ in an ovum 37 mm. in diameter, Figure 6.1). It lies in the center of a thickened mass of cytoplasm which, according to *Olsen*, is about ten times the diameter of the nucleus. As seen from the surface, the germinal vesicle appears spherical, but in cross-section it is actually flattened against the vitelline membrane. The margins of the lenticular nucleus are folded into the interior in such a way that sections give an effect of rod-shaped bodies springing from the membrane which were doubtfully interpreted as chromosomes by *Holl*. The real chromosomes are however in the center in the form of double rods. The germinal vesicle and its surrounding cytoplasm constitute the germinal disc.

Fig. 6.1.   Vertical section of germinal vesicle of hen's egg. Size of egg 37 mm. in diameter; size of nucleus 455μ x 72μ.

The process of maturation in the hen's egg has been described by *Olsen* as follows: approximately twenty-four hours before ovulation, vacuoles appear in the upper wall of the germinal vesicle, and the membrane starts to break down. The contents of the vesicle then spread out laterally and flatten to form a thin sheet under the vitelline membrane. The chromatin remains in a very limited area near the center of the distintegrating vesicle. The first maturation spindle arises with its long axis at right angles to the surface of the egg, and the first polar body is extruded by a least one hour before ovulation. The first polar body in freshly extruded ova measures approximately 18 μ x 8 μ and lies in a small depression near the center of the germinal disc and just beneath the vitelline membrane. At the time of ovulation the second maturation spindle has formed. Whether the entry of sperm is necessary for the completion of maturation is not clearly demonstrated, but *Olsen* is of the opinion that the second polar body is not extruded until after fertilization.

The wall of the germinal vesicle begins to break down in ovarian eggs of about 18.75 mm. diameter, the full size of the egg of the pigeon being about 25 mm. Part of the fluid contents of the germinal vesicle flows out and forma layer outside the distintegrating wall. The chromosomes and nucleoli form a group near the center of the upper plane surface of the germinal vesicle. The first maturation spindle is formed before ovulation, containing eight quadruple chromosomes (tetrads).The spindle is still in the equatorial plane stage when the ovum is grasped by the mouth of the oviduct. The bulk of the substance of the germinal vesicle soon forms a yolk-free cone extending from the maturation spindle deep into the superficial yolk. The outer end of the spindle is in almost immediate contact with the surface of the ovum. In the late stages of formation of the first polar body each tetrad, or quadruple chromosome, separates into two dyads or double chromosomes, and the members of each pair of dyads separate and approach opposite ends of the spindle (anaphase).

Thus at each end of the spindle there are eight dyads. Those at the outer end then enter a little bud of protoplasm projecting above the surface of the blastodisc, and this bud with the dyads is cut off as the first polar body, which lies in a depression of the blastodisc beneath the vitelline membrane. Eight dyads, therefore, remain within the blastodisc.

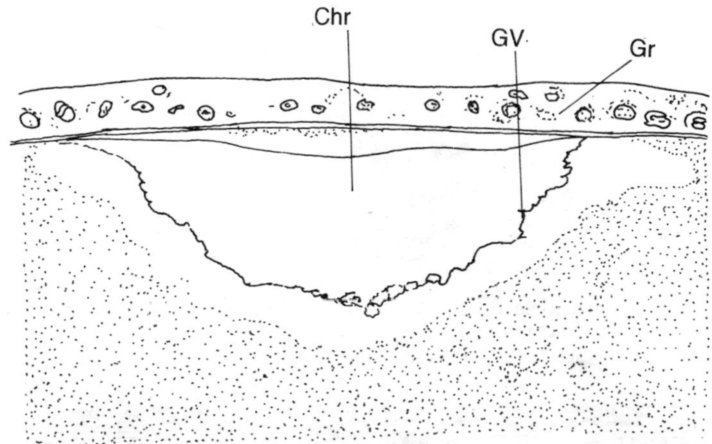

Fig. 6.2.   Vertical section of the germinal vesicle and part of the germinal disc of on ovarian ovum 3/4 inch in diameter; pigeon. Chr., Chromosomes. Gr., Granulosa, G.V., Wall of germinal vesicle.

A second maturation spindle is then formed almost immediately, apparently without the intervention of a resting stage of the nucleus, and takes a radial position similar to that occupied by the first, with the dyads forming an equatorial plate. Each dyad then divides along the preformed plane of division, and the daughter chromosomes diverge towards opposite poles of the spindle. The outer end of the second maturation spindle then enters a superficial bud of the protoplasm of the blastodisc similar to that of the first maturation spindle; and this bud together with the contained chromosomes becomes cut off as the second polar body.

The result of these processes of maturation is the formation of three cells, the two polar bodies and the mature egg. The polar bodies are relatively very minute and soon degenerate completely.

After the formation of the second polar body there remain in the egg eight chromosomes, each of which represents one quarter of an original tetrad. These form a small resting nucleus known as the egg nucleus or female pronucleus. It is many

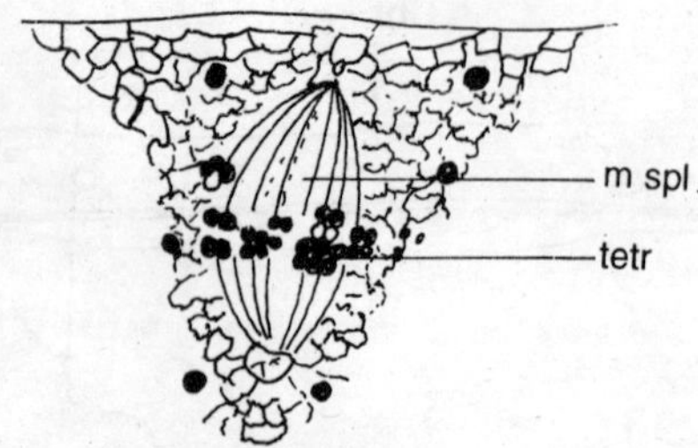

Fig. 6.3.   Vertical section of the blastodisc of the pigeon's egg showing the first maturation spindle.
m. Sp. 1, First maturation spindle. Tetr., Tetrad.

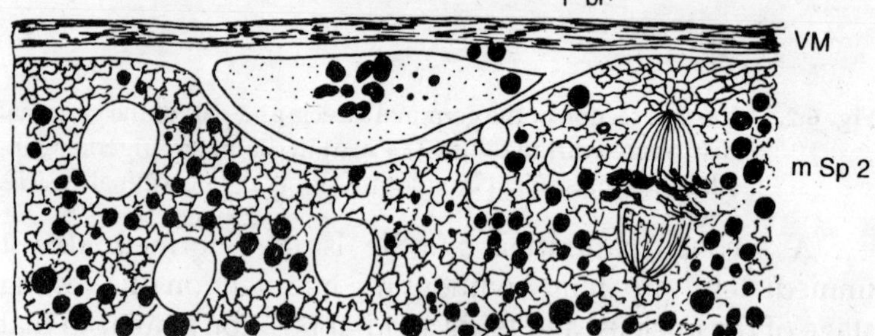

Fig. 6.4.   Second maturation spindle and first polar body of the pigeon's egg; a combination of two sections.
m. Sp. 2, Second maturation spindle. p.b. 1, First polar body. V.M., Vitelline membrane.

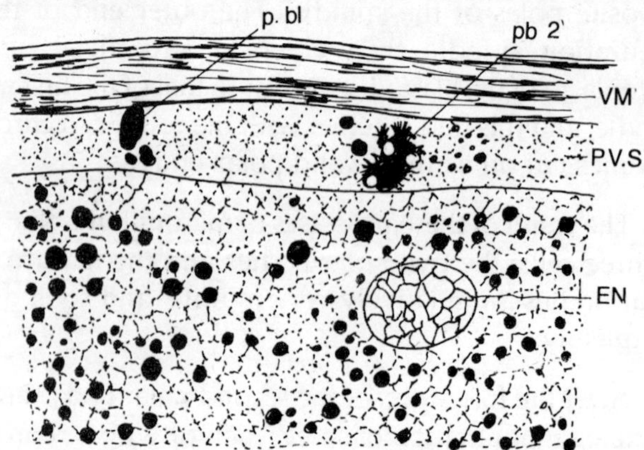

Fig. 6.5.   Egg nucleus (female pronucleus) and polar bodies of the pigeon's egg.
E.N., Egg nucleus. p.b. 1. First polar body. p.b. 2, Second polar body. p.v.s., Perivitelline space. V.M., Vitelline membrane.

times smaller than the original germinal vesicle and it rapidly withdraws from the surface of the egg to a deeper position near the center of the blastodisc.

FERTILIZATION

The spermatozoa traverse the entire length of the oviduct and are found in the uppermost portion in a fertile hen. The period of life of the spermatozoa within the oviduct is considerable, as proved by the fact that hens may continue to lay fertile eggs for a period of at least three weeks after isolation from the cock. After the end of the third week the vitality of the spermatozoa is apparently reduced; as eggs laid during the fourth and fifth weeks may exhibit, at the most, abnormal cleavage, which soon ceases. Eggs laid forty days after isolation are certainly unfertilized, and do not develop.

Parthenogenetic cleavage occurs in about 15 per cent of such unfertilized eggs, but lit is of short duration. Evidence of abnormality such as multipolar mitosis is common. The mitotic nuclei apparently contain the diploid number of chromosomes, which suggests that the second polar body may have been retained because of lack of the stimulus of fertilization. By the time that the egg is laid, most of the parthenogenetic cleavage nuclei have become necrotic. Parthenogenetic cleavage also occurs in the pigeon but becomes abortive as soon as the blastoderm reaches late stages of cleavage and before there is any subgerminal cavity.

The ovum enters the infundibulum within fifteen minutes after ovulation and is surrounded there by a fluid containing spermatozoa. According to *Olsen*, the second polar body, comparable in size and appearance to the first, is formed after sperm penetrate the egg. Although probably correct, this conclusion should be accepted with caution, since it is based on inference from the observation that the newly ovulated ovum has only a single polar body whereas ova taken from the infundibulum have two. As many as three to five sperm may enter the ovum, but only one unites with the female pronucleus. The male pronucleus, which is formed from the head and middle piece of the sperm, is similar both in ovoid appearance and size (11.3 µ x 5.6 µ) to the female pronucleus.

The egg of the pigeon varies somewhat from the above account: After the ovum enters the infundibulum, from twelve to twenty-five spermatozoa immediately penetrate the egg membrane (presumably by enzymic action, by analogy with other species; cf. Tyler) and enter the blastodisc, within which the heads, which represent the nuclei of the spermatozoa, enlarge and become transformed into sperm nuclei. The fate of the middle piece and tail of the spermatozoon is not known. At the time of entrance of the spermatozoa the first maturation spindle is in process of formation; it lies in the center of a group of granules at the surface of the egg, which is bounded by a non-granular zone of protoplasm, called by Harper the polar ring, in which the sperm nuclei accumulate. External to the polar ring the protoplasm is granular again. The sperm nuclei remain quiescent until maturation is completed, and, when the egg nucleus is reconstituted, one of them, which may be called the male pronucleus or primary sperm nucleus, moves inwards and comes into contact with the egg nucleus. The opposed faces of the conjugating nuclei become flattened together, until the contours form a single sphere, the first segmentation nucleus, in which a partition separates the original components, the sperm and egg nucleus. The partition apparently disappears. Each gametic nucleus forms an independent group of chromosomes of the same (haploid) number and of homologous types with the exception of the "X" or sex-determining chromosome which is absent in approximately half of the egg nuclei. The union of the gametic nuclei restores the diploid number of chromosomes typical of the species.

Shortly after its formation, the first segmentation nucleus prepares for division in the usual mitotic way. The first segmentation (or cleavage) spindle thus formed lies near the center of the blastodisc a short distance beneath the surface and its axis is tangential to the surface, or, in other words, at right angles to the axis of the ovum. The fertilization may be considered to be completed at this stage.

The entrance of several spermatozoa appears to be characteristic of vertebrates with large ova; thus for instance, it has been described in selachians, some amphibians, reptiles and birds. Such a condition is known as *polyspermy;* it is normal in the forms mentioned, but occurs only under abnormal conditions in the great majority of animals. Harper observed that the number of sperm nuclei formed in the pigeon varied from twelve to twenty-five in different cases. Only one of these serves as a functional sperm nucleus; the remainder or supernumerary sperm nuclei migrate, as though repelled, from the center towards the margins and deeper portions of the blastodisc, where they become temporarily active, dividing and furnishing a secondary area of small cells (accessory cleavage) surrounding the true cleavage cells produced by division of the central portion of the disc around the descendants of the segmentation nucleus. It has been supposed by some authors who studied the selachians that the descendants of the supernumerary sperm nuclei form functional nuclei of the so-called periblastic, but this view has been disproved for the hen and for the pigeon in which it can be demonstrated that the supernumerary sperm nuclei have but a brief period of activity, and then degenerate.

CLEAVAGE OF THE OVUM

The fertilized ovum is morphologically a single cell, with a single nucleus, the first segmentation nucleus. The living protoplasm is aggregated in the blastodisc, and the remainder of the ovum is an inert mass of food material destined to be assimilated by the embryo which arises from the blastodisc. The first step in the development is a series of cell divisions of the usual mitotic type, restricted to the blastodisc, which rapidly becomes multicellular. As the early divisions take place nearly synchronously in all the cells, there is a tendency for the number of the cells to increase in geometrical progression, furnishing 2, 4, 8 and 16 etc., celled stages; but sooner or later the divisions cease to be synchronous. All of the cells of the body are derived from the blastodisc, and the nuclei of all cells trace their lineage back to the first segmentation nucleus. The supernumerary sperm nuclei do not take part in the formation of the embryo.

Cell division is the most conspicuous part of the early development; hence this period is known as the cleavage, or segmentation period. But it should be remembered that cell division is as constant a process in all later embryonic stages.

The type of cleavage exhibited by the bird's egg is known as meroblastic, for the reason that only a part of the ovum is concerned, the germinal disc. This is obviously due to the great amount of yolk.

To understand the form and significance of the cleavage of the bird's egg, it is necessary first of all to gain a clear idea of the structure of the blastodisc and its relations to the yolk. At the time of the first cleavage the blastodisc is round in surface view and about 3 mm. in diameter; the center is white and is surrounded by a darker margin about 0.5 mm. wide. These two zones have been compared to the pellucid and opaque areas of later stages. We shall call the outer zone the *periblastic zone*, or simply *periblast*. In section, the blastodisc is biconvex, but the outer surface which conforms to the contour of the entire egg is much less arched than the inner surface. The disc is everywhere separated from the yellow yolk by a layer of white yolk; on the other hand, there is no sharp separation between the disc and the white yolk. The granules of the latter are largest in the deeper layers and there is a gradual transition from them to the smaller yolk granules with which the disc is thickly charged. It is practically impossible in a section to say where the protoplasm of the disc ceases; it is indeed probable that it extends some distance into the white yolk both beneath and around the margins of the disc. Thus a cone, apparently of protoplasm, extends into the neck of the latebra a considerable distance. In other cases it does not extend so far.

The Hen's Egg

The form of cleavage of the hen's egg is illustrated in Figure A.E. The first cleavage appears in surface view as a narrow furrow extending part way across the blastodisc. It occurs just as the egg is entering the isthmus which is about three hours after the estimated time of fertilization. Approximately twenty minutes later, while the ends of the first cleavage furrow are still extending towards the periblast, the second division begins. It is a vertical division in each cell like the first, and the two furrows meet the first cleavage furrow at right angles. They may meet the first furrow at about the same point, in which case they form an approximately straight line, or more commonly, they may meet the first cleavage furrow at separate points, in which case the intervening part of the first furrow becomes bent at an angle, forming a cross furrow. The third set of cleavage planes are vertical like the preceding planes, but they tend to be variable otherwise.

Before describing the later cleavage stages, we should note certain important relations of the first four or eight cells: First, these are not complete cells in the sense that they are separate from one another. They are, indeed, areas with separate nuclei marked out by cleavage furrows in a continuous mass of protoplasm. The furrows do not cut through the entire depth of the blastodisc and the cells are therefore connected below by the deeper layer of the protoplasm; nor do the furrows extend into the periblast, and all the cells are therefore united at their margins by the unsegmented ring of periblast. Second, according to several observers, the center of the cleavage, i.e., the place where the first two cleavage furrows cross, is sometimes excentric. It was believed by those who emphasized this point, that the displacement is towards the posterior end of the blastoderm; but *Coste,* for instance, failed to note any

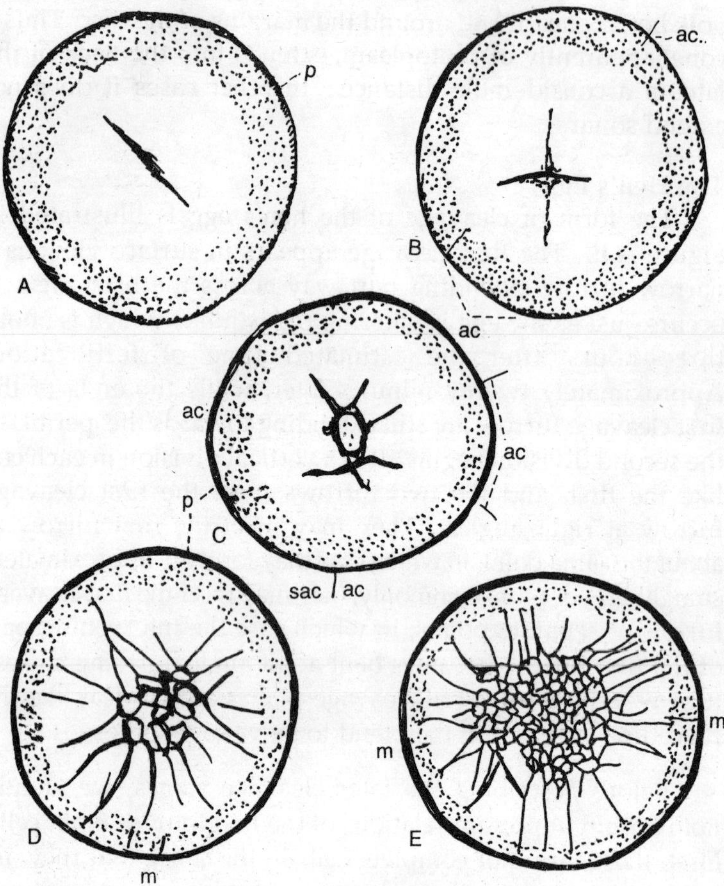

Fig. 6.6. Five stages in the cleavage of the hen's egg. Surface views of the blastoderm and adjacent periblast.

ac, Accessory cleavage furrow, m; Radial furrows. p., Inner part of marginal periblast, sac., Small cell formed by the accessary cleavage.

excentricity, and Patterson noticed both conditions, and showed that the displacement might even be towards the anterior end of the blastoderm. In the pigeon, according to Blount's observations recorded below, excentricity appears to be exceptional: moreover, the excentric area may bear any relation whatever to the future hind end of the embryo, so that in the pigeon it will not bear the interpretation that has been placed on it in the hen's egg.

After the third cleavage the ovum leaves the isthmus, and, within four hours after entering the uterus, progresses from the eight to approximately the 256 celled stage. These later cleavages in the hen's egg are very irregular, but two classes of furrows may be distinguished in surface view: (1) those that cut off the inner ends of the cells, and (2) those that run in a radial direction. The furrows of the first class produce a group of cells that are bounded on all sides in surface view, but these are, at first, still connected below by the deeper protoplasm. They may be called the central cells. These are bounded by cells that are united in the marginal periblast, and thus lack marginal boundaries as well as deep boundaries; they may be called the marginal cells. The distinction between central and marginal cells is one of great importance which should be clearly grasped.

In the surface views of later cleavages the following points should be noted: (1) the group of central cells increases by the addition of cells cut off from the inner ends of the marginal cells, and by the multiplication of the central cells themselves: (2) the marginal cells increase by the formation of new radial furrows, thus constantly expanding the diameter of the whole blastodisc, which may properly be called the blastoderm now that it is divided into cells. The increase of the central cells is much more rapid than that of the marginal cells, and the cells themselves are much smaller than the marginal cells, both because of their mode of origin and also because of their more rapid multiplication. The area of the central cells is also constantly increasing, with consequent reduction of the marginal zone. Emphasis has been laid by several authors on the excentric position of the smallest cells, and the inference has been drawn that these represent the hinder end of the blastoderm. Similar excentricity in the pigeon's egg is without reference to the future embryonic axis.

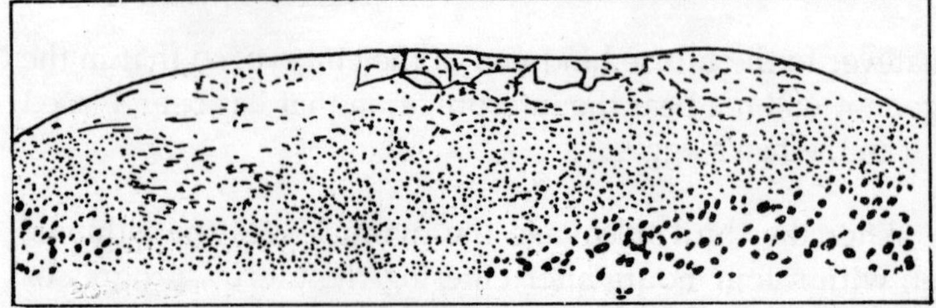

Fig. 6.7.   Median section of a blastoderm of the hen's egg which showed about 64 cells in surface view.

s.c., Subgerminal cavity.

But the surface views do not show what is going on in the deeper parts of the blastoderm. At the eight-celled stage a narrow space appears in the depth of the central portion of the blastoderm approximately between protoplasm and yolk; this is the subgerminal cavity which furnishes a lower boundary to the central cells. It is an area of progressive liquefaction of the yolk, presumably produced by enzymes from the overlying cells, and is probably not homologous to the blastocoele of other species. In later stages it extends peripherally to the inner margin of the periblast, and thus all of the central cells become completely bounded. A new class of cleavage planes then forms in these cells after the thirty-two celled stage, horizontal or parallel to the surface; in this way the central part of the blastoderm becomes two cell layer deep, and later several layers deep. The subgerminal cavity never cuts under the marginal cells, which remain united below and at their margins by the periblast.

In the older accounts of the horizontal cleavages by *Kolliker*, *Duval*, and others these are represented as forming before the subgerminal cavity, thus leaving the deeper cells in continuity with the yolk. Such cells are then supposed to continue budding off cells from their upper surfaces. But this view has been shown to be incorrect _ by the observations of Blount on the pigeon described below and by Patterson on the hen included above.

The Pigeon's Egg

The cleavage of the pigeon's egg, as worked out in detail by Blount, is made the basis of the description of the formation of the germinal wall in the absence of any consistent account for the hen's egg. The fundamental features of the cleavage are the same as in the hen's egg, so that the description need not be repeated.

The feature to be particularly emphasized in the cleavage of the pigeon's egg is the occurrence of a secondary or accessory cleavage in the marginal zone or periblast. When the origin of these cells is traced it is found that they arise around the supernumerary sperm nuclei, which accumulate and multiply in the periblast. Another interesting point illustrated by the figures is that the marginal cells have a peripheral wall wherever the accessory cleavage occurs, but between the groups of accessory cleavage cells the marginal cells are continuous with the periblast, as they are everywhere in the hen's egg. In a section of a germinal disc, showing the accessory cleavage it is seen that the peripheral boundary of the marginal cells cuts under the margin for a considerable distance.

The accessory cleavage becomes manifest at the time of appearance of the first cleavage plane, and increases in amount up to about the 32 celled stage, and thereafter gradually decreases until it completely disappears. The peripheral boundaries of the marginal cells disappear at equal pace, and, when the accuracy accessory cleavage is finally wiped out, the marginal cells are everywhere continuous with the periblast, as in the hen's egg. In some eggs the accessory cleavage is much more extensive than in others; indeed, in some it appears to be entirely absent, but this is relatively rare. In the stage shown in Figure for instance, there is usually considerable accessory cleavage; but in this egg there is none. The variation is obviously due to the variable number of supernumerary spermatozoa which enter the ovum.

The question arises whether the disappearance of the cell

walls around the sperm nuclei is caused by degeneration of the latter, or is simply a later syncytial condition in the periblast in which the sperm nuclei are embedded. There can be little doubt that the former alternative is correct. While in the stages of the accessory cleavage, sperm nuclei are readily found both in the accessory cleavage cells and also in the unsegmented periblast, they decrease in number as the accessory cleavage planes disappear, and when the latter are entirely lost the periblast is absolutely devoid of nuclei. Fragmentation of the sperm nuclei is a frequent accompaniment of their disappearance.

Thus the accessory cleavage is a secondary and transient feature of the cleavage of the pigeon's egg due to polyspermy. After it has passed, the ovum is in precisely the same condition as the hen's ovum of the same stage of development. In the hen's egg there is a very limited and inconspicuous accessory cleavage around the fewer supernumerary sperm nuclei that occur (Patterson, Olsen). But most of these nuclei in the hen tend to pass into deeper portions of the disc and there undergo complete fragmentation, producing superifical furrows only rarely.

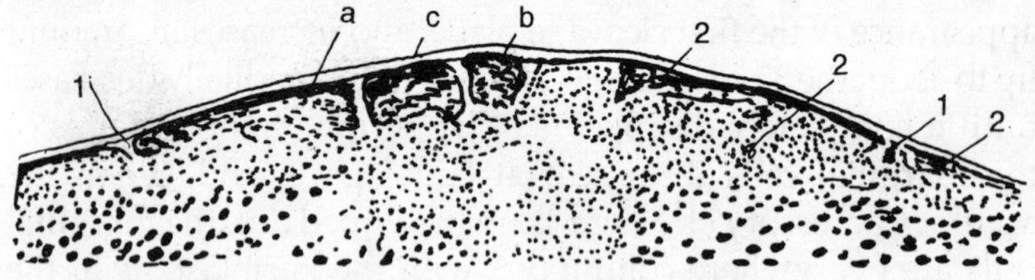

Fig. 6.8. Transverse section of the blastoderm of a pigeon's egg about
8³/₄ hours after fertilization.

1, Accessory cleavage.

2, Migrating sperm nuclei. a, b, c, d, Cells of primary cleavage

Another feature brought out by these photographs requires emphasis. The periblastic ring shows no definite outer margin, but beyond the zone of the accessory cleavage there may occur two or three concentric circles variously indicated.

Vacuoles, appearing black in the photographs, are very common in the outer zones. These appearances indicate that the periblastic protoplasm extends farther out in the superficial white yolk than is usually believed to be the case; and this suggests an interesting comparison with the teleostean ovum, where the periblastic protoplasm surrounds the entire yolk as a very thin layer. Sections confirm the idea that the periblastic protoplasm has an extension beyond the so-called margin of the blastodisc. Some eggs show a more definite margin than others; it may be that there is a periodic heaping of the periblast at the margins, for which again an analogy may be found in teleosts.

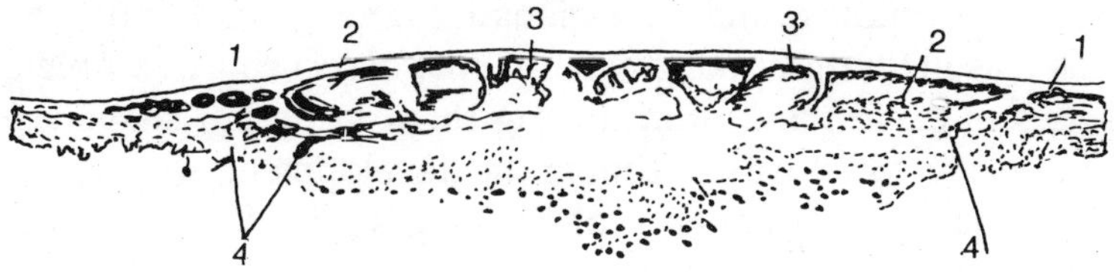

Fig. 6.9.   Transverse section of the blastoderm of a pigeon's egg at the end of the period of multiplication of sperm nuclei, about 10 hours after fertilization.

1, Accessory cleavage around the sperm nuclei. 2, Marginal cells; sharply separated from the sperm nuclei. 3, Central cells. 4, Sperm nuclei.

Although the smallest cells may be more or less excentric in the segmented blastoderm of the pigeon, their position bears no constant relation to the future embryonic axis. They may lie in this axis in front of or behind the middle, or to the right or left of it.

At the eight-celled stage a horizontal fissure begins to appear beneath the central cells. This marks the full depth of the blastoderm at all stages, and the several-layered condition arises by horizontal cleavages between this and the surface. Comparison of Figures drawn at the same magnification, shows that the depth does not increase by addition of cells cut off from below, as was once supposed to be the case in the bird's ovum. The horizontal fissure not only marks the full depth of the blastoderm, but it also indicates the site of the subgerminal cavity which arises gradually by accumulation of fluid and liquefied yolk between the cells and the underlying unsegmented protoplasm and yolk. The subgerminal cavity gradually extends towards the margin of the blastoderm, but it is bounded peripherally by the zone of junction between the marginal cells and the periblast.

Origin of the Periblastic Nuclei, Formation of the Germ Wall

Our knowledge of this part of the subject in the hen's egg is very incomplete, and the various accounts are contradictory. The reason for this is the great difficulty of securing a complete series of stages, and of arranging them in proper sequence. There is no exact way of timing the development, so that one has to judge the sequence of the stages, all of which come from the uterus, by reference to the time of laying of the preceding egg, by the degree of formation of the shell, by the size of the cells, and by the appearance of the sections. This is at best only approximate, and the securing of any given stage is largely a matter of chance. In the pigeon, which provides the basis for the following account, the time since the laying of the first egg is a fairly exact criterion of the stages of development of the second egg, so that a complete series may be collected.

The periblastic ring is entirely devoid of nuclei after the supernumerary sperm nuclei have degenerated. The marginal cells become greatly-reduced in size owing to multiplication and continuous production of central cells, and their nuclei thus approach more and more closely to the periblastic ring. The scene then changes; the marginal cells cease to produce central cells; when their nuclei divide the peripheral daughter-nuclei move out into the periblast, which is thus converted into a nucleated syncytium.' The periblastic nuclei multiply rapidly and invade all portions of the periblastic ring, which maintains its original connection with the white yolk. Not only do the periblastic nuclei invade the periblastic ring, but some of the also migrate centrally into the protoplasm forming the floor of the subgerminal cavity. They do not, however, reach the center, but leave a nonnucleated subgerminal area, corresponding approximately to the nucleus of Pander, free from nuclei. The subgerminal syncytium may be known as the central periblast to distinguish it from the marginal periblast. They are, of course, continuous. In sections one has the appearance of nuclei in the yolk, for there is no sharp boundary between periblast and yolk. The syncytium, which has received its nuclei from the marginal cells, is the primordium of the germ wall. The periblast probably aids in the liquefaction of the yolk, thus increasing the depth and extent of the subgerminal cavity.

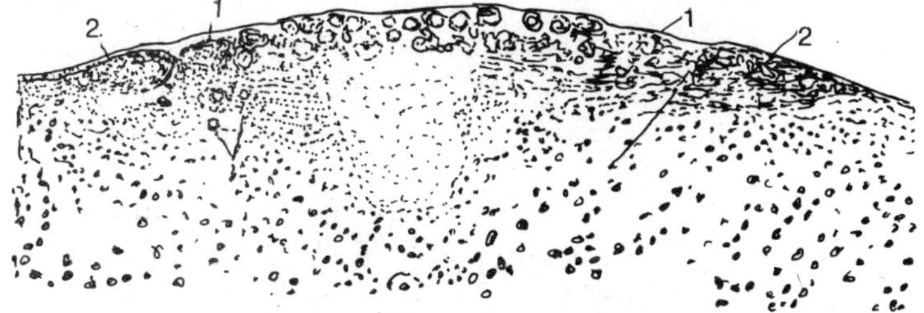

Fig. 6.10.   Transverse section through the center of the blastoderm of a pigeon's egg, 14 $^1/_2$ hours after fertilization.

1, Marginal cells. 2, Marginal periblast. 3, Nuclei of the subgerminal periblast.

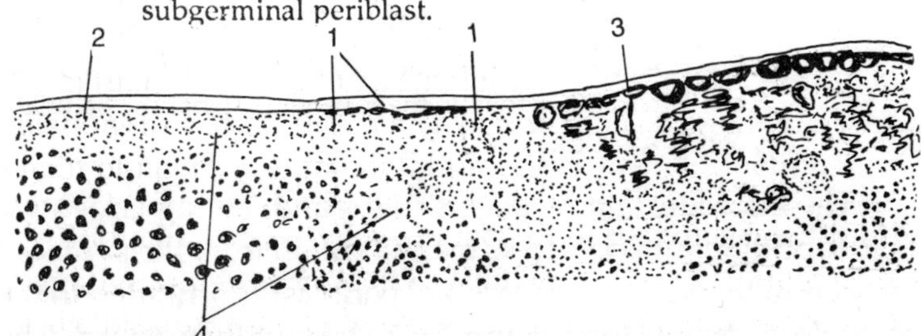

Fig. 6.11.   Posterior end of a longitudinal section through the blastoderm of a pigeon's egg about 25 hours after fertilization.

1, Nests of periblastic nuclei. 2, Periblastic nucleus in marginal position. 3, Syncytial mass derived presumably from the periblast, in process of organization into cells. 4, Vacuoles.

There is a sharp contrast between the segmented blastoderm and the syncytial periblast not only in structure but also as regards fate. The marginal cells constitute a zone of junction between blastoderm and periblast. Thus it will be observed that the large marginal cells on each side are continuous with the periblast, and nuclei are found in the periblast both central and peripheral to the zone of junction. The latter forms a ring around the blastoderm. It persists during the expansion of the blastoderm over the surface of the yolk.

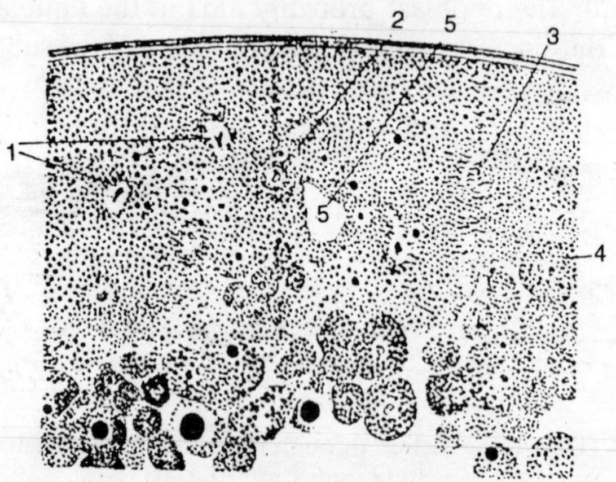

Fig. 6.12.   Part of the margin of a horizontal section through the blastoderm of a pigeon's egg about 25 hours after fertilization.

1, Periblastic nuclei. 2, 3, Cells organized in the periblast. 4, A cell apparently added to the blastoderm from the periblast. 5, Vacuoles.

The blastoderm now begins to expand, owing largely, at first, to additions of cells to its margin cut off from the periblast. The central as well as the marginal periblast contributes to the blastoderm, but he former appears to be rapidly used up. The marginal periblast, which is commonly called the germ wall from this stage, on the other hand grows at its periphery while it adds cells to the blastoderm centrally, and thus it moves out in the white yolk, building up the margin of the blastoderm at the same time. The original group of central cells appears to correspond approximately to the pellucid area; the additions from the germ wall would thus constitute the opaque area.

Some phases of these processes are illustrated. In the vertical section, the surface of the germ will next to the blastoderm is indented as though for the formation of superficial cells. Along the steep central margin of the germ wall groups of cells are apparently being cut off and added to the cellular blastoderm. In the horizontal section, the process of cellularization at the central margin of the germ wall is apparently proceeding rapidly.

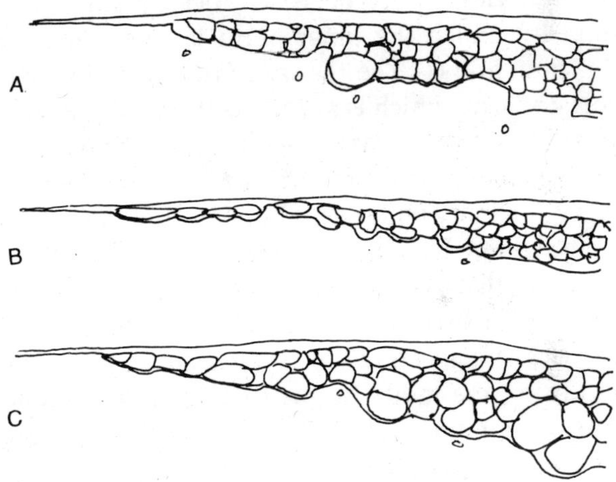

Fig. 6.13. Outlines of the margins of transverse sections of the blastoderm of pigeon's eggs; 26 (a), 28 (B) and 32 (C) hours after fertilization. (After Blount)

The superficial cells thus added to the margin of the cellular blastoderm become continuous with the ectoderm, and the deeper layers later form the yolk sac entoderm which becomes continuous with the embryonic entoderm secondarily. We can thus distinguish a syncytial, more peripheral, and a cellular, more central, portion of the germ wall.

In later stages the central margin of the syncytial part of the germ wall becomes much less steep, owing apparently to active proliferation of cells. This is illustrated. Still later, the external margin extends out peripherally and forms a short projecting shelf, appearing wedge-shaped in section. This we shall call the margin of overgrowth.

Thus we may distinguish the following zones: (1) margin of overgrowth; (2) zone of junction with the yolk (syncytial germ wall); (3) the inner zone of the germ wall, and (4) the original cellular blastoderm (pellucid area).

Origin of the Entoderm

Four main views have been advanced historically as to the method by which the inner germ layer or entoderm is formed.' These are: (1) The theory of delamination, namely, that the blastoderm, which is several cells thick at the close of cleavage, splits into two layers — an outer epiblast (which will later segregate into ectoderm and mesoderm) and an inner sheet of primary entoderm (hypoblast); this is the oldest view, promulgated by *Ollacher* (1869). (2) The theory of invagination, i.e., that the primary entoderm arises as an inrolling of the margin of the blastoderm; this view, which was supported by *Haeckel, Goette, Rauber, Duval Patterson* and others, was an attempt to bring the mode of gastrulation in birds into line with lower vertebrates. (3) The theory of multiple ingrowth, advanced by Merbach; according to this view, the epiblast was thrown into many irregular folds, from each of which the entoderm arose by inwandering of single cells or many isolated groups of cells. (4) The view of *Nowack* that the primary entoderm arises as an ingrowth of cells from the germ wall, more particularly from the posterior portion, was adopted in substance by *O. Hertwig* and combined, in part, by Jacobson with the blastoporal feature of the second theory above.

The account of the origin of the entoderm which follows is that which seems most probable in the light of recent work. It is essentially a modernized version of the theory of delamination with emphasis upon the properties and activities of individual cells.

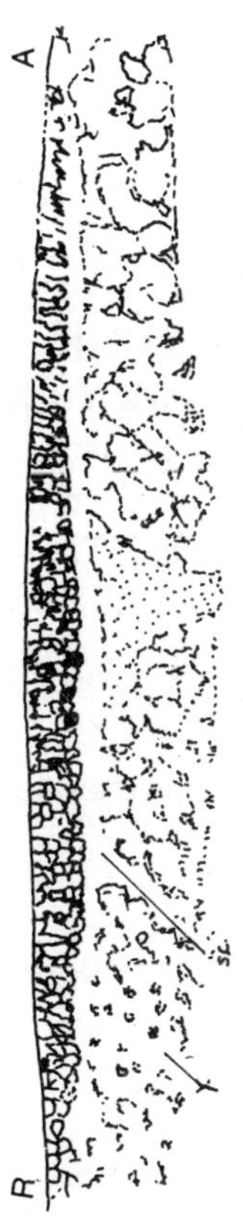

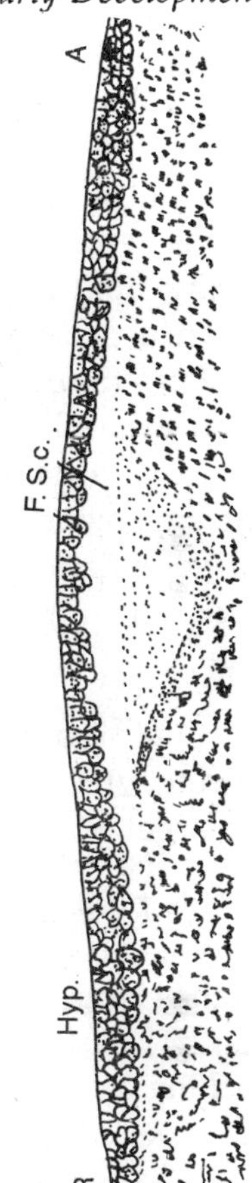

Fig. 6.14.   Median segittal section of a blastoderm is an advanced morula stage. Note that there are differences between the blastomeres in size and content of yolk, but that no assortment of the cells according to these differences has yet occurred.

A., P., Anterior and posterior ends of the blastoderm. s.c., Subgerminal cavity. Y., yolk.

Fig. 6.15.   Segregation of the hypoblast from the epiblast. Note the increased expanse and thickness of the blastoderm, the down-wandering of large yolky cells toward its lower surface beginning posteriorly, and the deeper subgerminal cavity. Flask-shaped cells (F.) are seen which still retain their contact with the surface coat.

Hyp., Hypoblastic cells.

The segmentation of the blastodisc during the period of cleavage results in a disc-shaped aggregate of rounded cells piled one on top of the other, four or five deep. Structurally, the disc is an extremely shallow dome, resting on the yolk at its edge (the germ wall) and separated centrally from the yolk by a narrow, fluid-filled, sub-germinal cavity. On the floor of the latter are occasional scattered cells which may have become detached from the overlying disc or may have been left behind by the ever-expanding germ wall. The cells of the blastoderm show some variation in size and yolk content, but there is no spatial arrangement with respect to these differences. The only visible differentiation is a thickening toward one side of the blastoderm (the future posterior end) where the cells are more numerous.

A gradual assortment of the blastodermal cells then takes place, during which the smaller yolk-poor cells are segregated from the larger yolk rich ones.

It seems probable that the segregation of the yolky cells from the others is a manifestation of negative" tissue affinity" between the ectodermal and entodermal cells as described by Holtfreter. The two kinds of cells may become incompatible after reaching a certain point in their differentiation and then mutually repel each other. The frequent occurrence of yolk-filled flask-shaped cells, with their small ends still in contact with the surface but with the body of the cell "squeezed" toward the interior, is suggestive of such a view.

Meanwhile the blastoderm expands over the yolk and becomes somewhat thinner and more transparent where it overlies the sub-germinal cavity. The central portion (area pellucida) is thus demarcated from the area opera (or germ wall) by virtue of the later's thickness and intimate contact with undigested yolk. The depth of the subgerminal cavity also increases as the blastoderm expands.

As the large yolk-rich hypoblastic cells accumulate toward

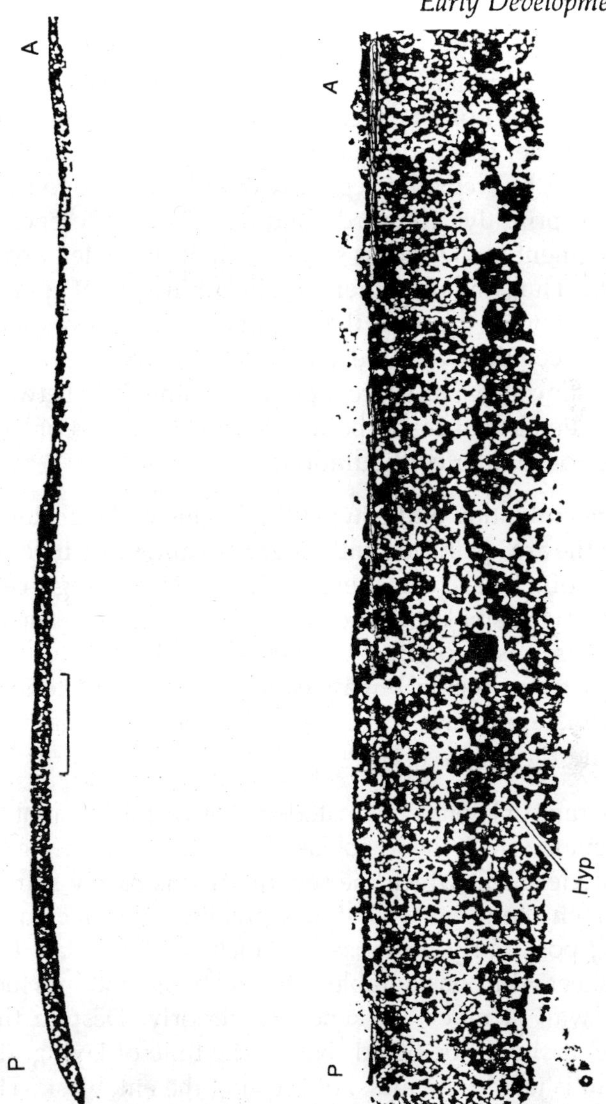

Fig. 6.16. Median sagittal section of the pre-streak blastoderm. Note the thickening of the posterior half of this blastoderm (the embryonic shield) and the presence of the hypoblast beneath the thickened area, discontinuous at its anterior edge.

Fig. 6.17. Details of the bracketed portion of the pre-streak blastoderm.

the lower surface of the blastoderm, the upper yolk-free cells become aligned (perhaps through the aid of a surface coat)' into an epithelium of single-cell thickness. The hypoblastic cells, on the other hand, do not organize into a coherent sheet for some time, but remain irregularly distributed one to three deep in the primitive rounded condition. This difference in the arrangement of the cells is sufficient in itself to demarcate two layers. Thus, in effect, there is a delamination of epi and entoblast, but not in the sense that a split or space is necessarily formed between the two. Due to the irregular arrangement of the entoblastic cells, small spaces appear here and there between the two layers, but at many points the separation is only in differences of cellular configuration.

The delamination of the hypoblast begins in the posterior portion of the area pellucida and progresses anteriorly in lesser amount; a point is reached where groups of cells or single cells separate from the epiblast at interrupted loci. Thus, the primary entoblast arises as a continuous layer of cells in the posterior part of the area pellucida, but appears "frayed-out" anteriorly and antero-laterally and occurs only as scattered islands of cells beyond the "frayed" edge.

At the time of laying, the blastoderm shows much variation in the extent to which the hypoblast has formed. The range is from blastoderms in which the separation has barely started and in which the germ wall shows practically no increased thickening posteriorly, to those in which the hypoblast is a separate sheet with considerable anterior spread and in which the germ wall is much thickened posteriorly. Despite this variability, it should be noted that, at the time of laying, the blastoderm is just in the midst of forming the entoblast. The process continues for many hours after incubation begins, but meanwhile other events supervene.

7

Development upto the Formation of the First Somite

Structure of the Blastoderm in Freshly Laid Eggs

There is more or less variation in the progress of development of the hypoblast at the time of laying. These differences, however, are apparent only in sectioned blastoderms; in surface view the blastoderms are similar in appearance, showing only an outer white ring (the area opaca) of even width and a central translucent area (the area pellucida) through which Pander's nucleus and scattered yolk globules can be distinguished. Beneath the pellucid area is the subgerminal cavity bounded marginally by the germ wall; the cavity is somewhat deeper at the posterior end of the blastoderm. The posterior part only of the pellucid area is two-layered. The lower layer or hypoblast (primary entoblast) terminates posteriorly at the germ wall but does not reach the wall either laterally or anteriorly. The hypoblast is a loosely coherent layer of rounded cells, uneven in thickness, continuous posteriorly, but grading out to discontinuous clusters and scattered single cells at its anterior and lateral edges.

The germ wall is slightly thicker in depth at the posterior than at the anterior margin. There is, however, no consistent difference in its breadth at the two regions. Beyond the zone of junction the blastoderm overlaps the yolk a short distance.

The epiblast is thicker in the area pellucida than in the area opaca. Its cells are more coherent than the underlying entoblast and have assumed for the most part the characteristics of a single-layered epithelium.

Changes Prior to the Formation of the Primitive Streak

With the beginning of incubation the blastoderm resumes development. The changes which take place are the result of (1) mitosis throughout the rapidly growing germ, and (2) orderly cellular movements, both of which occur concurrently. These processes, of course, have been in operation since the first cleavages. The part played by cellular movements, however, becomes increasingly important from this time onward.

The most important change during the first few hours is the continued separation and further differentiation of the *hypoblast*. The cells lose their primitive rounded shape, flatten, and unite to form a continuous layer except at the "frayed-out" anterior edge. These changes in shape, along with a migration forward of the cells, extend the hypoblast toward the germ wall. The process begins at the posterior edge of the hypoblast, progress anteriorly, and is still continuing during stages of the primitive streak. The hypoblastic epithelium is very thin. The *epiblast*, on the other hand, has now become a definite columnar epithelium with taller cells posteriorly than anteriorly.

The combination of thickened epiblast plus the hypoblast beneath it causes a visible opacity in the posterior portion of the *area pellucida*. In surface view it is seen as an indistinct circular or half-moon shaped area with one edge next to the area opaca. This is the *embryonic shield*. It is the first gross indication of the polarity of the blastoderm, for a line bisecting it will, in general, coincide with the future antero-posterior axis. The embryonic shield is not ordinarily seen in freshly laid eggs but is common in many so-called unincubated blastoderms and appears distinctly after four or five hours of incubation (Spratt). The entoderm at this stage has the same general crescentic shape and anterior extent as the thickened epiblast which overlies it.

In addition to the flattening and creeping forward of the hypoblast, there are shiftings of cells in the epiblast which are a prelude to the appearance of the primitive streak. If the pellucid area is marked at different points with carbon particles is extensive postero-medial movement of lateral epiblastic cells toward the median posterior quadrant. The cells of the posterior border of the pellucid area move medially, and, if already in the mid-line, move slightly forward.

The Primitive Streak

Our knowledge of the mode of origin of the primitive streak progressed slowly during the descriptive era of embryology until suitable experimental techniques could be evolved. Inevitably, concepts arose in the interim (e.g., concrescence) which, although speculative in nature, were widely adopted and even somewhat reluctantly given up as the technique of vital marking revealed the true dynamics of the process. In many cases experiments with vital dyes yielded equivocal results due to the tendency of the dyes to diffuse and fade and the difficulty of preserving the color when blastoderms were prepared for histological examination. Recently, these technical difficulties have been circumvented by marking the blastoderm with particles of carbon (Jacobson, Spratt), and the results thus obtained, in conjunction with the data of Pasteels and Wetzel, provide the basis for the following account.

The cellular movements in the epiblast, described above, immediately precede and lead to the formation of the primitive streak. After about six to seven hours of incubation the steak appears in the line of bisection of the embryonic shield as an elongated, slightly opaque band. It occupies the posterior third or two fifths of the circular pellucid area and, in a more definite way than the shield, marks the antero-posterior axis of the future embryo. It is relatively narrow in front and widens posteriorly where it is also less dense.

In sagittal section it is seen that the epiblast has become

nearly twice as thick in the region of the streak as it is elsewhere. Here it has lost its epithelial organization, and cells are rapidly moving inward from the surface to form an intermediate layer between the epiblast and hypoblast. This new layer spreads anteriorly and laterally from the streak, and may tentatively be called mesoblast, although it is probable that some of its cells migrate down into the hypoblast to form entoderm.

The streak then increases in length. According to *Spratt*, this occurs " as a wave of proliferation and invagination (i.e., inwandering) of epiblast cells in the mid-line." From stage 2 (initial streak) to stage 3 (intermediate streak), 56 per cent of the increase is due to additions to the front end of the streak, 19 per cent to additions to its posterior end, and 25 per cent to a "stretching" of the initial streak. These facts have been revealed by the movements of carbon marks. It will be noted that a mark placed at the anterior end of the initial streak is not pushed forward by the elongating streak any more than a floating cork is displaced by a passing wave front; the streak merely continues its formation in front of the node. Some of the particles of carbon are carried inward by the inwandering of cells as the streak forms, and they then spread anteriorly and laterally with the newly formed mesoblast. At an intermediate stage the streak has reached the center of the pellucid area where the chorda and medullary plate-forming areas of epiblast are located. At this point its anterior end becomes club-shaped and eventually develops a pit, the primitive pit, which corresponds to the neurenteric canal of other vertebrates.

The neurenteric canal is a canal that connects the posterior end of the central canal of the neural tube with the intestine. It arises from the anterior end of the blastopore, and is typically developed in Selachia, Amphibia, reptiles, some birds (e.g., duck, goose, Sterna, etc.). It begins in the primitive pit and extends forward into the head-process. Subsequently the primitive pit becomes surrounded by the medullary folds, and thus opens into the neural canal. An opening is later formed through the entoderm so that the definitive canal connects neural tube and hind gut. In the chick the neurenteric canal is never typically developed. Usually it is represented only by the primitive pit. In exceptional cases, traces of it are found in the head-process (Lillie).

At first the surface of the primitive streak is even, but, as it is elongates, a groove appears down its center where the inward migration of cells is taking place. This groove is known as the primitive groove; it is bounded by the primitive folds and terminates abruptly in the primitive pit. The primitive streak ends anteriorly in front of the pit in a cluster of cells called the *primitive knot (Hensen's node).*

The appearance of the intermediate streak in transverse sections is shown in Figure 7.1 In the region of the primitive knot (A) the epiblast is somewhat thickened, forming a projection above and below. Cells become detached from the lower surface of the epiblast, and are converted into migratory cells between the two primary layers. Immediately behind the primitive knot the primitive groove begins abruptly; it is the site of active inward migration from the epiblast, and the cells spread out laterally beneath the surface forming wings of cells which do not, however, reach the area opaca. Conditions are very similar along the entire length of the primitive streak at this time; but near the posterior end a few cells of the mesoderm reach the area opaca and begin to insinuate themselves between the epiblast and the germ wall. There is no evidence at any place that any of the mesodermal cells are derived from the entoderm, but there is good evidence that at least some of the cells which pass through the streak find their way into the entodermal layer. The axial thickening of the primitive groove comes in contact with the entoderm and appears in places fused to it.

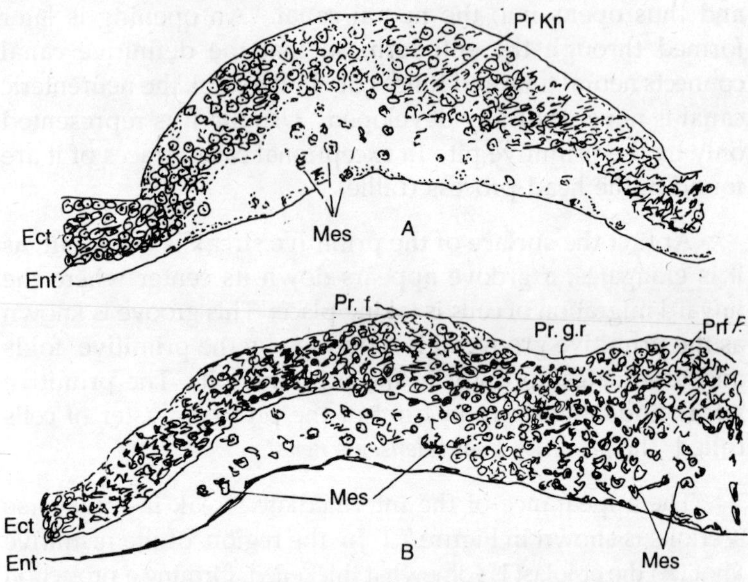

Fig. 7.1.   Transverse sections through an intermediate primitive streak of the chick. Incubated 17 $^1/_2$ hours.

A. Through the anterior end of the primitive streak (primitive knot). Mesoblastic cells are moving inward from the epiblastic (ectodermal thickening: some are scattered between the two primary germ layers. The hypoblast (entoderm) shows no proliferation, though some mesoblastic cells are adhering to it.

B. Fourteen sections posterior to A. (Entire length of the primitive streak is 80 sections.) The mesoblastic wings are forming; the primitive groove and primitive folds are indicated.

Ect., Ectoderm (epiblast). Ent., Entoderm (hypoblast. Mes., Mesoblast. pr. f., Primitive fold. pr. gr., Primitive groove. pr. kn., Primitive knot.

Further elongation of the streak from stage 3 to its definitive length (stage 4) occurs "solely by an intussusceptive or stretching process" which is localized mainly in its posterior portion. "This accounts for 46.7 per cent of the total length of the definitive streak; 26.6 per cent is contributed by the original length of the beginning [initial] streak; 19.79 per cent is the result of addition to the anterior end of the original streak; and 6.91 per cent was added to the original posterior end."

The fully formed streak is shown in stage 4. Its posterior end terminates in an expansion which is not very obvious in surface view, and hence is not usually described; it may be called the primitive plate. In some cases the primitive streak and groove are bifurcated at the posterior end but this condition is by no means typical. Simultaneous with the postero-medial movement of the epiblastic sheet and the "stretching" of the streak, the shape of the pellucid area is itself altered from a circular to pear-shape. The inward migration of the converged epiblastic cells through the primitive groove continues as before and occurs throughout the entire length of the streak including the primitive pit and Hensen's mode.

Formation of Entoderm During Streak Stages

It will be recalled that the hypoblast was in the act of organizing into a flattened and coherent epithelium and was creeping forward from the posterior part of the pellucid area at the time that the egg was laid. These changes continue during the first few hours of incubation while the primitive streak is forming in the epiblast; in fact, the forward movement of the hypoblast just precedes the anterior differentiation of the streak.

Spratt cites several lines of evidence which suggest a casual influence of the hypoblast upon the development of the streak: (1) the two structures develop at equal pace; (2) when the orientation of the hypoblastic cells is changed experimentally, the direction of the streak is correspondingly changed; (3) if the hypoblast is removed no streak is formed, etc. Further work is needed, however, to prove whether the suspected relationship is valid.

As the hypoblast forward, some of its cells tend to pile up in a thicker crescent-shaped area (the *Entodermhof* of Wetzel) located behind its fenestrated free edge and ahead of the developing streak. As Rudnick has pointed out, this thickened, almost mesenchymatous, crescent corresponds in general to the germinal crescent which gives rise to the so-called primordial germ cells. The entodermal sheet is not complete throughout the area pellucida until about the twelfth hour of incubation, at which time the streak has almost reached its definitive length.

Although the experimental evidence is meager, it seems probable that the hypoblast receives contributions of cells from the epiblast during the entire period of existence of the primitive streak. When the surface of the epiblast is marked with a vital dye or with carbon, some of the marked cells are found eventually in the entoderm. In view of the absence of any boundary between the mesoderm and the hypoblast at the level of the streak, the most reasonable assumption is the cells from the surface pass into both underlying layers. However, the extent of the contribution to the hypoblast is not known. From the available experimental evidence and that furnished by comparative embryology, it appears, therefore, that the entoderm of the chick has a double source : (1) primary entoderm which arises by delamination from the epiblast before the appearance of the primitive streak, and (2) secondary entoderm which arises by migration of epiblastic cells through the streak.

The present author would like to venture an opinion that the initial delamination and later inward migration described in the literature are merely phases of one and the same process. It is obvious that in pre-streak stages the yolk-rich cells of the epiblast must move inward before they can be *"delaminated."* The migration and downwandering of cells merely become more rapid and more localized, leading to the appearance of the streak. The proportions of cells going into the underlying layers would reach a balance according to their respective affinities or compatibilities.

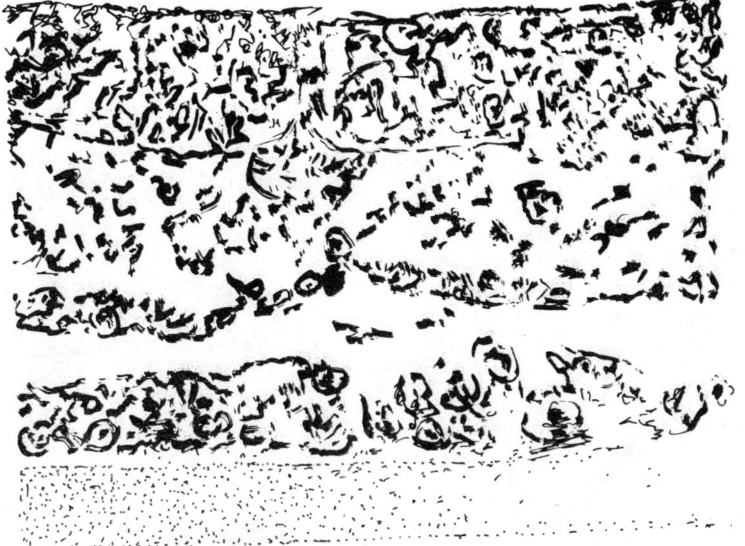

Fig. 7.2.   Transverse section through a blastoderm with a long head-process, showing particles of carbon localized in the notochord, lateral mesoderm and entoderm.

Since the epiblast contributes cells to the mesoblast (and probably also to the hypoblast), it seems best to use the non-committal terms, *"epiblast,"* and *"hypoblast,"* until the germ layers are clearly established in front of the streak as indicated by the appearance of the embryo. Therefore, we shall arbitrarily use the more general terminology until the head fold appears (stage 6), after which the epiblast and hypoblast anterior to the node will be called ectoderm and entoderm, respectively. At the level of the streak (and later in the tail bud), however, the three germ layers will still be in the act of forming for several days.

Regression of the Primitive Streak

Formation of the Head-process: Soon after the streak reaches its definitive length, the first sign of the embryo proper appears as a short mesodermal rod (ca. 0.1 mm. wide x 0.16 mm. long) which extends forward from the anterior border of the primitive knot. This is the head-process or notochord. It is formed by

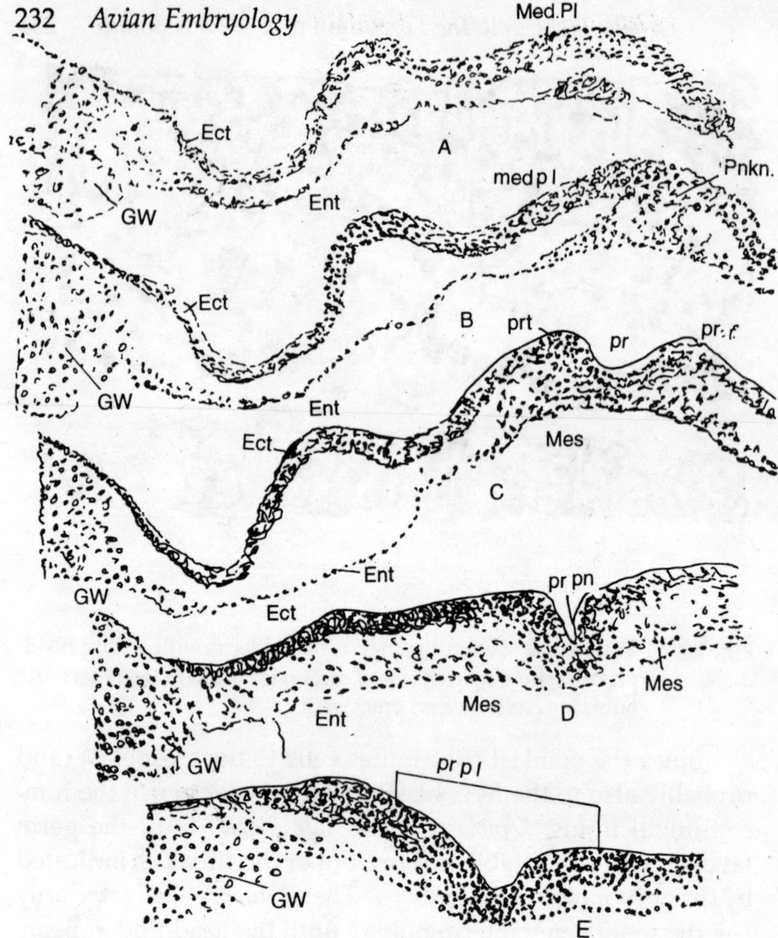

Fig. 7.3.   Five sections through the head-process and primitive streak of a chick embryo. The head-process is very short.

A. Through the head-process, now fused to the entoderm.

B. Through the primitive knot.

C. Through the anterior end of the primitive groove.

D. A little behind the center of the primitive streak.

E. Through the primitive plate.

Ect., Ectoderm. Ent., Entoderm. G.W., Germ wall. H. Pr., Head-process. med. pl., Medullary plate. Mes., Mesoblast. pr. f., Primitive fold. pr. gr., Primitive groove. pr. kn., Primitive knot. pr. pl., Primitive plate.

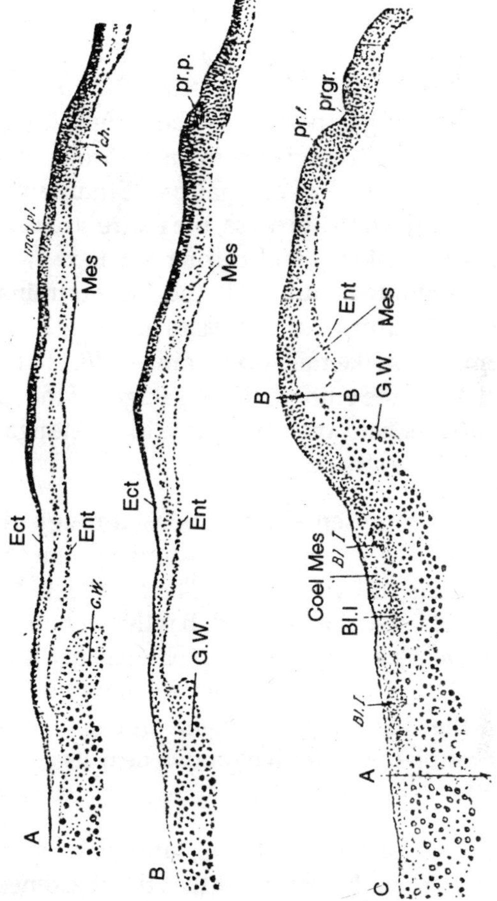

Fig. 7.4.   Three transverse sections of a late stage (corresponding to about Figure 7.3), through the head-process and primitive streak of a chick embryo.

A. Near the hind end of the head-process.

B. Through the primitive pit.

C. A short distance behind the center of the primitive streak. The region between the lines A-A and B-B is represented under a high magnification.

Bl. 1., Blood island. Coel. Mes., Coelomic mesoblast. Ect., Ectoderm. Ent., Entoderm. G.W., Germ wall. med. pl., Medullary plate. Mes., Mesoderm. N'ch., Notochord. pr. f., Primitive fold. pr. gr., Primitive groove. pr. p., Primitive pit.

an axial concentration of the chorda-mesoblast just anterior to the node and is continuous anteriorly and laterally with the other cephalic mesoblast which underlies the anterior part of the epiblast. Marking experiments show that the cells from which it is formed have migrated inward through the streak during the latter's period of elongation. Simultaneous with the appearance of the head-process, the entire streak and node begin to move posteriorly, and during the next twenty-four hours the streak shortens until only the node refrains. At the same time the head-process increases in length and other parts of the embryo make their appearance. We shall consider first the structure of the embryo during stages of the formation of the head-process, and then the mode of its elongation and the regression of the streak.

Fig. 7.3. A-E represent five sections through the head-process and primitive streak of a chick embryo at a time when the head-process is still very short. The first section through the head-process is described beyond. B is through the primitive knot; there is a greater accumulation of cells than in the preceding stage and it will be observed that they are now fused with the entoderm, so that the latter no longer appears as a distinct layer. C is through the primitive groove near its anterior end. D is behind the

Center of the primitive groove, and E is through the primitive plate. It will be observed that the thickened area in the primitive plate where cells are migrating inward is very wide.

It will be noted that the mesoblast is thickest at its apparent place of origin (the streak) and thins out laterally as it spreads towards the periphery. Mesoblastic cells also creep into the opaque area, and thus it produced a three-layered portion of the latter which corresponds to the future vascular area. The mesoblast migrates out not only from the sides and front of the streak but also from its hind end, that is from the primitive plate, so that the mesoderm extends into the opaque area behind the embryo at a very early stage.

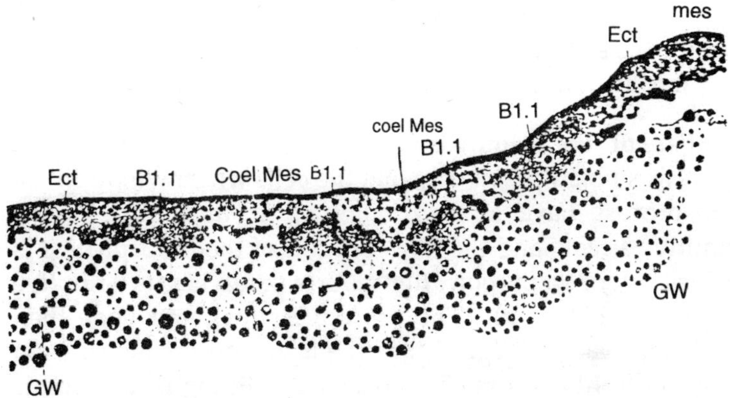

Fig. 7.5.   The part of the section shown in Figure 7.4 C. between A-A and B-B more highly magnified. Abbreviations same as Figure 7.4.

The head-process itself, consists of a thicker central mass of cells with lateral wings; the central part, or primordium of the notochord, is continuous posteriorly with the axis of the primitive streak. The head-process becomes inseparably fused with the entoderm in the mid-line immediately after its formation, and this fusion is continued back along the axis of the primitive streak. The fusion is particularly intimate and persistent at the extreme anterior end of the head-process; behind this point the notochord and entoderm soon separate again in the course of development. But the anterior end of the notochord remains attached to the entoderm for a considerable period after the formation of the head fold. A longitudinal section shows the head-process as a mesodermal appendage to the anterior end of the primitive streak, or the primitive knot.

The embryo arises in front of it around the head-process as a center; the ectoderm of the anterior portion of the blastoderm, under which the head-process forms a median axis, becomes the thickened medullary plate from which the neural folds are formed the anterior end of the primitive streak marks the hind end of the differentiated portion of the embryo. As the embryo grows in length, the primitive streak decreases and finally disappears, except for remnants which enter with the primitive knot into the tail bud, at the time that the tail fold is formed.

The Fate of the Primitive Streak

It was noted previously that the cells which will form the head-process have already migrated inward through the primitive streak, and the first portion of the head-process is molded out of them soon after the streak reaches its definitive length (ca. 1.88 mm). The primitive streak and node then move backwards ("regress") as the head-process increases in length in front of them. This posterior shifting is real, and is most rapid in the anterior half of the streak, as can be shown by marking the node and various parts of the streak and nothing their movement with reference to stationary marks. At the same time the head-process grows in length, its rate and extent of growth paralleling the rate of regression of the streak. Careful measurements show that the tip of the head-process moves forward only slightly with reference to stationary marks, and that nearly all of the increase in length occurs by additions to the posterior end of the early head-process. It was formerly thought that the head-process was formed out of the cells of the regressing node but Spratt's experiments show that this is not true. The increase in length of the head-process is largely due to proliferation and "stretching" of the cells immediately in front of the node. If the anterior half of the blastoderm, including the node, is served from the posterior half, the notochord, neural plate, and node extend by growth back from the cut surface to form a tail-like appendage. These findings have led *Spratt* to suggest that the cause of regression of the streak is a "pushing" of the node and streak backwards by the growth of the head-process. There is some question, however, whether the node is entirely passive during regression, for *Spratt* also reports that no regression occurs if the node is excised from the blastoderm. Further study is needed to provide an adequate explanation of this complicated process.

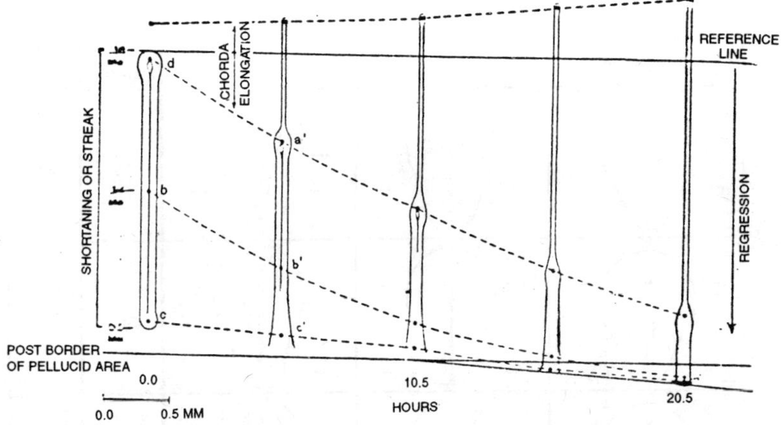

Fig. 7.6.   Regression of the primitive streak. Note that the node moves backwards with respect to the stationary line, and that the shortening takes place first at the posterior end of the streak. Note also that the primitive groove disappears first at the posterior end.

Changes also occur in the epiblast which are correlated with the growth of the head-process and regression of the streak. The whole portion of the epiblast anterior to the level of the node grows at a faster rate than the remainder, thus keeping pace with the increase in length of the head-process. Back of the node there is a continued movement of lateral areas of epiblast toward the streak and inward migration of the cells to form mesoblast, but the extent of these movements is much less. There is no movement of cells from the surface of the node or through the primitive pit to the interior during regression; the node remains intact until it becomes the major part of the tail bud. Not only the epiblast, but also the mesoblast and hypoblast are apparently displaced back-wards by the regression of the streak.

As the node moves backwards, the primitive streak decreases in actual length. The shortening occurs mainly at the posterior end, where making experiments show that its cells spread out laterally and posteriorly to form embryonic

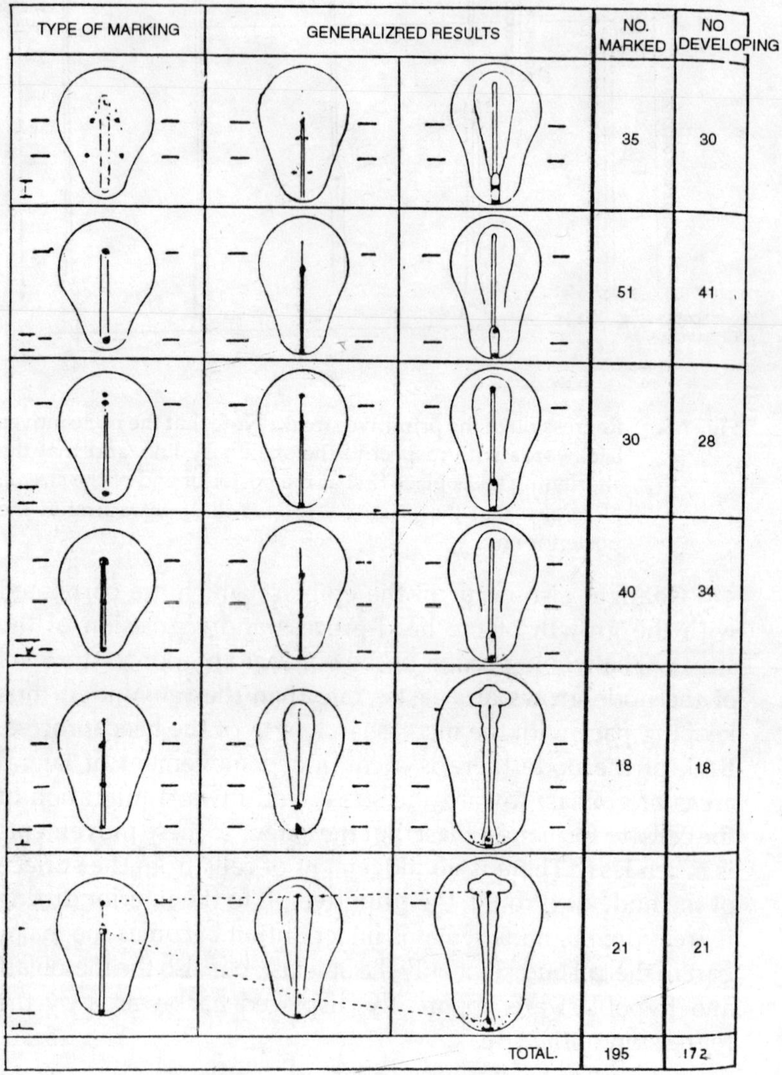

TYPE OF MARKING	GENERALIZRED RESULTS	NO. MARKED	NO DEVELOPING
		35	30
		51	41
		30	28
		40	34
		18	18
		21	21
	TOTAL.	195	172

Fig. 7.7.   Summary of the movements of epiblastic cells and the manner of increase of the head-process and decrease of the primitive streak as revealed by carbon-marking experiments.

and extraembryonic ectoderm and mesoderm.   With the dispersion of the cells, the primitive groove disappears first, and then the streak, at progressively more anterior levels. The shortening continues until only the node and a bit of the streak behind it remain, at which point the tail fold is formed and the primitive knot is incorporated in the tail bud. There is evidence that some of the cells in the streak are consolidated with the node to form the tail bud, since the latter is larger than the original node.

Although evidence is not yet available or the mechanism of shortening of the streak it seems probable that it is the result of the slowing and cessation of the medial movements and inward migration of the epiblast. The sequence of events (i.e., disappearance of the groove followed by dispersion of the streak itself) favours this view.

Maps of the Blastoderm During Stages of the Primitive Streak

Organ-presumptive Areas. The active shifting of cells in the epiblast and their migration through the streak to a new location raise the question of their fate in normal developments : to what will each cells or group of cells in the blastoderm give rise, i.e., what is their *prospective value?* Our knowledge of this subject is largely obtained by marking the blastoderm with vital dyes or carbon particles, as described above, and following the movements of the marked areas to their ultimate positions in the body of the embryo. Maps may then be constructed to show the approximate area in the surface of the blastoderm which are destined to take part in the formation of specific organs.

Figure 7.8 is a diagram of the epiblastic areas of the pre-streak blastoderm (stage 1), showing the boundaries of the streak at initial and definite stages on the right side, and the extent of the area which is to be moved inward through the streak. On the left side is shown the probable location of prospective organ-forming areas at stage 1 as inferred from marking experiments on blastoderms at the stage of the definite streak. The actual shapes of the areas and boundaries between them are hypothetical, therefore, in the absence of direct marking experiments at the pre-streak stage. It will be noted that thee prospective mesodermal areas are located with respect to the mid-line in the order that they will be moved inward.

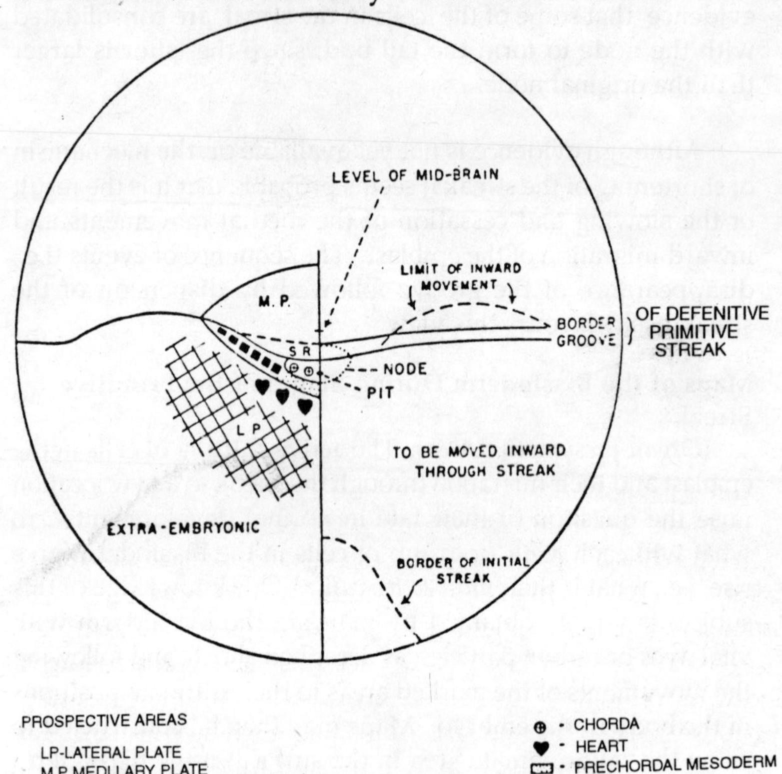

Fig. 7.8.   Diagram of prospective organ-forming areas in the pre-streak blastoderm as inferred from their positions at the stage of the definitive streak. Right side: boundaries of the streak at stages 2 and 4, and extent of the area to be moved inward. Left side: probable distribution of prospective embryonic areas.

A composite map of the prospective organ-forming areas at the stage of the definitive streak. The shapes and sizes of the areas are ever changing as some of them converge and migrate through the streak and others spread out correspondingly. It will be noted that the principal embryonic areas are concentrated around the node and fore part of the streak. It is this region which under-goes the most growth and elongation and gives rise to the major part of the embryonic axis during the regression of the streak.

Histogenetic Potency

In addition to the prospective value of any unit area of the early blastoderm, each part has a far greater capacity for development than is normally expressed. It is as though there were an untapped reserve potential of ability which is expressed only in case of some unexpected deficiency. For example, developmental areas at certain critical periods may lead to complete or partial twinning. Or, if a lesion is produced experimentally, then adjacent areas of the blastoderm are able to reconstitute the deficiency so that development may continue normally. This power to develop or *histogenetic potency* can be revealed by removing portions of the blastoderm to other sites for growth such as the chorio-allantoic membrane, the coelomic cavity, or plasma clots, where the cells may differentiate apart from organismic influences (e.g., the effects of neighbouring cells and embryonic fields).

The distribution of organ-forming potencies in the blastoderm at the stage of the definitive streak. When this map is compared with that of prospective values at the same stage it is noted that the ectodermal potencies show the same general axial seriation but that the mesodermal and entodermal potencies are much more closely grouped around the node than in the map of values. The size of any particular area of histogenetic potency, however, is usually greater than the corresponding area of prospective value (e.g., compare the heart-forming area in the two maps). With increasing age the potency of a given area becomes progressively more limited until it is equivalent to the prospective value.

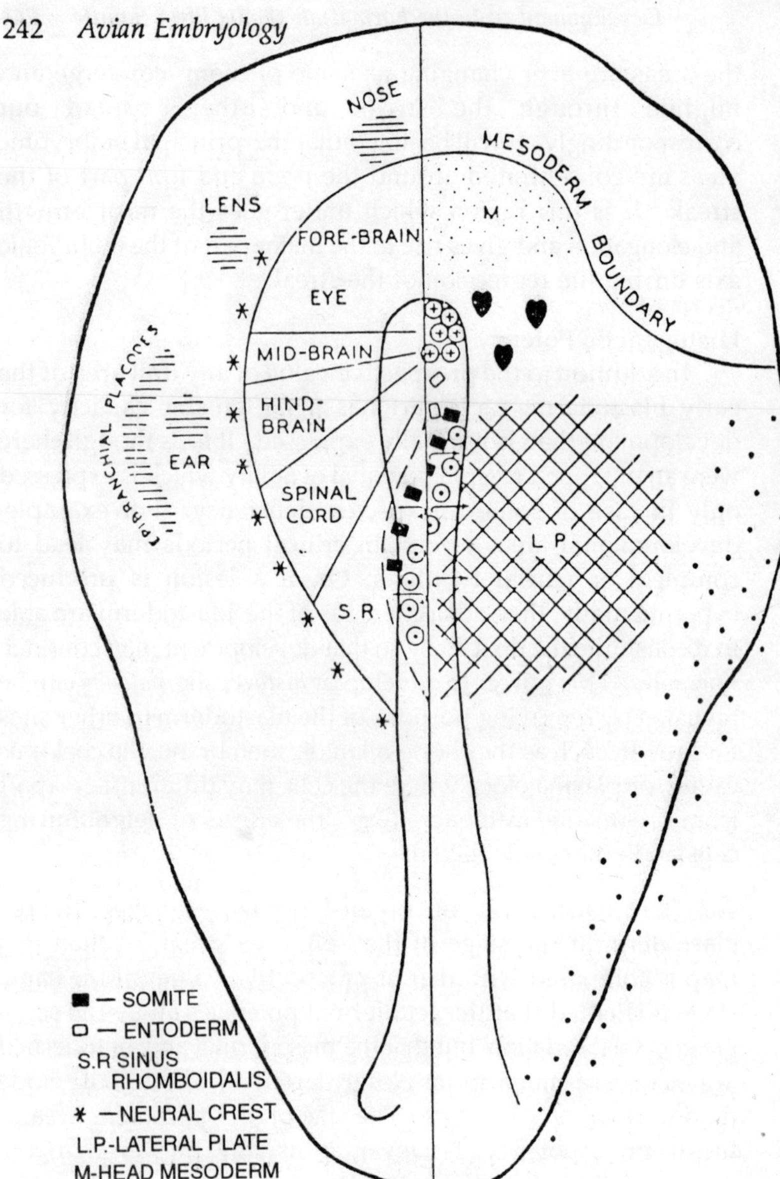

Fig. 7.9. A tentative map of prospective organ-forming areas in the blastoderm of the stage of the definitive primitive streak. Superficial layer shown on left; inwards-moved material on right. Boundary between lateral plate and blood-forming area is very uncertain.

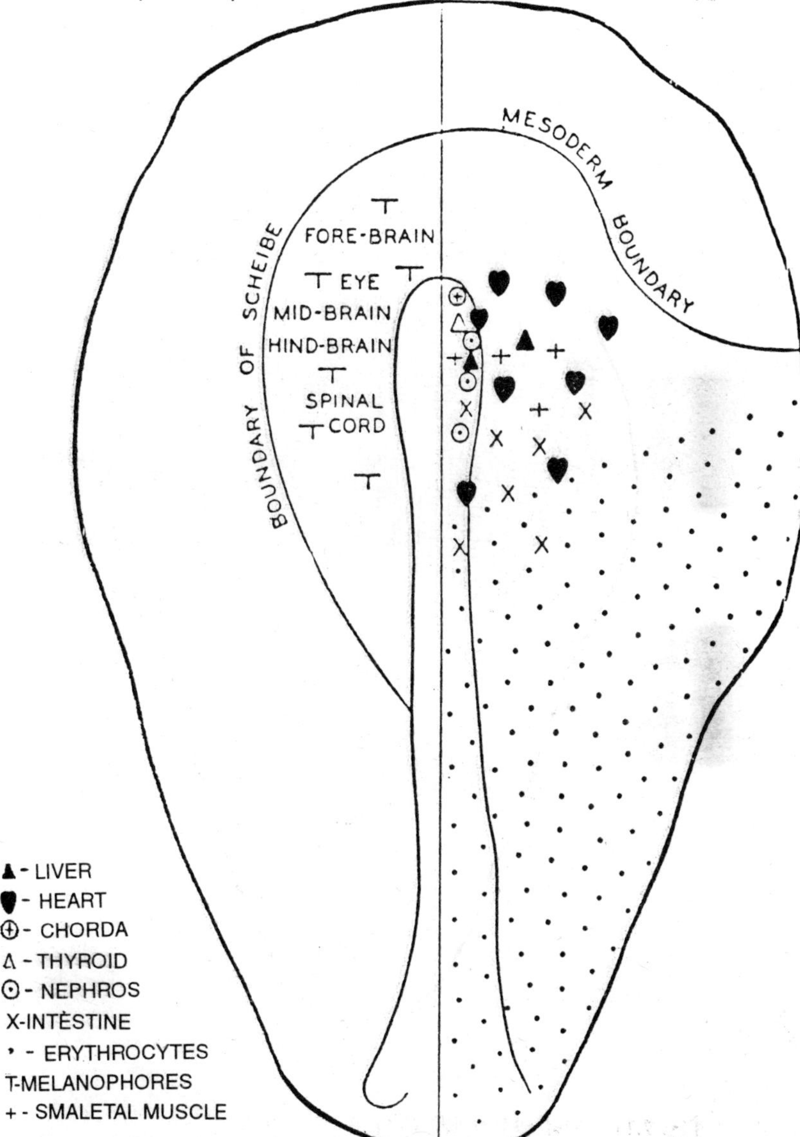

Fig. 7.10. Distribution of histogenetic potencies at the stage of the definitive primitive stage. Ectodermal potencies shown on the left, mesodermal and entodermal on right; these have not been tested separately. Posterior and lateral extent of mesodermal potencies is not known for certain.

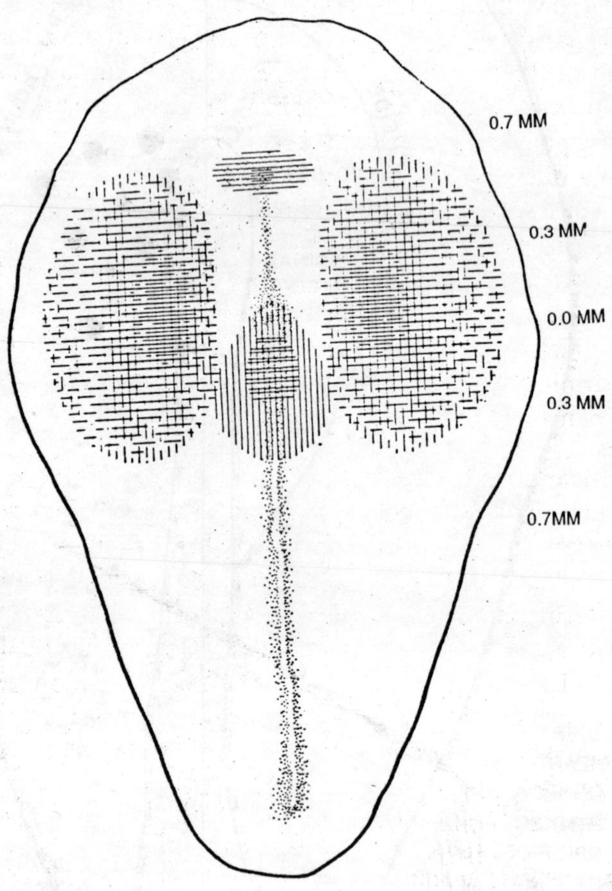

Fig. 7.11.   Map of histogenetic potencies at the stage of the head-process. (After Rawles). The eye-forming area is at the tip of the head-process, mesonephros-forming area around the anterior end of the streak, heart-forming area laterally. The density of shading indicates in a general way the intensity of histogenetic potency within each area; in the mesonephric area it also delimits the area of gonad-adrenal-forming potency.

By using the method of isolation of parts of the blastoderm, the prospective potencies of the head-process blastoderm of the chick have been thoroughly mapped when this map is compared with that of the potencies of the definitive steak it is noted that there has been considerable lateral and posterior spread of the mesodermal areas during the formation of the head-process. Some areas, such as those of heart and liver-forming potency, which were previously located medially have become bilateral. Each area of potency is an ever changing field which varies in extent and shape throughout development. It intensity is highest at one point and diminishes in varying degree toward an indefinite edge where it may overlap other areas. There is a tendency for the potencies of transverse areas to be greater toward the left side of the blastoderm (e.g., the eye-forming area, which is probably correlated with a general dominance of the left side during development.

Significance of the Primitive Streak

It is apparent from the foregoing that during gastrulation the embryo is converted from the one-layered to the three-layered state, and that the cells which eventually become head-mesoderm, notochord, somites, etc., are originally localized in definite areas on the surface of the blastoderm. These areas of the epiblast move into the primitive streak and through it to the interior of the germ: thus, organ-presumptive cells which are at first distributed in a single layer are brought into spatial relations to one another which permit interactions and a whole chain of new developmental processes (e.g., inductions).

Inductive activity has, in fact, been demonstrated for the primitive streak of the chick. *Hunt* found by grafting portions of the blastoderm to the chorio-allantois that only the nodal level was capable of forming axial structures and suggested that the latter were "organized" by the node. Waddington and his co-workers obtained direct evidence for such inductive activity by grafting portions of the primitive streak beneath the epiblasts of blastoderm *in vitro*. Not only was the primitive streak capable of organizing the overlying epiblast of the host, but there was a gradient of organizing activity with its high point at the anterior (nodal) end of the primitive streak. These relationships are akin to those which are well-known in the amphibian gastrula.

The primitive streak of the chick is comparable in *function*, therefore to the amphibian blastopore and the germ ring of the teleosts. All three are regions of high activity where outlying areas of cells on the surface are converged and moved inward to form the germ layers and future organs of the embryo. Such homologization was suggested as early as 1876 by Rauber, who considered the primitive streak to be an elongated blastopore. Insofar as *structure* is concerned, there are wide differences : e.g., at no time is there an opening in the primitive streak leading into an archenteron; the primitive streak more nearly resembles the closing amphibian blastopore with apposed lateral lips. However, the differences in morphology are secondary, and are imposed apparently by differences in the amount of yolk within the eggs of different forms.

The node and primitive pit of the definitive streak, being the area of highest morphogenetic activity and the site of inflow of the prospective axial mesoderm, are comparable to the dorsal lip of the amphibian blastopore and the teleostean germ ring at the level of the embryonic shield. If the regressing streak is homologous to a closing blastopore, one would expect that the less active portion (ventral lip) would close first of all, whereas the most active portion (Hensen's node = dorsal lip) would remain functional until the last. spratt's demonstration that the regressing streak is shortened at its posterior end supports such an interpretation. At the close of gastrulation the last of the cells to be moved inward are piled up with the node to form the caudal knot or tail bud from which the remainder of the embryonic axis is completed. The transition from primitive streak to tail bud is almost imperceptible, and the two structures are merely two morphological manifestations of the same phenomenon— the putting in place of prospective organ-forming cells.

The Mesoderm of the Opaque Area

We have seen that the prospective mesodermal cells move inward from the surface through the primitive streak, and spread out between the ectoderm and the entoderm to the margin of the pellucid area; the mesoderm then begins to overlap the opaque area at first behind, later at the sides, appearing between the ectoderm and the germ wall. Its peripheral extension; at first it spreads rapidly behind the embryo, but soon extends with equal speed opposite the primitive streak, and thus a considerable portion of the area opaca becomes three-layered, consisting of ectoderm, mesoderm, and germ wall. The contour of the anterior margin of the mesoderm is at first rounded convex anteriorly. Then the antero-lateral angles of the mesoderm begin to extend forward so that the anterior boundary becomes concave; the lateral horns thus established continue to grow forward and ultimately meet in front of the head they thus bound a mesoderm-free area in front of beneath the head, known as the proamnion, into which the mesoderm does not penetrate until a relatively late stage of development.

Blood islands develop early in the three-layered part of the opaque area; appearing first behind the embryo, they rapidly differentiate forward opposite the sides of the embryo and follow the expansion of the mesoderm. This three-layered portion of the opaque area is known as the vascular area (area vasculosa) after the appearance of the blood islands. It soon acquires a very definite peripheral boundary by the formation of the vena (sinus) terminalis at its margin. The two-layered peripheral portion of the opaque area is known as the vitelline area (area vitellina), and here again we distinguish two zones, an outer including the zone of junction, and an inner one.

The first blood islands are masses of cells lying on the germ wall behind the embryo; the first blood cells (*erythrocytes*)

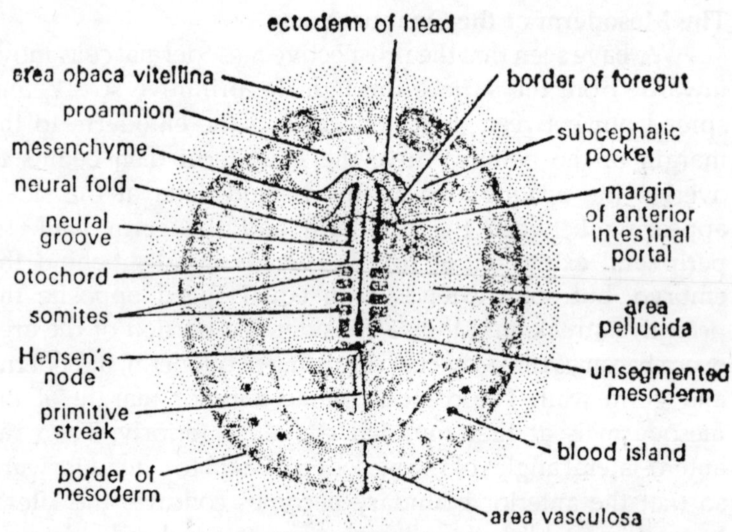

Fig. 7.12.   Chick embryo-24 hours (W.M.)

and blood vessels arise from them, hence their name. Soon after their origin the blood islands appear red owing to the formation of hemoglobin. Between the blood islands and the ectoderm is a layer of the mesoderm. If the blood islands be reckoned as mesoderm we must distinguish two layers of the latter, a deep or vascular (hemangioblastic) layer lying next to the germ wall, and an upper layer next to the ectoderm, which may be called the coelomic mesoderm, inasmuch as the body cavity (*coelom*) develops within it later.

There are two sharply contrasted views concerning the origin of the mesoderm in the area opaca. According to the one point of view it is simply a peripheral extension of the primitive streak mesoderm with which as a matter of fact it is continuous. According to the other point of view, it is split off from the germ wall. One thing is perfectly clear, namely, that the mesoderm of the opaque area arises in continuity with the primitive streak mesoderm; the second view would therefore be better expressed as *Ruckert* states it that the primitive streak mesoderm grows in the region of the area opaca at the expense of elements of the germinal wall.

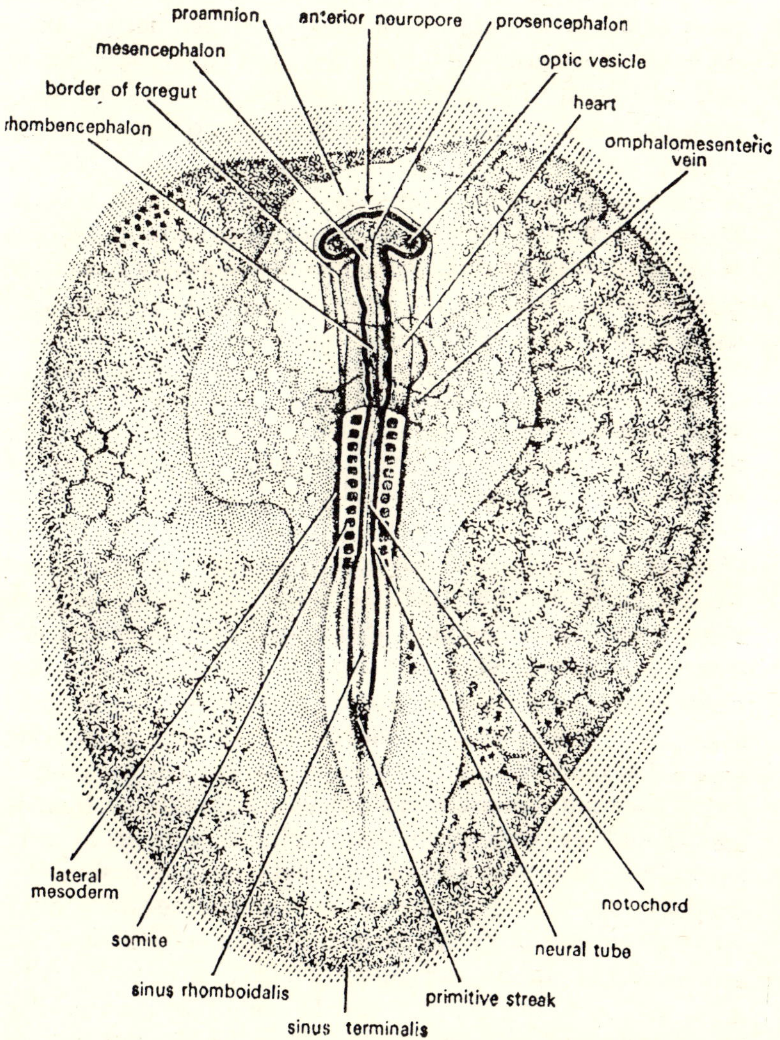

Fig. 7.13.   Blastoderm and embryo at the stage of 33 hrs. of incubation.

If the cells of the primitive streak mesoderm are compared with the cells of the forming blood islands a sharp contrast will be observed: the mesodermal cells of the area pellucida are devoid of yolk granules; young blood islands on the other hand contain yolk granules of precisely the same character as those of the germ wall, which must have been derived from the latter. If the origin of the blood islands be carefully traced, they are found to be rooted in the protoplasm of the germ wall; and prior to the appearance of the blood islands proper , protoplasm and nuclei of the germ wall aggregate superficially in a manner that appears to foreshadow the blood islands. Therefore, either the blood islands are derived from the cells of the germ wall, or cells of the mesoderm growing over the germ wall burrow into the latter, engulf yolk spheres, and reappear in masses as blood islands.

The second alternative is probably right in principle, for several reasons: Patterson and Wetzel prevented the mesoderm of the primitive streak from reaching the germ wall and found that blood vessels did not develop in the absence of streak mesoderm. In addition, when the head fold and anterior end of the embryos were isolated before the appearance of blood islands in the opaque area, the isolate developed its own blood system independently of the blood islands of the opaca (Reagan). This latter observation is supplemented by the multitude of experiments which led to the mapping of histogenetic potencies. It is clear that not only the opaque area but also a large share of the area pellucida and primitive streak itself can give rise to erythrocytes. Furthermore, there is considerable evidence from tissue cultures of pieces of early blastoderms that the area of erythrocytic potency first moves from a lateral position in the epiblast toward the streak, after which it spreads out toward the lateral border of the blastoderm and localizes in the area vasculosa. Thus, it can be stated with some assurance that a major part (if not all) of the vascular tissue is derived from mesoderm which has moved inward from the epiblast by way of the primitive streak. Whether the germ wall also contributes to the hemangioblastic layer is an open question.

Another question concerns the origin of the layer of coelomic mesoderm that overlies the blood islands: is it derived from the primitive streak mesoderm, or is it split off from the blood islands? When the latter first appear, in the periphery of the vascular area at least, there is no coelomic mesoderm above them. It appears later, at first not as a coherent layer, but as scattered cells that rapidly unite to form a layer. In many places the microscopical appearances indicate strongly that the cells are split off from the surface of the blood islands; but, as they are usually not far from the edge of the advancing coelomic mesoderm, it may be that they are derived from the latter. Ruckert states, however, that, in the case of some isolated blood islands behind the embryo, a layer of mesoderm is formed over them while they are still isolated. This would render the derivation from the blood islands probable in such cases. It is possible, therefore, that the coelomic mesoderm grows partly, at least, at the expense of the superficial cells of blood islands.

As rapidly as they are formed the various blood islands connect and anastomose with one another forming a vascular network lying between the coelomic mesoderm and the germ wall. This network spreads throughout the vascular area, and appears later in the pellucid area, and communicates with the blood vessels of the embryo. In the next chapter we shall consider the manner in which the extension takes place, and the origin of the blood vessels and blood cells.

The Germ Wall

The germ wall arises, as we have seen, through infiltration of the superficial white yolk by the periblast. These cells multiply and anastomose and form a multinucleated syncytium with the yolk granules in its meshes. By degrees the protoplasm itself takes up the yolk granules, which are gradually digested, and the germ wall thus becomes organized as a coherent layer. It then separates from the underlying yolk. The next period in the history of the germ wall is its differentiation, which takes place in the vascular are concomitantly with the formation of the blood islands: the vascular mesoderm is formed probably from the streak mesoblast with contributions of yolk globules (if not cells) from the germ wall. The remainder of the germ wall then differentiates into the characteristic entodermal epithelium of the the opaque area, which is known as the yolk sac epithelium (entoderm) because it is destined to form the lining of the yolk sac.

After the formation of the vascular area the term germ wall must be restricted to the lower layer of the vitelline area, because with in the vascular area the mesoderm and yolk sac entoderm have already differentiated. The development of the germ wall takes place in a centripetal direction; at any period during the overgrowth of the yolk the three stages of the germ wall may be found in the con-centric zones. The first stage, that of periblast, is found in the zone of junction (area vitellina externa); the second stage, that of organization of the germ wall, is found in the area vitellina interna; and the third stage, that of differentiation, is found at the margin of the area vasculosa. Within the latter area the differentiation is completed.

8

Advanced Development of Chick

Twenty-four Hours of Incubation

In the average college embryological laboratory it is customary to study chick embryos at definite intervals of incubation, overbridging the gaps between these stages by collateral reading. The 24-hour chick embryo is usually one of the first embryos to be examined in total and serially sectioned preparations. It is therefore desirable to devote some time to the discussion of the chick embryo at that age, and in the following chapters embryos of such ages as are usually preferred for general study will be discussed more thoroughly than others.

The Neural Folds

In the 24-hour chick embryo, the neural folds have approached each other to a variable degree in different regions of the body. Posteriorly they are widely separated, flaring out as they approach Hensen's node and getting lost in the superficial ectoderm on both sides of the anterior portion of the primitive streak. In the head region the neural folds diverge to a lesser degree, but they are well separated and the neural groove is deeper here than in the central and posterior portions. The folds approach each other closest in the area located immediately behind the head and anterior to the first somities, and it is here that they will fuse first to form the neural tube. Viewed from the dorsal aspect in a total preparation, the neural folds appear as two bands, diverging slightly anteriorly and more so posteriorly. As a matter of fact, the neural folds on the anterior tip of the head are curved forward and downward, giving the head a notched appearance.

The Enteron

The enteron is composed of the pocket-like fore-gut and the open mid and hind-gut. Examined from the dorsal aspect, the opening into the fore gut (*anterior intestinal portal*) can be located behind the crescentic head fold. From the anterior intestinal portal the fore-gut juts forward into the head as a broad flat pocket with single cellular sides, floor and roof.

The Mesoderm

The mesoderm has spread out considerably between the ectoderm and entoderm and extends far out into the blastodisc, except in the mid-interior region where its absence in front of the head is indicated by the proamnion. In the 24-hour chick the mesoderm has formed from four to five paired somities, located to the right and left of the notochord in the central portion of the embryo. Their organization and relationship to the remainder of the mesoderm can best be observed in transverse sections. The mesoderm can be conveniently divided into three district parts. These are the dorsal or segmental mesoderm represented by the somities; secondly, the non-segmented intermediate mesoderm; and thirdly, the lateral mesoderm consisting of an upper somatic and a lower splanchnic layer. In the early embryo, the somatic layer is always apposed to the ectoderm, and the two in connection are referred to as the *somatopleure*. The splanchnic mesoderm, when in close apposition to the entoderm, forms with it a double layer which is called the *splanchnopleure*. In the peripheral portion of the circular blastodisc, the lateral mesoderm has not as yet differentiated into the somatopleure, and splanchnopleure. Its structure will be discussed in connection with the formation of the blood vascular system.

In addition to these three well-defined divisions of the mesoderm, one may include the mesenchyme, consisting of loose cells which are budded off from the somites in all parts of the body, especially in the head region.

The dorsal or segmental mesoderm produces the mesodermal somites. From the developmental point of view, it is the most important differentiation of the mesoderm, because the somites give rise to several fundamental organic differentiations. They make their appearance as blocks of cells containing a transitory central cavity referred to as the *myocoel*.

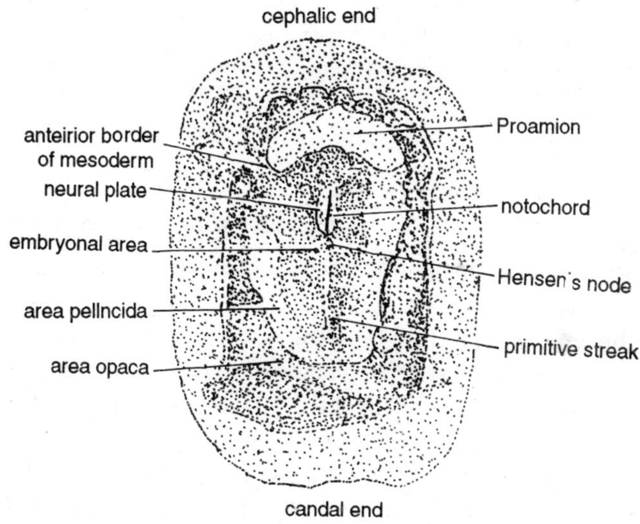

Fig. 8.1.   Chick embryo-18 hours (W.M.)

The intermediate mesoderm connects the dorsal with the lateral mesoderm. It is not segmented, though it gives rise to the segmented *nephrotomes*, and later to the segmented *mesonephros*.

The lateral mesoderm is unsegmented, agreeing in this respect with that of all other vertebrate embryos. It is composed

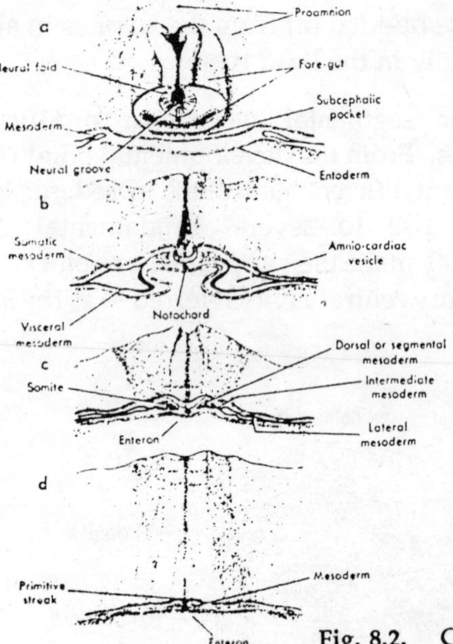

Fig. 8.2.   Chick embryo-24 hours

of the somatic and splanchnic layers which enclose the coelomic cavity. Since these two sheets of the lateral mesoderm extend from the embryo far out into the extraembryonal area, the coelome must necessarily be very extensive. For descriptive purposes the coelome within the embryo is therefore referred to as the *intra-embryonic coelome*, and that found in the extraembryonic area as the *extraembryonic* coelome. In early embryos, as for example, those studied here, there is no sharp demarkation between the two coelomes.

Near the anterior intestinal portal the paired sheets of the splanchni, mesoderm show conspicuous thickenings as they branch off from the intermediate mesoderm. Moreover, because of the elevation of the gut at the intestinal portal, the two mesodermal sheets are well separated, thus forming a very spacious intraembryonic coelome. The thickened portions of the splanchnic mesoderm on either side of the gut initiate the formation of the heart, which will be discussed in more detail in the next chapter. It will be completed and functioning within an additional twenty-four hours of incubation. The spacious part of the coelome, called the *amniocardiac vesicle*, is destined to form the pericardial cavity.

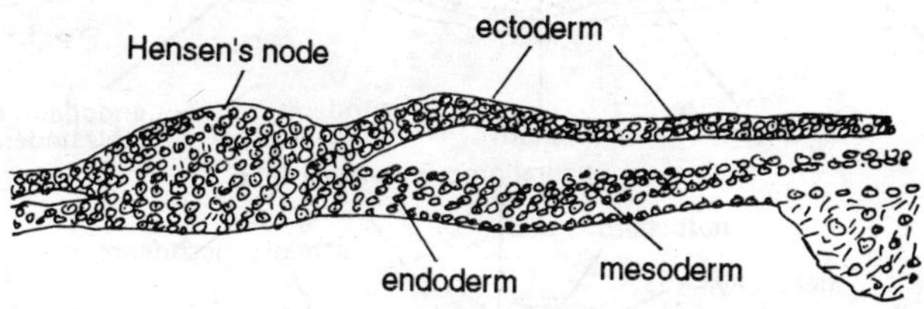

Fig. 8.3   T.S. of 18-hrs. Chick Embryo through Heusen's node.

The Area Opaca Vasculosa and Vitellina

The total 24-hour chick embryo shows that the area opaca and area pellucida have retained the same fundamental relationship that they had in earlier stages. However, the embryonic area and the growth of the mesoderm have partly obscured the earlier definite outlines of the area pellucida. This is true to a much greater extent of the area opaca. In its posterior half, closest to the area pellucida and encroaching upon it, it has developed a mottled and blotchy appearance and has differentiated into the so-called *area opaca vasculosa*, which is destined to give rise to a large part of the extra-embryonic vascular system. In further development the area opaca vasculosa; or simply *area* vasculosa, tends to grow forward, thus encircling the entire embryo in later stages (48 hour check). The area vasculosa does not occupy the entire area opaca. In early embryos it is limited to the portion adjacent to the area pellucida. Beyond the area vasculosa the area opaca does not have a mottled appearance, and this peripheral region is known as the *area opaca vitellina*. A clearer picture of the structure and relationship of these two areas is obtained from the study of transverse sections which extend far out into the extra-embryonal area.

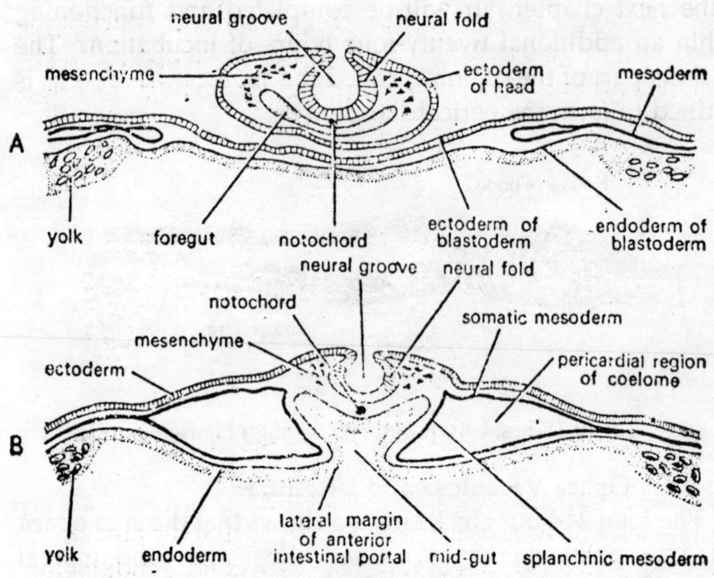

Fig. 8.4.   T.S. of 24 hrs chick Embryo
      A. T.S. through head
      B. T.S. through mid-body.

The Area Vasculosa

Following one of the paired mesodermal sheets as it passes out between the ectoderm and entoderm in the posterior region of the embryo, we find that it traverses the area pellucida and enters the area opaca. It becomes apparent here that the underlying entoderm is gradually losing its definite sheetlike structure, and in tracing it further outward we observe that its cells become disjoined and eventually disappear entirely, well within the peripheral limits of the blastodisc. The mesoderm remains intact for a longer distance, but even this sheet becomes disorganized before reaching the periphery of the blastodisc. In sections of embryos of twenty hours of incubation, the mottled appearance of the area vasculosa observed in total preparations is seen to be due to localized mesodermal thickenings referred to as blood islands. In later development, when the undifferentiated mesoderm gives rise to the somatic and the splanchnic layers, the blood islands are embedded in the latter, which at that time rest on the expanding thin yolk entoderm.

In their earliest stages the blood islands are merely large compact clusters of cells within the mesodermal sheet, resting above the entoderm of the yolk sac. Within ten hours of further development they evolve cavities isolating the central cells from the peripheral ones. The former will be transformed into blood cells containing hemoglobin, while the latter will form a wall of flattened endothelial cells around the blood cells. In the meanwhile the latter have multiplied into such large numbers that they touch and anastomose with each other, thus forming a vascular capillary network containing blood corpuscles. Finally, the cavities within the blood islands become filled with a liquid (plasma) acting as a vehicle for the corpuscles when circulation is later established. In the discussion of the development of the heart in the following chapter, it will be seen how this capillary system becomes connected with the embryo.

The Area Opaca Vitellina

The area opaca vitellina is located beyond the area vasculosa in the peripheral region of the blastodisc. It contains chiefly a surface layer of ectoderm and underlying undifferentiated cells which are more numerous near the area vasculosa than they are in the periphery. It consists of the margin of overgrowth and the germ wall where the cytoplasm of many cells is continuous ventrally with the yolk, and is therefore identical in its general structure with the peripheral layers of the blastula and gastrula, which have been discussed before.

9

Thirty-Three Hours of Incubation

During the period of incubation from twenty-four to thirty-three hours, the chick embryo undergoes extensive changes in which several permanent anatomical structures begin their differentiation. In a total preparation of the 33-hour chick the most obvious transformation seems to occur in the head region. In the 27-hour chick the neural folds begin to fuse near the future myelencephalon, and within six hours the fundamentals of the vertebrate brain have been laid down.Furthermore, the entire head region has elongated and grown forward over a deep subcephalic pocket, and anterior to the latter the superficial ectoderm of the blastodisc is elevated in a crescentic fold *(amniotic head fold)*. Within the head region the fore-gut has kept pace with this growth and has elongated considerably, so that the anterior intestinal portal comes to lie relatively further backward. Close to the latter, two blood vessels have been formed which unite anterior to the first somite to form the tubular heart, and postero-laterally they flare out to become continuous with the area vasculosa. Lastly, the fact that the embryo has grown considerably in length is evident by the increase in the number of mesodermal somites.

The Neural Tube

In the 33-hour chick the neural tube is differentiated into the anterior, more or less dilated portion and the central and posterior, uniformly tubular part, the former giving rise to the brain and the latter to the spinal cord. During an interval of nine hours, from the 24- to the 33-hour embryo, the neural folds unites dorsally, and, proceeding with their fusion forward as well as backward, form the neutral tube. The neural folds do not close stimultaneously throughout their entire length; anteriorly they are delayed for a few hours and posteriorly even longer. The opening at its anterior tip is called the *anterior neuropore;* this will close eventually, leaving a scar that is still visible in the 33-hour chick embryo. Posteriorly the fusion of the neural folds is delayed much longer, thus forming a posterior opening to the neural tube referred to as the *posterior neuropore.* Behind the last mesodermal somite the two neural folds diverge, and after having passed the ever-shortening primitive streak, they gradually fade into the superficial ectoderm. In this manner the neural folds encompass a shallow area, called the *sinus rhomboidalis,* which represents the most posterior portion of the materials which will complete the neural tube. Its posterior limit is therefore indicative of the posterior tip of the future spinal cord.

Since the disappearing primitive streak extends into the posterior tip of the sinus rhomboidalis, it has been assumed that the primitive pit may represent an abortive attempt to form a neurenteric canal. If one may assume that the neural folds will close over the area of the sinus rhomboidalis and thus form a roof over the primitive pit, and if then the latter would perforate into the enteron, the neurocoel would become continuous with it.

The manner in which the neural folds fuse is similar to that described for the frog embryo. Each fold consists of a single-layered superficial (non-nervous) and a deeper, many-layered (nervous) ectoderm, embodying the bulk of the former neural plate. By the union of the folds the superficial ectoderm unites over the neural tube while the latter detaches itself and sinks into the tissue.

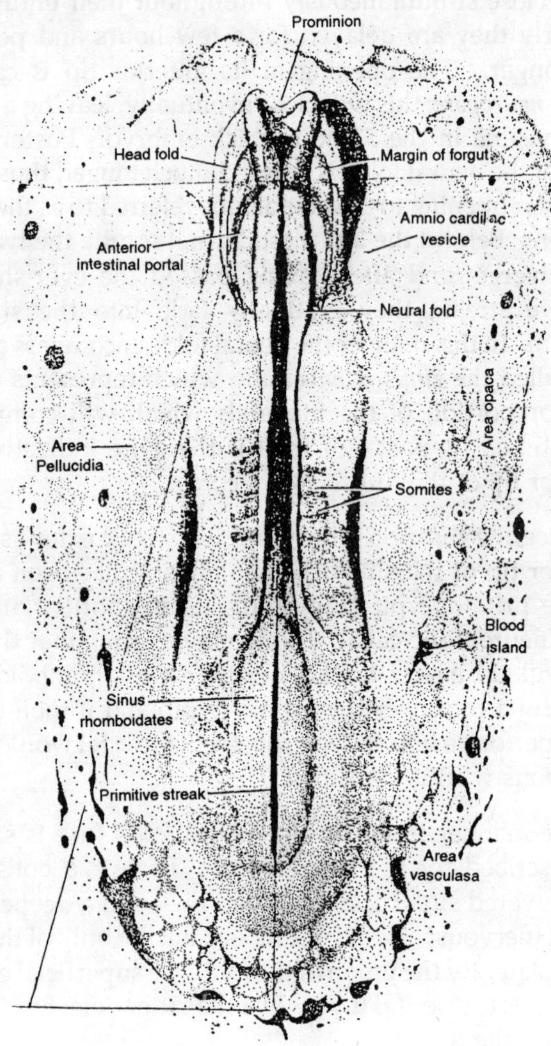

Fig. 9.1.   The 25-hour thick embryo with five somites, dorsal aspect. Total view.

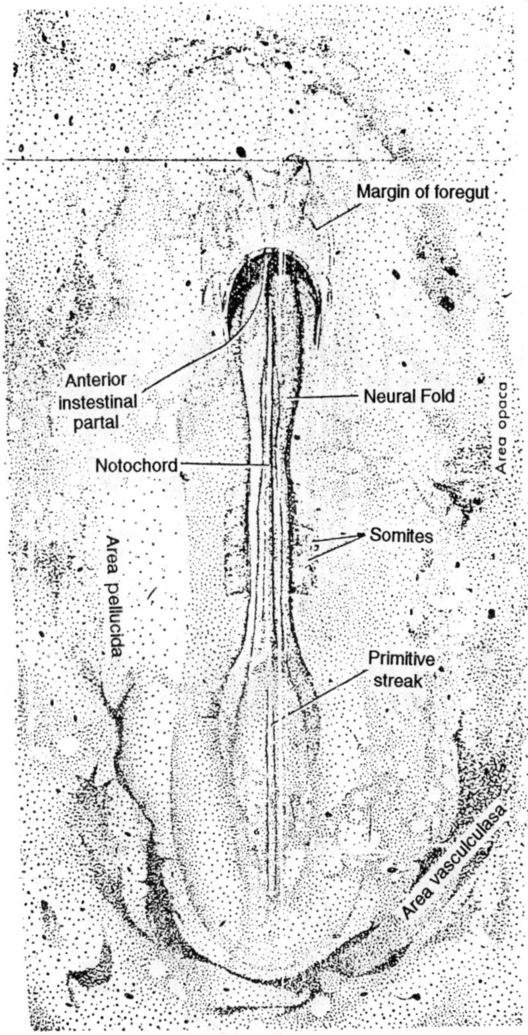

Fig. 9.2.   The 25-hour chick embryo with five somites, ventral aspect. Total view.

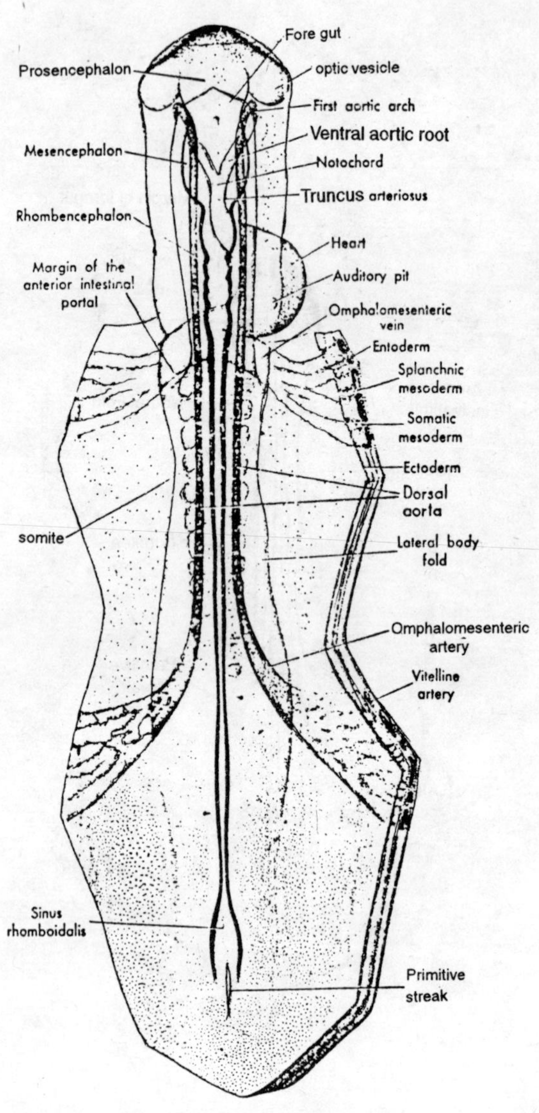

Fig. 9.3.   The 33-hour chick embryo with 13 somites, dorsal aspect.

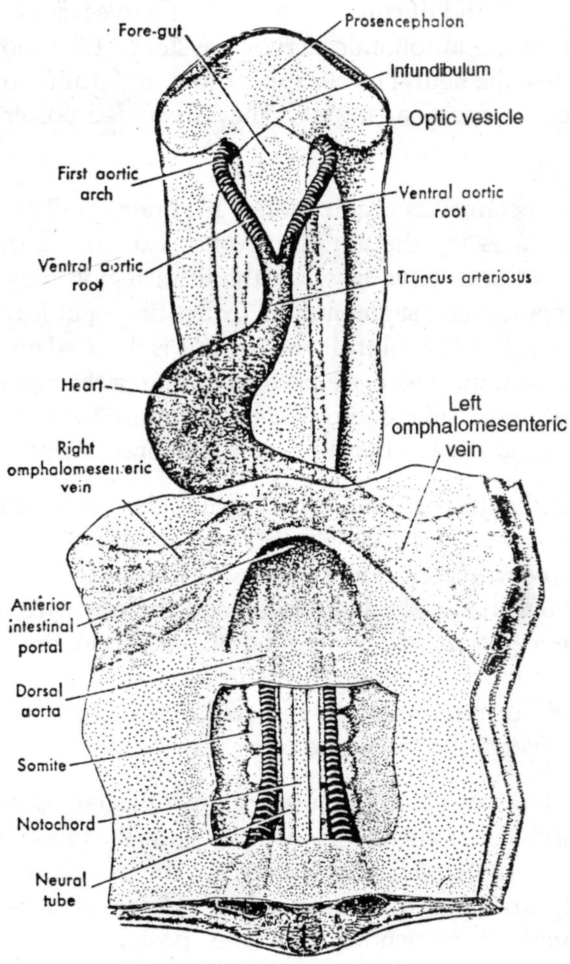

Fig. 9.4.   The 33-hour chick embryo.

The Neural Crests

When the two neural folds unite in the mid-dorsal line,

loose cells appear between the two ectodermal components of each fold. These cells form two bands (neural crests) on the dorsal right and left sides of the forming neural tube. They are segmented and form the primordia of the dorsal root ganglia of the spinal and of some of the cranial nerves, as well as the ganglia of the autonomic nervous system. Like most other structures, the neural crests appear first in the anterior region of the embryo, and as it grows they are added posteriorly.

The Brain

The metameric organization of the brain is indicated early by *neuromeres* in the anterior portions of the neural folds. Before the latter have fused in the head region they show a vague neuromeric segmentation, consisting of at least eleven neuromeres. It is doubtful whether this is the total number for the entire brain. There are probably more in the anterior part that are not definite enough to be recognized and a number are added to the hind part of the rhombencephalon.

Shortly after fusion of the neural folds, the three primary brain divisions of the vertebrate brain are differentiated from specific neuromeres. The prosencephalon, of forebrain, takes up the first three, the mesencephalon the next two, and the rhombencephalon the remaining six. Beyond the eleventh neuromere the metameric organization of the neural tube becomes so indefinite that the actual posterior limit of the rhombencephalon cannot be recognized.

With the differentiation of the three primary divisions of the brain, all neuromeric constrictions within the prosencephalon and mesencephalon disappear except those separating these primary divisions from each other. However, most of the neuromeric constrictions in the posterior part of the rhombencephalon remain visible for a longer period — at least for more than three days.

With the sharp, demarcation of the primary divisions of the brain, the neurocoel has also been divided into certain definite regions. The cavities of the forebrain, midbrain, and hindbrain are the *prosocoel, mesocoel* and *rhombocoel* respectively. The last is continuous with the neural canal of the spinal cord at the posterior part of the rhombencephalon.

The prosencephalon becomes expanded by the lower lateral outgrowths of the sac-like optic vesicles. In the 33-hour chick embryo they are the most conspicuous differentiations of the entire brain, extending outward from the right and left sides of the prosencephalon toward the ectoderm of the head. Their cavities (*opticoels*), which are in free communication with the prosocoel, are actually divisions of it. On the floor of the prosencephalon, immediately posterior to the optic vesicles, is a median depression which forms the infundibulum, a part of which eventually forms the posterior lobe of the pituitary.

The mesencephalon containing the mesocoel, as well as the rhombencephalon with its rhombocoel, shows no other differentiation in the 33-hour chick embryo besides those referred to above.

The Enteron

The fore-gut of the 33-hour chick is slightly less than one millimeter in length. It has elongated at the expense of the open mid-gut. The anterior intestinal portal is continually receding posteriorly as a result of the ventral fusion of the lateral walls of the open gut. In this manner the open gut behind the anterior intestinal portal is slowly converted into a closed gut. At the same time the amnio-cardiac vesicles, which were located at either side of the intestinal portal in the 24-hour chick, become located anterior to it in the 33-hour embryo. This fact must be kept in mind in the following discussion of the development of the heart.

Restricted and small as the fore-gut appears to be, it is still possible to distinguish certain features on it. For example, at its anterior ventral side it is directly apposed to the superficial ventral ectoderm of the outer covering of the embryo without any intervening mesoderm. This is the oral plate, which will rupture in proper time to form an opening for the mouth. The oral plate in the chick embryo is homologous to that of the frog embryo, and it is advisable to recall its formation in the latter. The anterior tip of the fore-gut in the 33-hour chick embryo is round in transverse section. It broadens and flattens over the anterior heart region and appears therefore more or less slitlike in transverse section. This part of the fore-gut is destined to become the pharynx. Over the posterior heart region and still further back the fore-gut becomes rounded again, and in due time this part will become the oesophagus.

The Mesoderm

Though the organization of the mesoderm has become increasingly more complex, it is not as striking in the structure of the somites as it is in the lateral mesoderm. The somites are bulkier and better defined as compared with those of the 24-hour chick, and in the anterior region they have almost lost their connection with the intermediate mesoderm. Not considering these minor changes, a cross section through the embryo at the level of the sixth or seventh somite will show no radical changes in its organization. The differentiations in the mesoderm of the 33-hour chick are chiefly concerned with the vascular system. Beginning in the 20-hour chick with the formation of the area vasculosa, this system culminates in its completion and function in embryos of forty to forty-four hours of incubation. Of course, there will be changes and additions to this complex system; in fact, an extensive circulatory system *(allantoic)* will be added at the end of the third day of incubation, but this does not alter the fact that a complete blood circulation has been established in the 44-hour chick embryo and that by means of it metabolism is carried on.

In the 33-hour chick two complete circulatory arcs are in process of formation, one within the embryo (intraembryonic) and the other outside of it (extraembryonic). The center of each is the heart.

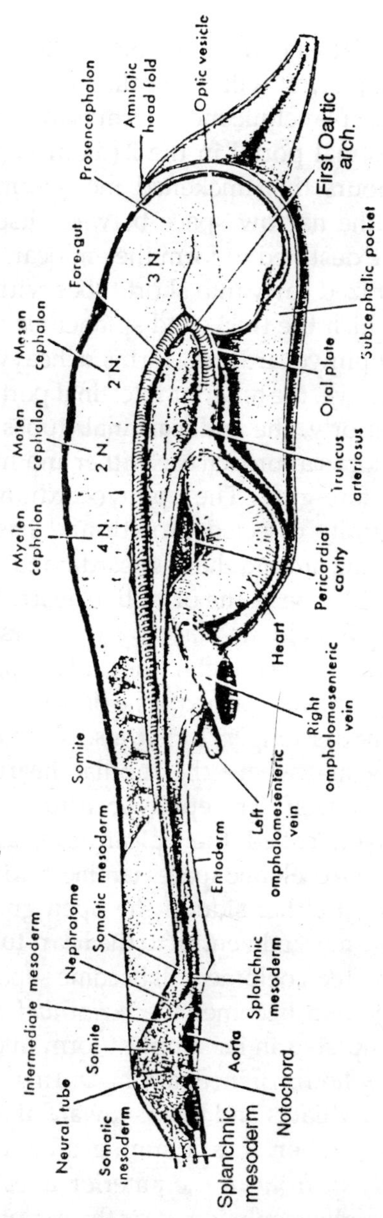

Fig. 9.5.   The 33-hour chick embryo, lateral view.

The Heart

The beginning of the heart formation has been alluded to in the preceding chapter. It will be recalled that it consists of the thickening of the splanchnic mesoderm near and posterior to the anterior intestinal portal in the 24-hour chick. In the succeeding three hours this thickened mesodermal portion buds off cells into the narrow space between itself and the entoderm. These are destined to form the endocardium of the heart, and are organized early into short tubes with walls one cell thick around which the thickened splanchnic mesoderm, referred to as the epimyocardium, forms a heavy covering. When, as outlined above, the anterior intestinal portal becomes pushed more posteriorly, the endocardinal tubes with their epimyocardial envelopes approach each other and fuse beneath the newly formed fore-gut. The epimyocardium has kept pace with this centrally directed growth and has formed a thick investing covering around the endocardium, thus forming a short, double-walled tube — the primitive heart. Dorsally as well as ventrally it is held in place by the dorsal and the ventral mesocardium respectively, the latter disappearing immediately after it has been formed. The dorsal mesocardium persists for a few hours longer, but it also vanishes, except near the sinus venosus, where the tubular heart begins to dilate and twist upon itself. The epimyocardium gives rise to a thinner outer layer, called the epicardium, and a thick myocardium. The large coelomic spaces (amino-cardiac vesicles) which were located at either side of the open gut in the 24-hour chick have now merged ventral to the heart tube. In later development this united coelome will become separated from the general coelome and becomes the *pericardial cavity*. The tubular heart is completed in its simplest form in embryos of twenty-nine to thirty hours of incubation. During the ensuing three or four hours it dilates and bends toward the right side. In these early stages when it is a simple tube, the heart is located immediately in front of the anterior intestinal portal and beneath the rhombencephalon, near the region where the auditory pits will presently form. At first the heart is very

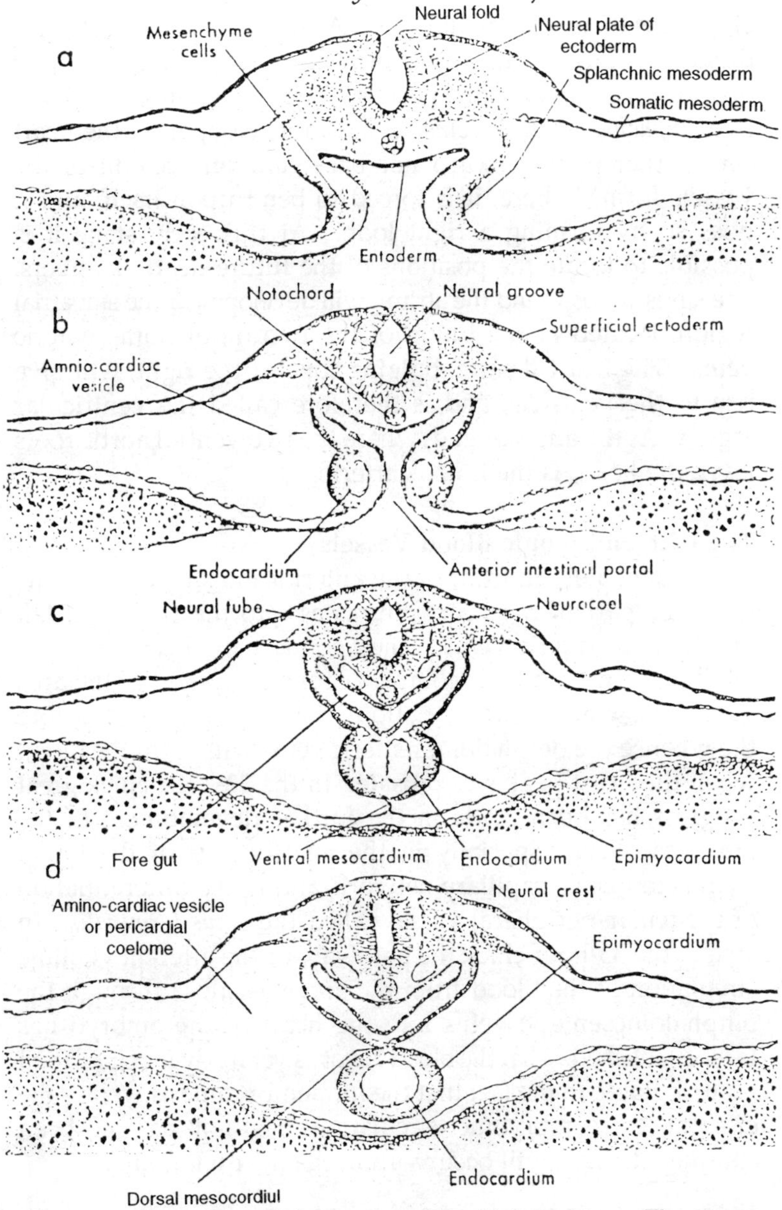

Fig. 9.6.   Formation of the heart in the chick embryo. The stages represented in this series are present in (a) the 25-hour embryo; (b) the 26-27 hour embryo; (c) the 27-28 hour embryo; and (d) the 29-hour embryo.

short, less than $^1/_2$ mm. in length. As it begins to elongate, it is divided anteriorly into two branches which form the two ventral aortic roots, while posteriorly it is continuous with the two *omphalomesenteric veins.* Since in its elongation the heart can neither push forward nor backward on account of the vessels formed there, it is forced to bend upon itself, and it does so by forming a right loop. At this early stage it is possible to locate the positions of the future heart chambers. The sinus venosus and the atrium will develop near the sinoatrial region, located at the union of the two omphalomesenteric veins. The heart flexure, bulging toward the right, will give rise to the ventricle, and is therefore called the ventricular region. At the anterior end, where the two ventral aortic roots leave the heart, is the bulbus arteriosus.

The Extraembryonic Blood Vessels

Immediately in front of the anterior intestinal portal, the posterior part of the heart bifurcates into the right and left omphalomesenteric veins. They connect the heart with the vitelline veins and their complex capillary system of the area vasculosa, whose differentiation has been described before the whence the omphalomesenteric veins will drain the blood when the heart begins to pulsate. In the 33-hour chick great progress has been made in the further development of the area vasculosa, especially in the anastomosis of the blood islands to form a capillary network, and in the differentiation of a circular peripheral vessel called the *sinus terminalis.* In fact, in the 33-hour chick the pathway for the afferent vitelline circulation of the blood from the area vasculosa through the omphalomesenteric veins into the heart of the embryo has been established. That the blood is not as yet ready for circulation is attributable to the fact that the intraembryonic as well as the efferent extraembryonic circulation has not as yet been completed. This will occur within the next ten hours.

The Intraembryonic Blood Vessels

The divided anterior end of the tubular heart becomes the

paired ventral aortae (ventral aortic roots), which pass beneath the anterolateral tip of the fore-gut, where they swing upward, and then pass backward over the fore-gut as the two dorsal aortae (dorsal aortic roots). They are well-defined and are easily identified in transverse sections through the head region of the embryo, but as they pass backward they become less conspicuous and finally become lost as fine capillaries in the tail region. In embryos of thirty-seven to thirty-nine hours of incubation, it is possible to follow these capillaries further until they lose themselves posteriorly in the area vasculosa.

The anterior and posterior cardinal veins are also in the process of development in the 33-hour chick, but they are less striking and less definite in their appearance than the aortae. The anterior cardinals are located on either side close to the brain, except in the heart region where they are wedged between the dorsal aortae and the body wall on their descent toward the posterior portion of the heart (sinoatrial region). The posterior cardinal veins, being narrower in diameter than the anterior ones, are still more difficult to identify. When present, they are located near the intermediate mesoderm and the ectoderm of the body wall.

While it is not at all difficult to determine that the splanchnic mesoderm gives rise to the heart, the blood and the extra-embryonic blood vessels in the area vasculosa, the source of the intraembryonic blood vessels is not so clear. Apparently they are derived from mesenchyme cells, and they appear in their proper places when needed, leading into the correct channels and making the right anastomoses.

FORTY-EIGHT HOURS OF INCUBATION

External Morphology

General Appearance. After forty-eight hours of incubation the chick embryo shows a striking difference in appearance from embryos of earlier stages. Because of twisting (*torsion*) of the anterior part of its body to the right side, it seems to have lost its bilateral symmetry. Furthermore, a sharp bending (*flexure*) of the anterior portion of the head has curved the prosencephalon at right angles to the rhombencephalon. These are the most apparent changes in the examination of the total embryo. However, there are many others : as for example, the increase in the number of somites, the expansion of the area vasculosa, the growth of the heart, and the functioning of the vitelline circulation. Torsion and flexure of the embryo renders the study of the developing organs more difficult. These processes should therefore be well understood and should be kept in mind in the identification and location of specific organs and organ systems.

Changes in Portion of the Embryo

Though the 33-hour chick appears to be almost perfectly bilaterally symmetrical and appears to be without any curvature in its longitudinal axis, this is not exactly true. Viewed from the side, the prosencephalon is bent slightly downward with the center of rotation located in the midbrain. This is the beginning of the *cranial flexure*, which increases as the embryo grows older. Since the head of the young embryo is raised only slightly above the underlying tissue, a downward flexure of the head would bury it in the yolk in the region of the proamnion. This is avoided by a lateral torsion which begins at the head region. The embryo is twisted to the right, so that the left side is now closet to the yolk and the right side is facing upward. Torsion of the body begins at the tip of the head in embryos of approximately thirty-eight hours of incubation and proceeds backward as far as the eight or ninth somite in the 48-hour chick. At that time the second flexure (*cervical flexure*) can be recognized in the posterior part of the head. It is not very conspicuous at that time, but in later embryos it will be a prominent anatomical feature.

INTERNAL MORPHOLOGY

The Nervous System

The Brain. On account of the curvature of the head of the

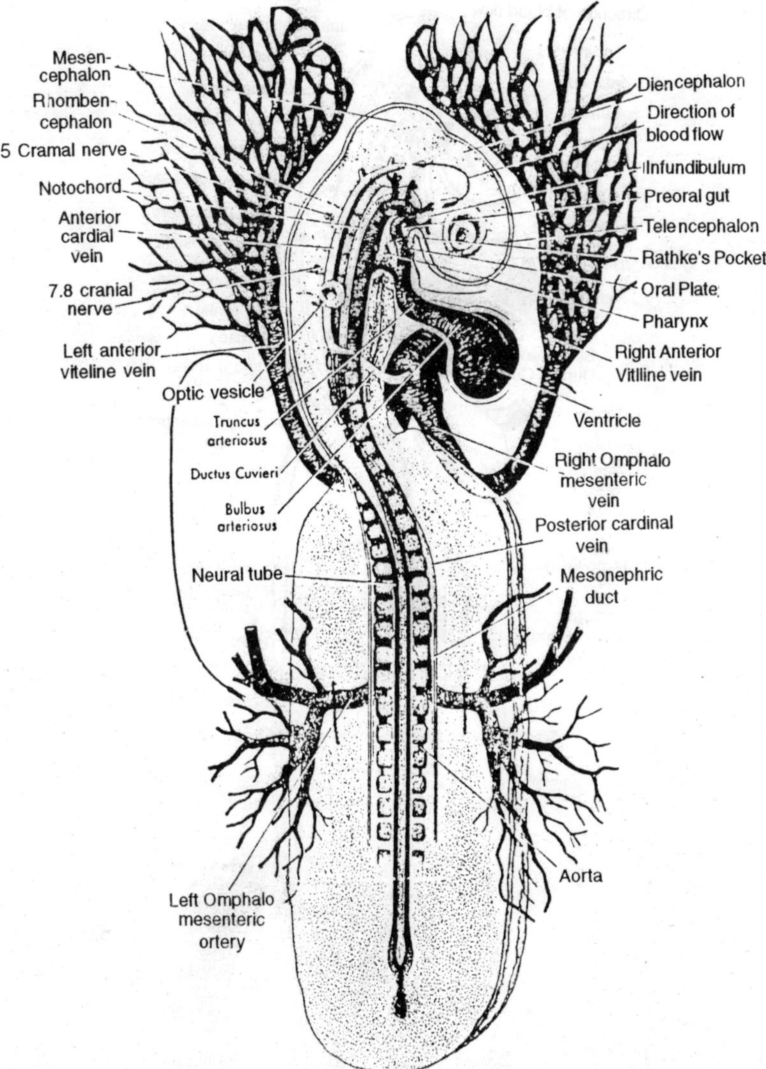

Fig. 9.7. The 48-hour chick embryo with 24 somites, dorsal view.

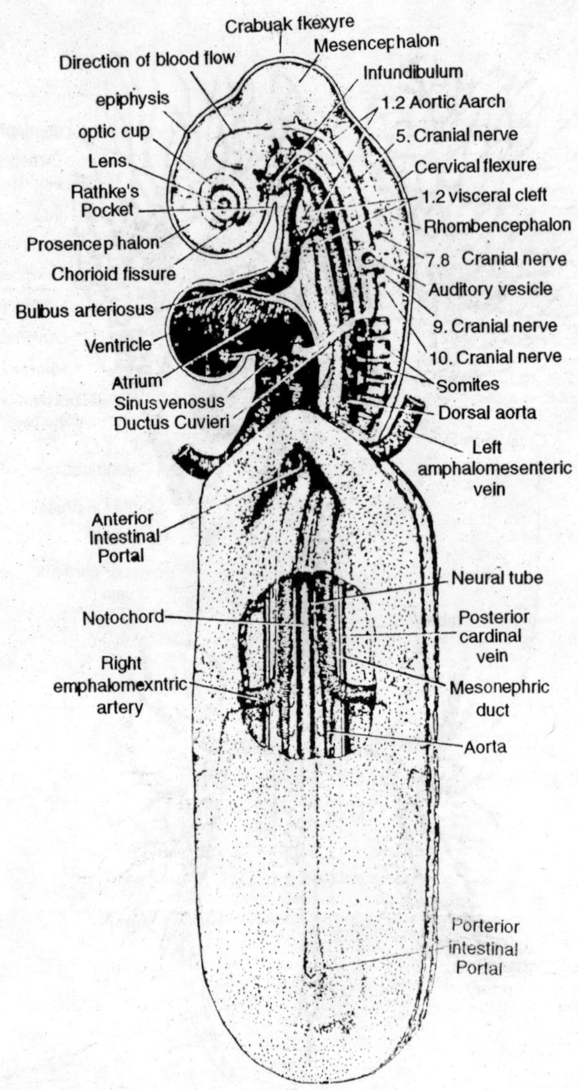

Fig. 9.8.   The 48-hour chick embryo, ventral view.

embryo, the terms "anterior" and "forward" no longer stand for the same spatial directions as they did in the 24-hour and the 33-hour embryo. In future discussions and descriptions these terms will be used in relation to the anatomical organization of the embryo. Therefore, the most anterior region of the 46-hour chick is the tip of the forebrain and not the center of the midbrain, though, on account of its flexure, the prosencephalon is actually closer by straight line to the tail than is the mesencephalon.

In an embryo of forty-eight hours of incubation the brain and the spinal cord stand out more prominently than other organs. In spite of the flexure of the head and the torsion of the body, no far reaching changes have occurred in the general organization of the brain and spinal cord. There are a number of minor differentiations which will give rise to various important organs in later development, but these are all in their incipient stages. The extent of the three primary brain divisions can easily be ascertained by the two grooves separating the mesencephalon anteriorly from the prosencephalon and posteriorly from the rhombencephalon. The latter is continuous with the spinal cord without showing any definite posterior boundary. The organization in the neurocoel is similar to that of the 33-hour chick, the brain cavities consisting of the prosocoel, mesocoel and rhombocoel being continuous with the central canal of the spinal cord. Posteriorly the neurocoel no longer opens to the outside through the posterior neuropore to the sinus rhomboidalis. It is completely enclosed, even in its most posterior aspect where its completion has eliminated every trace of the primitive streak.

A sagittal section through the brain of the 48-hour chick indicates the relative thickness of the different parts. The sides and the floor of the prosencephalon are slightly thicker than its roof. The mesencephalon is uniformly thick throughout its entire length. The rhombencephalon shows are greatest differentiation in that it has acquired a very thin roof, while its floor and especially its sides are thick.

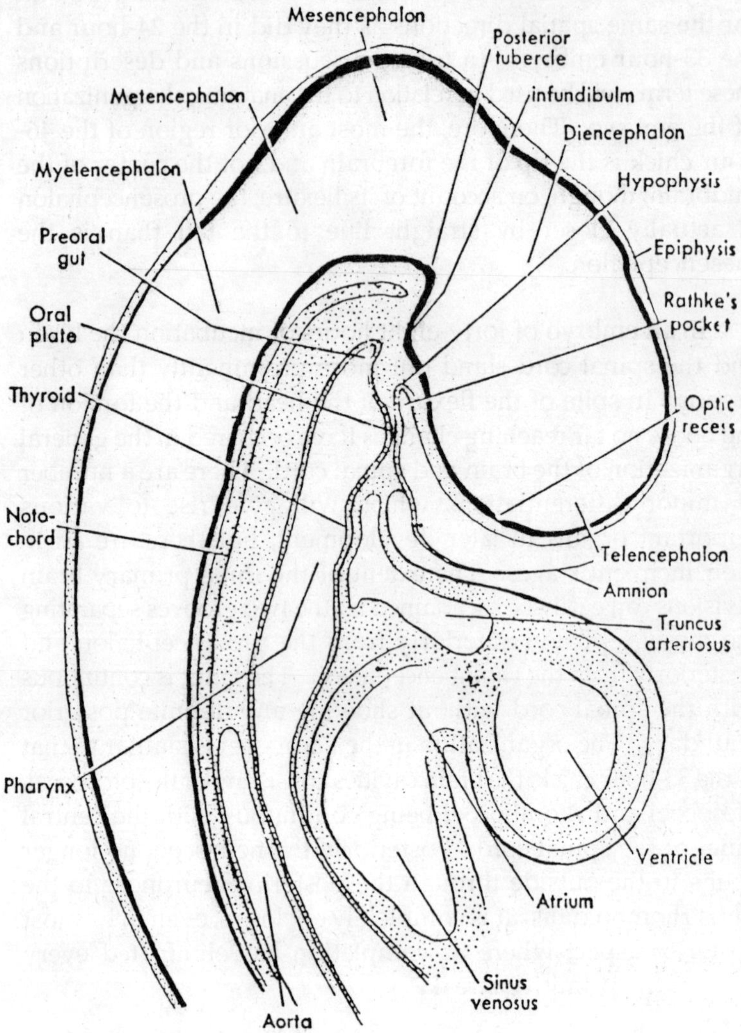

Fig. 9.9.   The 48-hour chick embryo. Sagittal section of anterior portion.

At this early stage the brain seems to be entirely out of proportion to the remainder of the neural tube or spinal cord. It comprises almost one half the entire neural tube. This condition prevails only in the early embryo, and the brain progressively becomes relatively smaller in its development to the adult condition by pushing the successive divisions of the brain closer together.

The Prosencephalon and its Differentiations

Beginning with the prosencephalon, one may observe a broad and shallow transverse groove appearing on its median dorsal surface, which, in the 48-hour embryo, has not cut far down toward the ventral side. Eventually it will divide the prosencephalon into the anteriorly located *telencephalon* and the posterior *diencephalon*. At the present stage the demarkation between the two is so faint that it is best to reserve their discussion for embryos of more advanced development.

The Eye

The last reference to the differentiation of the paired eyes was made in the discussion of the 33-hour chick. At that time they were hollow, lateral outgrowths (optic vesicles) from the prosencephalon. Within a few hours (37-hour embryo) each optic vesicle constricts at its base where it grows out from the ventro-lateral sides of the prosencephalon and forms the optic stalk. The vesicles proper continue to push out further in a lateral direction toward the outer ectoderm, and where they approach it closest it thickens to form the primordia of the crystalline lenses. Each optic vesicle begins now to invaginate, pushing inward at the lower portion where it is attached to the optic stalk and proceeding upward until the entire optic vesicle is transformed into the optic cup. It resembles a shallow bowl from which a piece of the rim has been broken. The gap thus formed is the choroid fissure, which is located at the lower side of the optic cup near the attachment of the optic stalk.

Concurrently, a similar process has taken place in the formation of the lens. Similar to the invagination of the optic vesicle, the ectodermal thickening in the skin facing the optic cup invaginates. In this way a thick-walled hollow sphere (*lens vesicle*) is formed, opening to the outside by a small aperture. The lens vesicle takes up its position within the optic cup where it is destined to form the cyrstalline lens. In the 48-hour embryo the lens vesicle is still open to the outside by a tiny pore in the outer ectoderm.

The Infundibulum

In chicks of thirty-three or a few more hours of incubation, a shallow depression appears in the mid-ventral line immediately posterior to the optic vesicles. This is the infundibulum. In the 48-hour embryo it is slightly deeper than in the 33-hour chick embryo and it is met by a mid-ventral inward growth of the outer ectoderm known as *Rathke's pocket*. In connection with ectodermal cells proliferated from Rathke's pocket, the infundibulum will form the *hypophysis* or *pituitary body* of the adult.

The Optic Chiasma and Optic Recess

Posterior to the infundibulum is a projection called the posterior tubercle (*tuberculum posterius*) which is located near the axial center of the cranial flexure. It projects from the floor of the posterior limit of the midbrain and represents the landmark separating the prosencephalon from the mesencephalon. In front of the infundibulum is the thickened *optic chiasma*, ending in a depression referred to as the *optic recess*. From here the optic stalks extend in a lateral direction into the optic cups. Anterior to the optic recess the thickness of the prosencephalon is increased, extending for a short distance anteriorly. It diminishes in the most anterior region of the forebrain, which is to give rise to the telencephalon.

The Epiphysis

The *epiphysis* cannot as yet be identified as a definite primordium. There is a large bulge in the mid-dorsar region of the prosencephaion, shaping itself later into a more restricted evagination which will eventually develop into the epiphysis or pineal body of the adult. In front of it is a faint transverse depression, referred to as the *velum transversum*, separating the prosencephalon into an anterior telencephalon and a posterior diencephalon.

The Mesencephalon

In the 48-hour chick the mesencephalon has undergone very little change when compared with that of the 33-hour embryo. Though there is no difficulty in identifying the midbrain by its central location, by its graceful curve, and by its sharp external demarkation from the forebrain and hindbrain, it exhibits no special organs or landmarks. Its roof, sides, and floor are uniformly thick and enclose the mesocoel which forms the *aqueduct of Sylvius* in the brain of the adult.

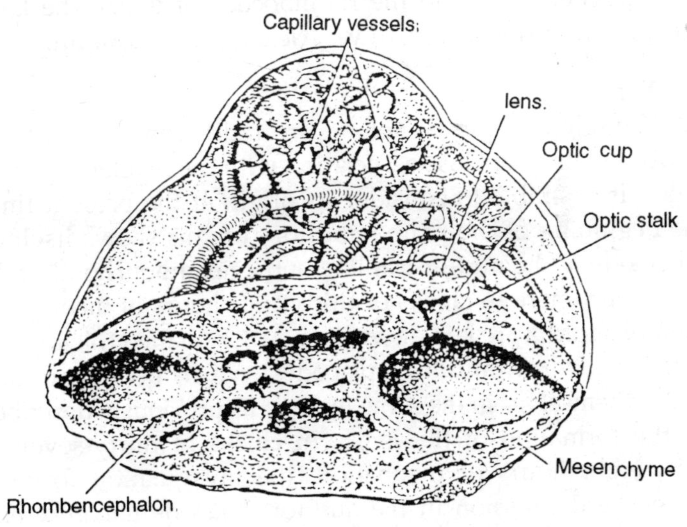

Fig. 9.10.   The 48-hour chick embryo. Section through the prosencephalon and optic cup. The profuse capillary network in the brain region is indicated here.

The Rhombencephalon

The hindbrain or rhombencephalon is the largest of the three primary brain divisions. Its anterior boundary is marked by a groove separating it from the mesencephalon, but its posterior limit is not at all definite merging gradually into the spinal cord. Judging from its later differentiation it is safe to assume that it extends as far back as the level of the second mesodermal somite. Where it joins the midbrain its roof, sides, and floor are uniformly thick, and it exhibits an oval outline in transverse section. Further back the roof becomes very thin. The short anterior portion of the hindbrin with its thick roof constitutes the metencephalon and the remainder the myelencephalon. The latter is further characterized by the persistence of neuromeres, which were a striking feature in the organization of earlier embryos. The spacious cavity of the mesocoel leads into the rhombocoel, of which the large portion will form the fourth ventricle of the hindbrain.

The Ear

Though the ear is not a direct differentiation of the hindbrain, its later close association with the myelencephalon makes it desirable to consider its origin at the present time. The beginning of the paired auditory organs shows itself on either side of the mid-dorsal line in the neighbourhood of the tenth neuromere in chicks of thirty-four hours of incubation. Their primordia are laid down as two disk-shaped thickenings in the ectoderm (auditory placodes) which invaginate and detach themselves in the same manner as previously described for the formation of the lens. In chicks of thirty-seven to thirty-eight hours of incubation, the invagination forms a saucerlike depression in the auditory placode (*auditory pit*). Pushing further into the tissue of the embryo, the auditory pits take on the shape of sacs (*auditory vesicles, otic capsules*), leading to the outside by slitlike openings. These begin to disappear almost entirely in chicks of forty-five hours of incubation and within three hours only a tiny upper pore is still visible, connecting the cavity of the auditory vesicle with

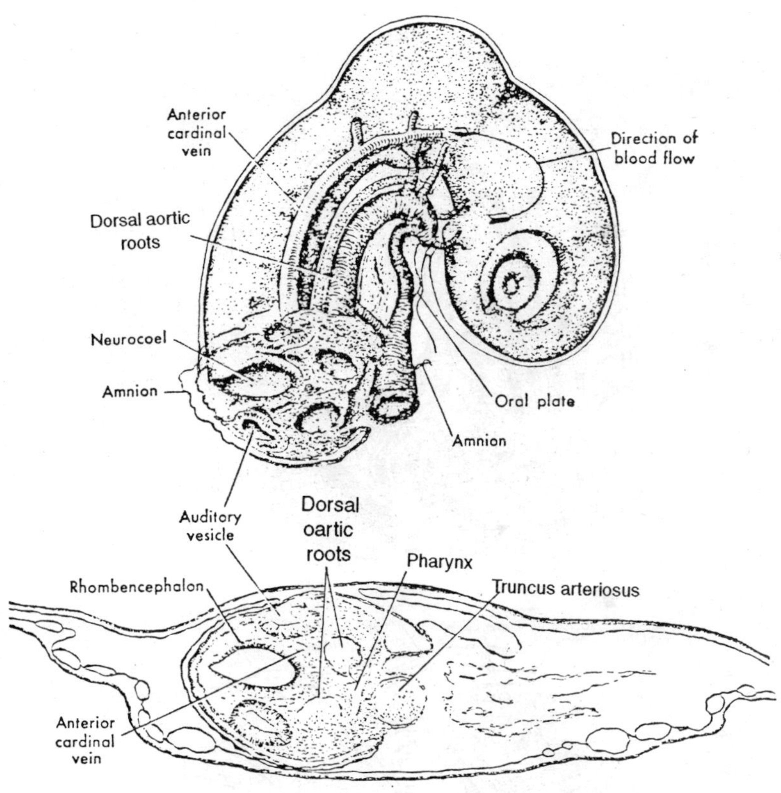

Fig. 9.11. The 48-hour chick embryo. Section through the rhombencephalon and auditory (otic) vesicle.

the outside. This is probably the old endolymphatic duct, which upon complete later disappearance reappears as a small evagination within the cavity of each vesicles and grows upward, giving rise to the endolymphatic sac in later embryos.

Fig. 9.12. The 48-hour chick embryo. Section through the heart

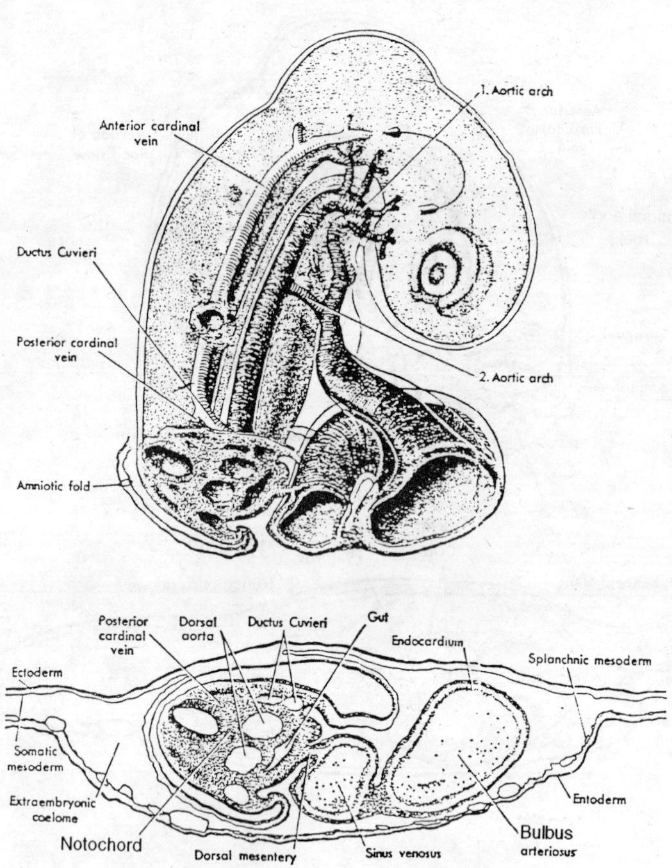

Fig. 9.12. The 48-hour chick embryo. Section through the heart

The Spinal Cord

Without any visible demarkation, the myelencephalon merges into the spinal cord. In its earlier stages, shortly after the fusion of the neural folds, the central canal is still an elongate oval in transverse section, but it soon becomes compressed laterally as the embryo grows older, assuming in the 48-hour chick the shape of a longitudinal slit. Concurrently, the sides of the cord thicken so that its roof and floor appear to be comparatively thin. Two types of cells can be distinguished in the tissue of the nerve cord of the 48-hour chick namely, the original *epithelial cells* augmented by the secondary *germinal cells*. The former have elongated considerably, but they still remain at one end in their original position bordering the central canal where they develop short processes extending into the canal. After this differentiation the epithelial cells are referred to as *ependymal cells* which have no nervous function. The smaller germinal cells are wedged between the ependymal cells and give rise to the *neuroblasts* and the *neuroglia cells*. The latter, with the ependymal cells, form the connective tissue of the nerve cord, while the neuroblasts differentiate into the various types of neurones.

The Neural Crests and Ganglia

There is little to be added to the description of the neural crests as given in the preceding chapter. The anterior mesodermal so mites are paralleled in their metameric arrangement by the neural crests. Anteriorly this differentiation is very distinct, but it loses in clearness as one proceeds toward the posterior portion of the embryo, where additional mesodermal somites are still in formation.

In front of the first pair of somites the neural crests extend into the head region, but their metameric regularity is obliterated there. It seems that some of them have fused to form large ganglia for the cranial nerves, of which the fifth, seventh, eighth, ninth, and tenth can be well distinguished in the 48-hour chick. The ganglion of the fifth cranial nerve (*trigeminal*) is located on the side of the anterior portion of the rhombencephalon, midway between the auditory vesicle and the furrow separating the hindbrain from the midbrain (*meso-metencephalic furrow*). It is usually referred to as the *Gasserian ganglion*. In front of the auditory vesicle, or auditory capsule, are two ganglia in close apposition so that they cannot be told apart in the 48-hour, chick. These are the *geniculate* and the *acoustic (auditory) ganglion* of the seventh, or facial, and the eighth, or acoustic (auditory), nerve respectively. The ganglia of the ninth (*glossopharyngeal*) and the tenth (*vagus, pneumogastric*) cranial nerve make their appearance in close union in the 36-hour chick immediately behind the auditory pit. At forty-eight hours of the incubation they are well-separated, the ninth being located close behind the autitory vesicle and the tenth a short distance further back.

The Alimentary System

The Fore-gut. In the 48-hour embryo the fore-gut has attained a length of $1\text{-}1/_2$ mm. It is only $1/_2$ min. longer than that of the 33-hour chick, but it has advanced considerably over the latter in its differentiations. The fore-gut resolves itself into three natural component parts — the preoral, the pharyngeal, and the oesophageal — of which the pharyngeal parts shows the greatest modifications.

The Preoral Gut

This portion occupies the most anterior part of the fore-gut, pointing toward the infundibulum and being continuous posteriorly with the spacious pharynx which it joins at the level of the oral plate. The significance and possible function of the preoral gut is not known. After seventy-two hours of incubation it gradually degenerates. For a time it persists in the roof of the pharynx of later embryos as a blind sac (*Seesel's pocket*), but even that vestige disappears and eliminates the preoral gut entirely from the alimentary canal.

The Stomodaeum and the Oral Plate

The first reference to the oral plate was made in the

discussion of the 33-hour embryo. Though its original structure has not undergone any changes, its position in the 48-hour chick has been altered by the cranial flexure and by the rapid growth of the mandibular arches. It should be recalled from the embryology of the frog that the paired *mandibular arches* (first visceral arches) are the primordia of the lower jaw. One should therefore expect that the stomodaeum and the future mouth will form above and between the two mandibular arches. That this is correct. Moreover, the cranial flexure, turning the lower head region inward, assists in the invagination of the outer ectoderm around the oral plate. Thus the stomodaeum, which will give rise to the mouth and oral cavity, is formed as a pocketlike in growth toward the pharynx, from which it is separated by the thin oral pate. Its subsequent rupture in the 72-hour chick will connect the pharynx with the outside.

The Pharynx

The pharynx is the largest and most spacious division of the fore-gut. It is of great interest to the comparative embryologist because its differentiations have a direct beating on the development of various organs and organ systems, as for example, the vascular and nervous systems, the lower part of the face, the ear, the neck, and certain endocrine glands. The pharynx is the longest division of the fore-gut, and it is wider and more spacious than either the preoral gut or the oesophagus.

The most conspicuous differentiations of the pharynx are the visceral clefts and visceral arches. In a general way they are the same as those of the frog embryo, but their development is slightly different and is modified because the visceral, clefts in bird embryos are never used for respiration as they are in the amphibians and fishes.

Visceral Clefts

It does not seem necessary to repeat here the formation of the visceral clefs, since they develop in the same manner. The external visceral furrows grow inward to meet corresponding pharyngeal pouches growing outward, and where the two

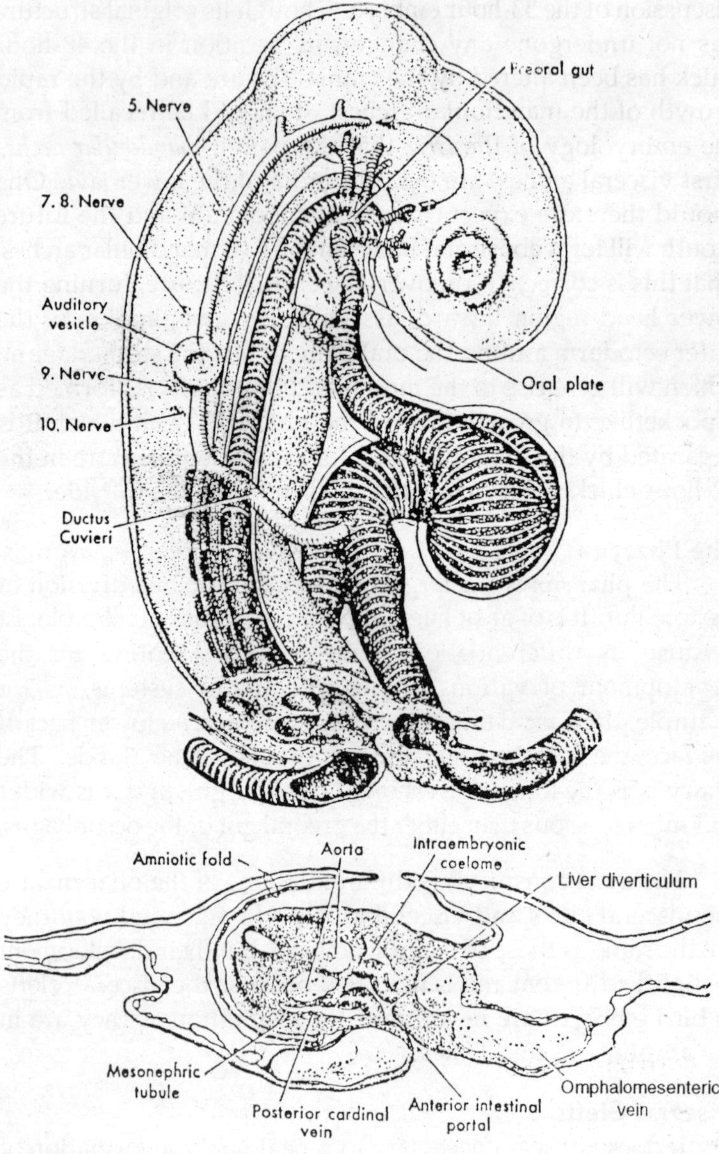

Fig. 9.13. The 48-hour chick embryo. Section through the anterior intestinal portal.

meet and perforate, a more or less open visceral cleft will be produced. If the term "cleft" is used in its strictest sense, then the chick embryo has only three of them, because the fourth one never breaks through but remains in the incomplete stage of fused visceral furrow and pharyngeal pouch.

The first pair of visceral furrows appears in embryos of thirty-four hours of incubation. Within an additional five or six hours of incubation, the second pair becomes visible posterior to the first. Concurrently, the corresponding pharyngeal pouches have evaginated from the pharynx, so that at the time when the heart begins to function (44-hour embryo), the first two pairs of visceral clefts are almost completed. The perforation of the first is accomplished in the 48-hour chick, but the opening of the second pair is delayed more than twelve hours. The third pair does not open until the fourth day of incubation, and, as stated above, the fourth does not open at all. The term *"cleft"* is really a misnomer for the first pair (hyomandibular clefts), because the opening is not at all slitlike or cleftlike, being confined to a small aperture on its upper portion. The second and third visceral clefts are more slitlike in structure, but even in these two the clefts do not open throughout their entire length. Indeed, they have two openings each, a small one above and a large one below, separated by a fusion of the clefts a little above their centers.

When it is recalled that the frog embryo has five visceral clefts, it seems that in the evolution of the birds and reptiles one of them has been lost. Evidence gathered from the changes occurring in the nervous and vascular systems during embryology of these types indicates that the fifth has been lost. This evidence is further supported by the presence of certain vestigial organs *(post-branchial bodies)* behind the fourth cleft. It is assumed that the post-branchial bodies represent the abortive pair of fifth pharyngeal pouches.

Moreover, the first visceral "cleft" never opens in the frog embryo, but does break through in the embryo of the chick.

Since the latter never uses its gill clefts for respiration, and since the birds have been evolved through reptiles from amphibians, one should expect that the frog embryo, not that of the chick, would have an open first visceral cleft. This is another striking example of an exception to the so-called biogenetic law.

The Thyroid Gland

The differentiation of the primordium of the thyroid gland begins in the 34- to 35-hour embryo. It consists of a thickening (*thyroid plate*) in the median ventral wall of the fore-gut. With the development of the pharyngeal pouches, this thickening is localized anterior to and between the second pair of pharyngeal pouches. From forty to forty-four hours of incubation, the thyroid plate changes into a saucer-shaped trough, developing later into a shallow pocket in the 48-hour chick. Its cells are elongate, almost columnar in shape, and are well differentiated from the surrounding entoderm cells.

The Laryngotracheal Groove and Oesophagus

The posterior portion of the pharynx is not so wide nor so strikingly compressed dorsoventrally as is its anterior division. Along the median region of its floor a faint groove appears in the 48-hour embryo. This central longitudinal trough contributes the primordium of the larynx, trachea, and the lungs, and is therefore referred to as the *laryngotracheal groove*. That portion of the fore-gut immediately abole the laryngotracheal groove will form the anterior portion of the oesophagus, while the fore-gut immediately posterior to it develop into the posterior oesophageal region.

The Mid-gut

The mid-gut is characterized by the lack of a cellular floor, and is usually called the open gut. It begins at the anterior intestinal portal, where it has well-defined lateral walls which diminish in size and tend to flatten out to become continuous with the roof of the open gut as one proceeds posteriorly toward the hind-gut. However, the lateral margins of the

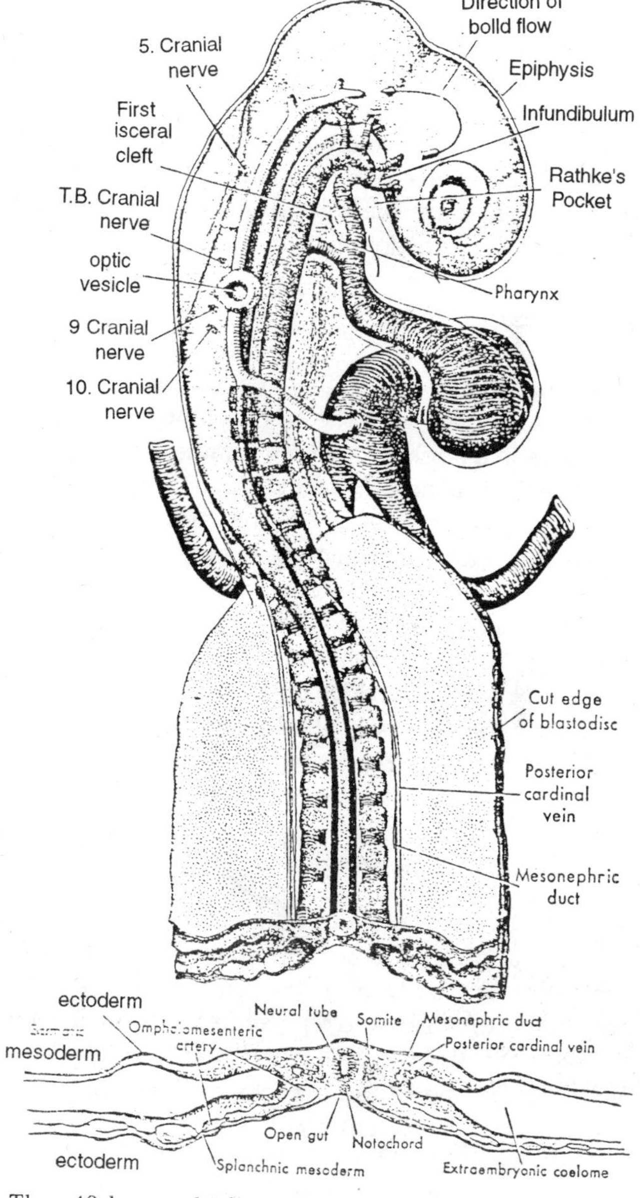

Fig. 9.14. The 48-hour chick embryo. Section through the omphalomesenteric artery.

closed fore-gut are continued in the mid-gut as lateral folds; they tug the lateral walls inward toward the mid-line, thus forecasting the future complete constriction of the gut from the underlying yolk. Near the anterior intestinal portal, above the junction of the two omphalo-mesenteric veins, two lateral evaginations form a pair of small pockets in the lateral wall of the gut. They are the liver diverticula, which will develop later into the functional liver in close connection with the omphalomesenteric veins by growing together and forming a single use.

The Hind-gut

As compared with the origin of the fore-gut, the hind-gut develops in a slightly different manner. It should be recalled that the fore-gut and the head fold developed simultaneously. This is not the case in the formation of the hind-gut, which makes it appearance in embryos of forty-five hours of incubation. The hind-gut begins its development without the external appearance of a tail fold. Only a faint posterior swelling, the beginning of the *tail bud*, is visible, containing within its tissue the saclike hind-gut which opens into the mid-gut through the *posterior intestinal portal*. The actual tail fold forms in slightly older embryos by the growth of the tail bud and its later projection over the outer ectoderm, much in the same manner as the head fold is formed by the projection of the head over the proamnion anteriorly.

At the most posterior portion of the hind-gut, the entoderm comes into direct contact with the superficial ectoderm in the formation of the *anal plate*. From its homology with that of the frog embryo, this region represents the posterior portion of the former primitive streak. It is placed dorsally at first — not ventrally, as one would expect — because of the delay in the formation of a projecting tail bud which is produced by the deepening of the tail fold, thus carrying the anal plate downward of the ventral side of the embryo.

The Vascular System

The uniformity of development processes in the vertebrates is shown more strikingly in the embryology of the vascular system than in any other organ system. The following account should be compared with parallel stages in the frog embryo, and illustrations of the two types should be carefully studied even if the attention of the reader is not specially called to it. It will be seen that the chick "recapitulates" the embryology of the frog very closely. In fact, some parts of the present account will appear to be mere repetition of the development of the circulatory system of the frog embryo and larva.

The Heart

In the discussion of the 33-hour chick embryo, the origin of the heart tube was traced from a splanchnic mesodermal tube to the thick-walled primordial heart, which had turned toward the right side in the form of a horseshoe curve. In the succeeding hours, ending with the 48-hour chick, the heart remains essentially a tube, having as yet no valves nor sharp demarcation of heart chambers. However, a great physiological advance has been made in that it has begun to function as a blood pump. Since there are no valves to check the blood from flowing back, it is propelled forward by peristaltic waves. This great change occurs at about forty to forty-four hours of incubation, shortly after the embryo has begun to turn over on its left side. In the 44-hour chick, or sometimes a little earlier, the heart has acquired its rhythmical beat, which was preceded by a two-hour period of spasmodic contractions, starting the heart on its life-long function.

Though the heart is still essentially a tube, it has become more complex in appearance than it was in the 33-hour chick embryo, and if at that time we could assign certain regions of the tube to particular later differentiations, we are able to do so much better now. In the 48-hour chick embryo the heart has grown considerably, and since it is held in place anteriorly by the two ventral aortic roots and posteriorly by the dorsal mesocardium and the omphalomesenteric veins, it increases the length of its curvature by twisting upon itself. In this manner it has formed a complete spiral turn. Posteriorly this turn begins with the sinus venosus, which is very close to the body of the embryo and is still held in place by a short dorsal mesocardium. The right turn spiral twist is formed chiefly by the thick heart tube, which embodies the atrium and the ventricle. The truncus arteriosus returns in its direction to the body of the embryo in completion of the turn, and is continued there as the paired ventral aortic roots (ventral aortae) within the embryo. The heart, having a right-handed spiral twist, must necessarily have the atrium on the left and the ventricle on the right side.

The Aortic Arches and the Aorta

Though the aortic arches as well as the visceral arches and clefts, are paired structures in the chick embryo, it is convenient in their description to refer to them in the singular form. This is done for the sake of clearness and ease of presentation and the singular form will be employed occasionally in the present discussion as well as later accounts. One should always be mindful that all such structures are paired and occur on both the right and left sides of the pharynx.

In the 33-hour chick embryo almost the entire systemic blood circulation was seen to be paired. The two ventral aortic roots were continuous with the two dorsal aortae by a pair of aortic arches. These two aortic arches are not only first in appearance, but also first in their morphological order. In the 48-hour chick the second aortic arch has been added and there may be even a vestige of a third. In this respect the chick embryo differs from that of the frog. In the latter, the first as well as the second aortic arch is never functional. In the chick embryo, functional vessels are formed in both of these arches, though at the end of the fourth day of incubation they will degenerate without leaving a trace.

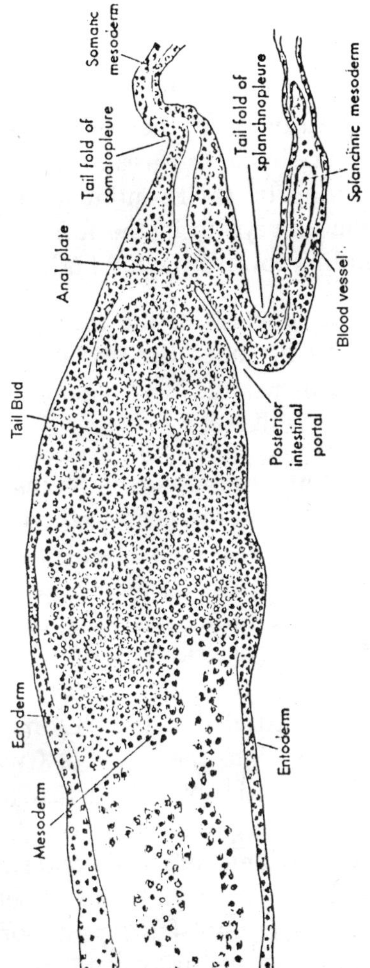

Fig. 9.15. Sagittal section through the posterior end of an embryo of about twenty-one somites.

The paired aortic arches become confluent with the dorsal aortic roots, located dorsolateral of the pharynx. Near the level of the sinus venosus they approach each other and unite into one centrally located dorsal aorta, but they separate again near the level of the posterior intestinal portal and continue backward as paired vessels. Near the eighteenth somite each member of the pair gives off a heavy branch, namely, the omphalomesenteric artery, which divides into the vitelline arteries, and these spread out over the yolk mass through a plexus of capillaries. Since by far the greater amount of blood is thus drained from the aortic, their continuation toward the posterior region of the embryo is in the form of a pair of diminutive vessels leading toward the tail bud.

Anteriorly, near the junction of the first pair of aortic arches and the dorsal aortic roots, the two internal carotid arteries are continued into the large head region to supply the rapidly developing brain with blood.

The Cardinal and the Omphalomesenteric Veins

The blood is returned from the head to the heart by the anterior cardinal veins which pass on the right and left side of the brain above the level of the aortae and then descend toward the heart, pouring their blood through the ducts of Cuvier into the sinus venosus. From the posterior part of the body of the embryo, the blood is returned through the posterior cardinal veins, which pass forward parallel to the mesonephric duct and empty their blood also into the ducts of Cuvier. The posterior cardinal veins are considerably thinner than the anterior cardinals, but near their entrance into the ductus Cuvieri they spread out, forming plexus of smaller vessels and sinuses (not shown in the figures). These plexus of blood vessels may be homologous to the branching-out of the posterior cardinal veins in the formation of the pronephric capsules in the region of the pronephric kidneys in the frog larva. The omphalomesenteric arteries and veins have also advanced in their development. They now lead the blood from and to the heart respectively. The transformation of the isolated blood islands into a complex capillary network of delicate blood vessels spread out over the yolk mass has been accomplished within twenty-four hours of development.

The Excretory System

The Pronephros. The pronephros, which has attained a high degree of prominence in the frog larva, is also present in the chick embryo, but it is doubtful whether it has any functional significance. The paired pronephric tubules are concentrated from the fifth to the sixteenth somite. As a matter of fact, the pronephros does not develop tubules but originates from the intermediate mesoderm as a series of cellular strands which grow upward and outward and then curve toward the posterior direction. These solid strands of cells connect with each other at their free ends and thus form the pronephric ducts. The latter become tubular and resemble that of the frog in structure and location. At the beginning of the third day the pronephros begins to degenerate. The development of the mesonephros has already begun and the pronephric duct will be utilized then as the mesonephric duct. Since the pronephros is never functional, the potential pronephric duct is usually referred to as the mesonephric duct at its very beginning

SEVENTY - TWO HOURS, AND CERTAIN LATER STAGES

External Morphology

The more interesting features of chick embryos beyond forty-eight hours of incubation are illustrated, which represent the 60- and 72-hour embryo respectively. Our interest will be centered chiefly on the latter, and intermediate development as far back as the 48-hour chick will be illustrated chiefly on the 60-hour chick embryo.

In a total preparation of the 72-hour chick, one's attention is immediately arrested by the increase of torsion as well as flexure. The slight turn to the right which began in the 37-hour embryo in the anterior part of the body has now involved more than half of the embryo. In fact, torsion has proceeded posteriorly almost to the level of the omphalomesenteric arteries. Only the tail region is still in its original position, and even that will complete its torsion within another twenty-four hours of incubation. The four-day-old chick embryo rests completely on its left side.

Another striking feature observed in the total embryo is the appearance of two additional flexures, one in the region of the tenth to the twelfth somite, known as the *dorsal* flexure, and the other in the tail region, referred to as the *caudal flexure*. In the 33-hour embryo the cranial flexure was observed as the first curvature of the brain. In the 48-hour chick the cervical flexure was added posterior to it, and in the 60-hour chick the dorsal flexure makes its appearance. In the 72-hour chick the caudal flexure has developed in the posterior end of the embryo. Generally speaking, it has now the outline imparting to it the general outline of the mirror image of a question mark.

The curvature of the dorsal flexure has increased considerably in length in the 72-hour chick and has there become continuous with the cervical flexure to form a semicircular outline of the anterior part of the body of the embryo. The effect of this change has transferred the otic capsules topographically more anteriorly than they were in earlier chicks. A transverse section through the level of the eye will also usually transect the ear in the 60-hour embryo, but in the 72-hour chick serial transverse sections, beginning an teriorly, will cut the ear first and the eye much later. Another outstanding characteristic is the mesodermal somites. They have become much more numerous, being approximately thirty-two in the 60-hour chick and thirty-five in the 72-hour chick. They are now much better defined than they were in younger embryos, and they have increased considerably in size in the anterior half of the body. They are apparently still in formation and are added posteriorly, a fact which is manifest in the tail region, where they are still small and well-separated from each other.

The heart is wedged between the lower part of the forebrain and the body wall. The aortic arches have increased from two in the 48-hour chick to three in the 60-hour embryo and to four

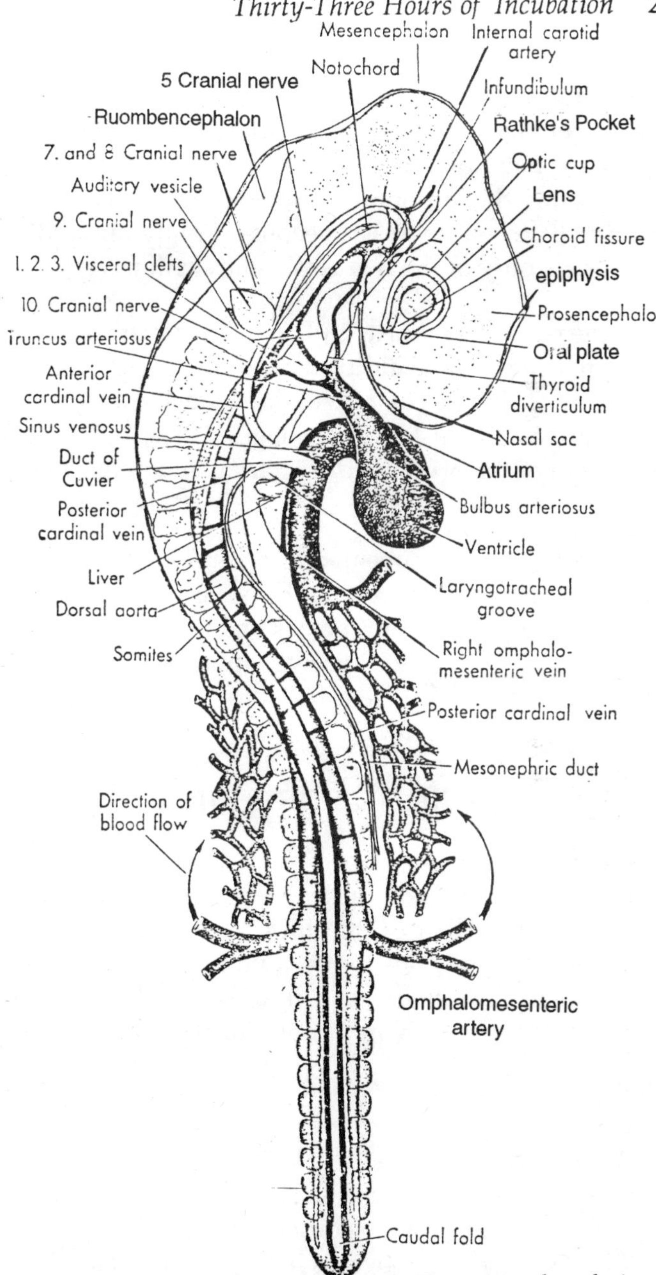

Fig. 9.16. The 60-hour chick embryo with 32 somites, dorsal view.

in the 72-hour chick. Moreover, the area vasculosa has expanded considerably, with the peripheral sinus terminalis far away from the embryo. The vitelline arteries and veins are well defined, the anterior, the posterior, and the united left vitelline veins forming the left omphalomesenteric vein, while on the right side its counterpart has become slightly reduced in size and consists only of the two two right vitelline veins.

In the region of the sixteenth to the twentieth, as well as in the twenty-fifth to the thirty-second somite, the limb buds have become visible. Their bases are wide, forming in slightly older embryos flipper-like appendages. Their form is strongly reminiscent of the origin of the pectoral and pelvic fins in the elasmobranch embryo.

The allantois has appeared as a small diverticulum from the alimentary canal just behind the posterior intestinal portal. It is very small in the 72-hour chick and is hidden between the two posterior limb buds. It grows rapidly and is soon visible as a baglike structure extending into the *seroamniotic cavity* (extraembryonic coelom).

INTERNAL MORPHOLOGY

The Nervous System

The Brain. The brain of the 72-hour chick has advanced considerably over that of the chick after forty-eight hours of incubation. Not only has it increased in size by growth of its three primary divisions, but it also has become greatly differentiated, especially in its anterior part.

The Prosencephalon

The beginning of the separation of the prosencephalon into the diencephalon and the telencephalon has already been pointed out in the 48-hour chick. The faint demarcation between these two parts can easily be observed in the 60-hour embryo, and it has become very obvious in the 72-hour chick. Not only has the telencephalon been separated from the diencephalon by a transverse constriction, but the telencephalon itself has become partly divided by a longitudinal furrow, forming in this manner two rounded anterolateral vesicles called the *telencephalic vesicles.* They are probably the most important differentiations of the higher vertebrate brain, since they will form the paired *cerebral hemispheres* of the adult. They increase considerably in size, so that they will extend backward over the diencephalon and mesencephalon. The cerebral hemispheres are chiefly the center of intelligence, by means of which the animal is able to learn and profit by past experience.

As the telencephalic vesicles continue to grow, they become progressively more constricted at the point of outgrowth, leaving finally only two small apertures through which the cavities of the left and right telencephalic vesicles (*telocoels*) communicate with the original prosocoel. These apertures are the *foramina of Monro (foramina interventricularia),* and the cavity of the left telencephalic vesicle becomes the *first ventricle* of the brain, while the right one becomes the *second ventricle.* These two cavities are usually referred to as the two *lateral ventricles.* Through the foramina interventricularia, the lateral ventricles communicate with the small remnant of the original prosocoel and the newly formed *diocoel* of the diencephalon, referred to as the *third ventricle* of the brain. Anteriorly the third ventricle is bounded by the *lamina terminalis* and posteriorly it opens into the cavity of the mesencephalon, called the mesocoel, or *aqueduct of Sylvius.* In its earlier stages of development the diencephalon and the telencephalon can be distinguished in the chick embryo as readily as in the frog larva. An imaginery line drawn from the posterior tubercle to the *mesodiencephalic constriction* separates the diencephalon from the mesencephalon. Another such line drawn from the optic recess to the *velum transversum (dien-telencephalic constriction)* separates the diencephalon from the telencephalon.

The Diencephalon

The differentiations of the diencephalon have been described in the 48-hour chick embryo, and little can be added to that account. Just posterior to the optic recess, the floor of the diencephalon thickens considerably and becomes the location of the optic chiasma. Posterior to the latter, the infundibulum has increased its evagination and is directly adjacent to *Rathke's pocket*, with which it is destined to form the *hypophysis*, or *pituitary body*, which functions later as a gland of internal secretion. The roof of the diencephalon remains thin throughout its entire development. Indeed, anterior to the epiphysis it differentiates into a membranous sheet which becomes richly supplied with blood vessels. In later development this membrane becomes convoluted and sinks into the cavity of the third ventricle, and is thus instrumental in furnishing the forebrain with a profuse blood supply. It is known as the *choroid plexus* of the third ventricle. In the 4-day chick the epiphysis changes from its earlier round and wart-like appearance into a slightly elongate outgrowth.

The Mesencephalon

The mesencephalon is characterized by its relatively thick walls, which gradually reduce its cavity, the *mesocoel*, to a narrow canal called the *aqueduct of Sylvius*. There is hardly any difference between the mesencephalon of the 48-hour and that of the 72-hour embryo. On its dorsal aspect are the *corpora bigemina*, though in certain Sauropsida (Ophidia) they have developed a transverse furrow and have become the *corpora quadrigemina*. They are the homologue of the corpora bigemina of the frog larva, except that in the case of the snake embryo the two additional bodies contain nerve fibres for hearing, while the two anterior ones have the same function as the corpora bigemina in the frog, namely, that of visual sensation. In later development the *crura cerebri* develop in the thickened floor of the midbrain. Their function in the chick is the same as that of the same structures in the frog larva.

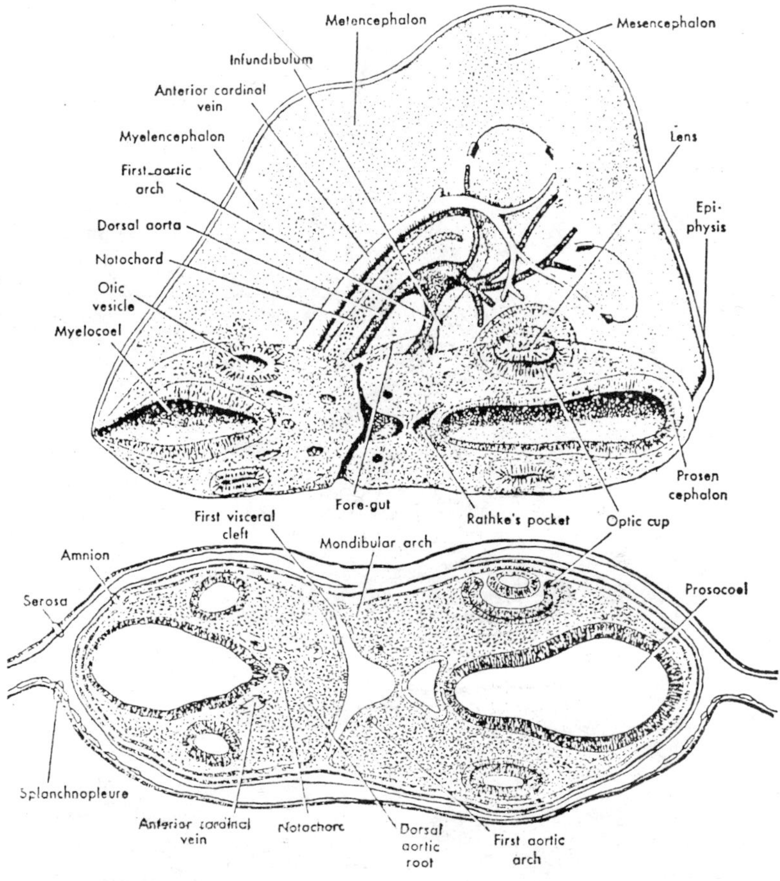

Fig. 9.17. The 60-hour chick embryo. Section through the optic cup and otic (auditory) vesicle.

The Rhombencephalon

Morphologically the rhombencephalon is divided into two parts, namely, the metencephalon and the myelencephalon.

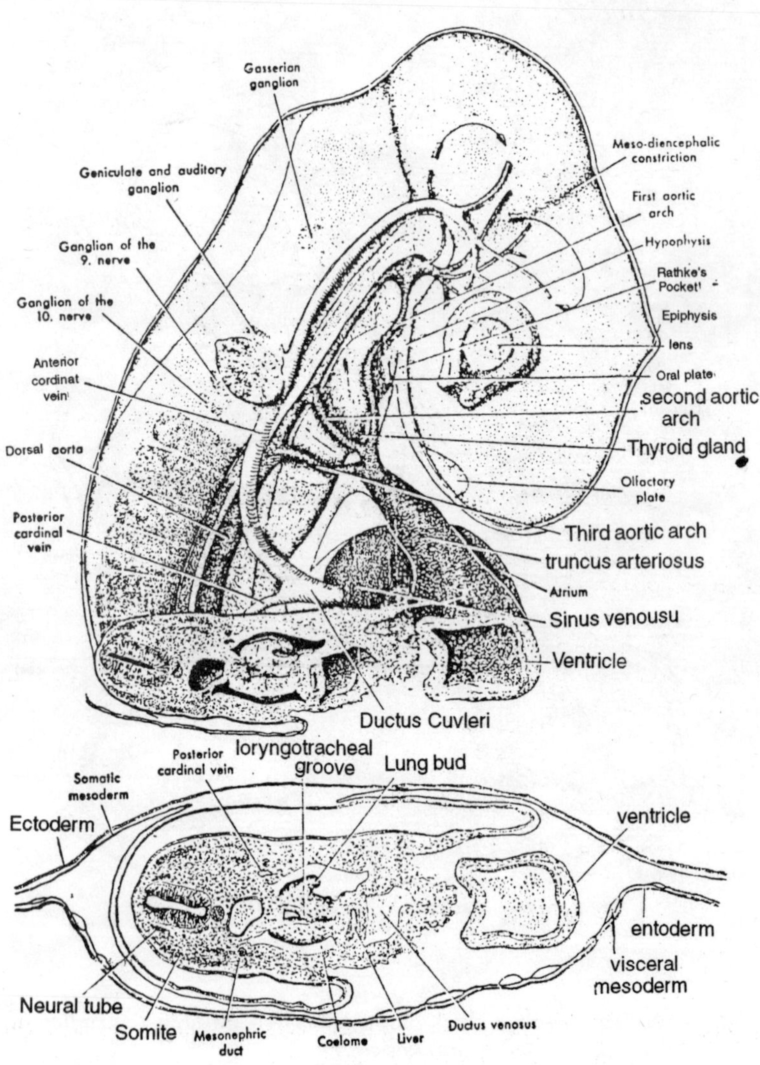

Fig. 9.18. The 60-hour chick embryo. Section through the ventricle and liver diverticulum.

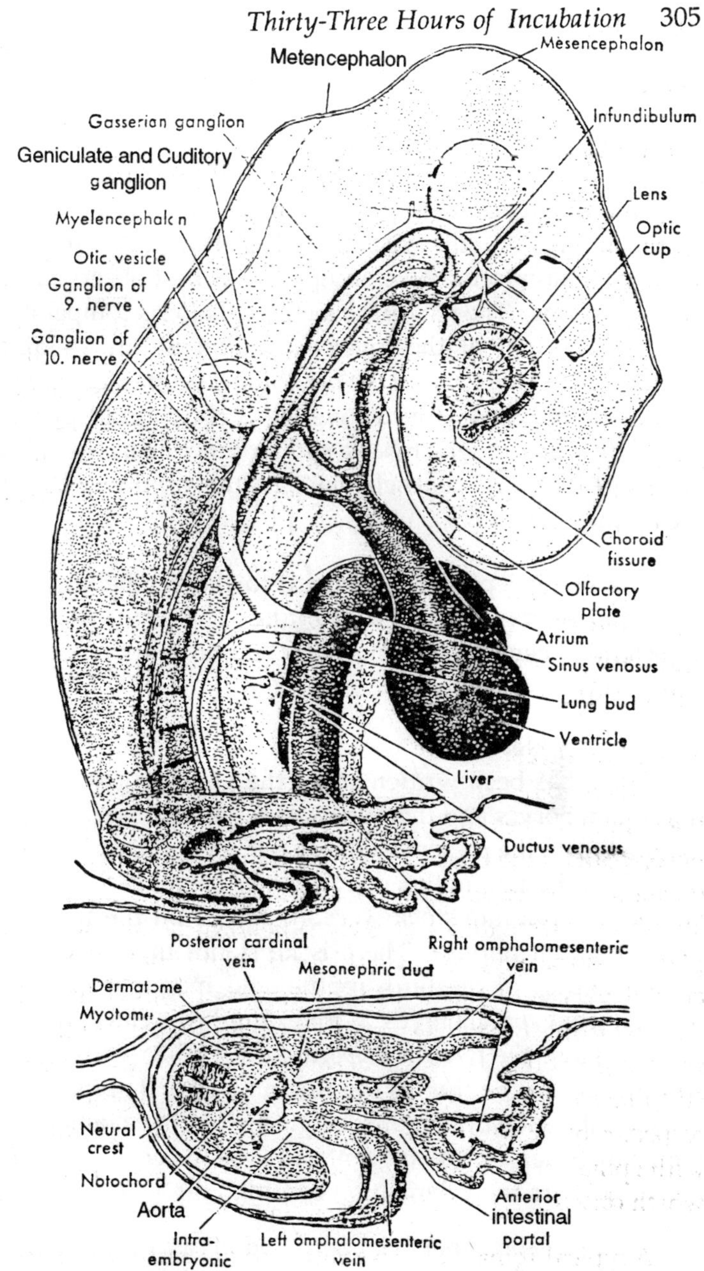

Fig. 9.19. The 60-hour chick embryo. Section through the anterior intestinal portal.

The former is separated from the mesencephalon by a deep dorsal and lateral constriction, referred to as the *mesometencephalic constriction*, which forms a narrow connection between these two parts of the primitive brain. Posteriorly it is continuous with the myelencephalon in which the original neuromeric segmentation is still visible. The transition from the metencephalon to the myelencephalon is best observed in its dorsal covering. The roof of the metencephalon is comparatively thick, while that of the myelencephalon is thin — almost membranous. This part of the hindbrain becomes the medulla oblongata, in whose dorsal covering is the choroid plexus of the *fourth ventricle*, which has a similar function as the choroid plexus of the third ventricle, namely, the supply of blood to the brain.

The metencephalon gives rise to the cerebellum. It is large and deeply furrowed in birds. The cerebellum is an association center for muscular co-ordination and for equilibrium.

The Cranial Nerves

Much has been written about the phylogenetic origin of the cranial nerves and their possible serial homology to spinal nerves, but all the present evidence is neither exact nor weighty enough for the assumption that all cranial nerves of the primitive ancestral vertebrate stock were organized on the metameric plan of spinal nerves. There is no doubt that some of the cranial nerves of the bird, reptile, and mammal have been derived from spinal nerves, as for example the eleventh (*spinal accessory*) and the twelfth (*hypoglossal*). These two nerves are still present in the frog as the first and second spinal nerves respectively. However, in homologizing the other cranial nerves with spinal nerves, we are confronted with certain objections which obscure their derivation considerably.

A typical spinal nerve is composed of two roots—a dorsal, which is an *afferent* or *sensory*, and a ventral, which is an *efferent* or *motor root*. These unite immediately outside the nerve cord to form a nerve cable containing both kinds of fibres. The dorsal root of each spinal nerve has a ganglion in which in proper time the ganglionic neuroblasts are developed. The neurones for the ventral root are formed in the ventral portion of the neural tube and are multipolar.

The *olfactory nerve* deviates greatly from the typical organization of the spinal nerve. It is purely sensory, and its neuroblasts develop from the epithelium in the center of the olfactory pit. They project their axones toward the ventral side of the forebrain and connect with the cerebral hemispheres there. For this reason it is often called "olfactory tract." The *optic nerve* conforms even less to the plan of the spinal nerves. Here it is the retina which develops the neuroblasts that send the axones toward the central portion of the diencephalon, and it was the retina which was part of the lateral wall of the prosencephalon prior to the formation of the optic vesicles. The retina is therefore truly a part of the brain, and its neurones are comparable to those found in the brain. The *auditory nerve* is also purely sensory in function.

The *oculomotor, trochlear* and *abducens* nerves, on the other hand, are purely motor or efferent in function, and no vestige of a sensory of afferent root is ever present in their formation. There is some evidence that these nerves may represent only the motor roots of former spinal nerves, because their centers of origin in the brain are in the gray matter which is continuous with that of the ventral horns of the spinal cord in which the motor neurones of the spinal cord in which the motor neurones of the spinal nerves are located. However, if one considers the exact origin of these nerves, as described below, the case is not quite as clear as it seems.

The *trigeminal facial, glossopharyngeal,* and *vagus* nerves agree more closely with the pattern of spinal nerves, since they are mixed in composition and are partly organized from ganglia. Upon more detailed examination this assumption is rather weighty because the large cranial ganglia are composed

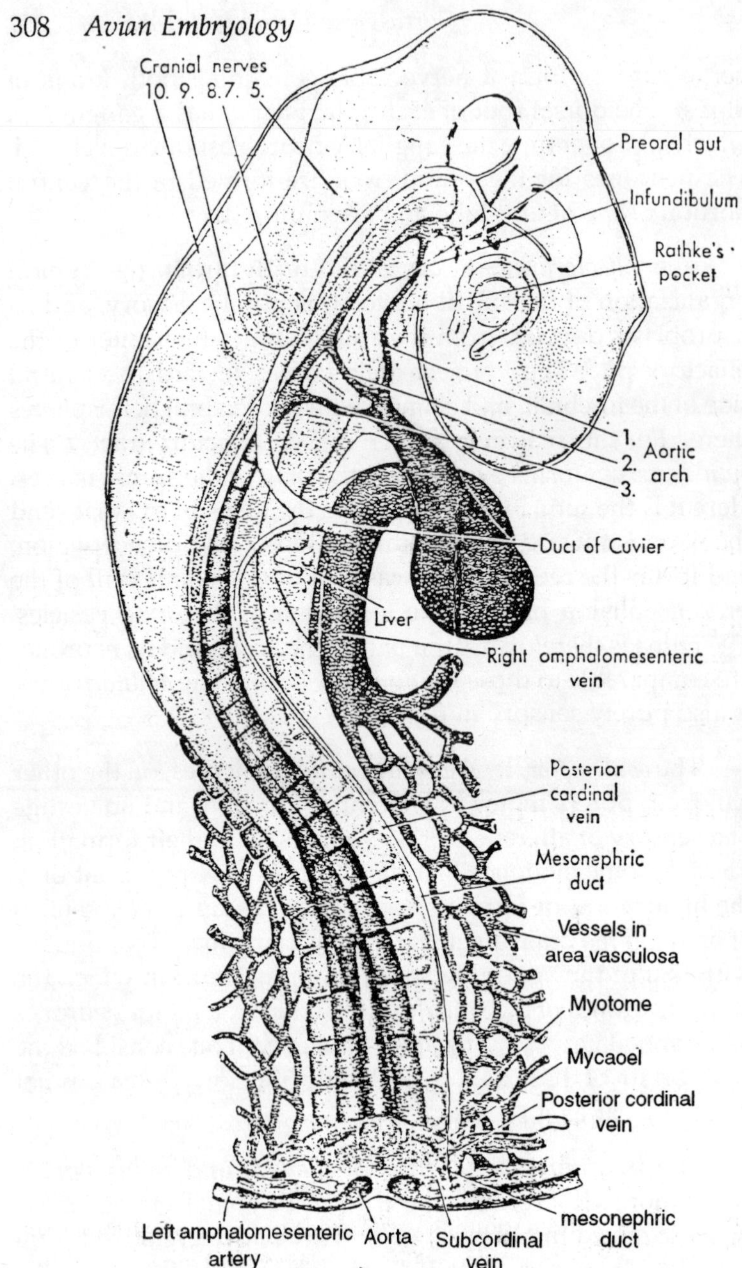

Fig. 9.20. The 60-hour chick embryo.   Section through the omphalomesenteric arteries.

of more than one unit, one being derived from neural crest material and the other from the ectoderm of the skin in the form of a placode. Since these four nerves supply the complex area of the visceral arches and the lateral line system in fishes and larval amphibians, it may be possible that their metamerism and fundamental organization have been greatly modified in the evolution from primitive ancestral chordates to modern birds and mammals. These nerves may have been metameric as are those of the present spinal nerves, but may have become aberrant and complex by mutual coalescence, conditioned by anatomical and physiological changes in the development of new organs and organ systems, such of gills, lungs, glands, musculature, skeletal structures, etc.

It will be profitable to review the development and make comparisons with the cranial nerves of the frog. A comparative study of the origin of the nervous system will not only be profitable from the evolutionary point of view, but it will also reveal many features of similarity and will illuminate certain apparent differences between the two types. All the cranial nerves are paired. For the sake of convenience and clearness in the following discussion, reference is usually made to only one member of the pair. The reader should be mindful of the fact that the discussion always applies to both.

The Olfactory Nerve and Olfactory Organ

The first indication of an olfactory organ in the chick becomes apparent at sixty hours of incubation. At that time the ectoderm on the anterolateral aspect of the head thickens to form paired olfactory plates. They are homologous to the olfactory placodes described in the development of the nose of the frog larva. The olfactory plates invaginate to form the *olfactory pits*, which by further growth develop into the *olfactory sacs*. At first their epithelial lining consists of uniform cells, but when in the 4-day chick the nesal sacs push further inward and become deeper, the centrally located cells grow large, while those near the periphery of the nasal sacs remain small

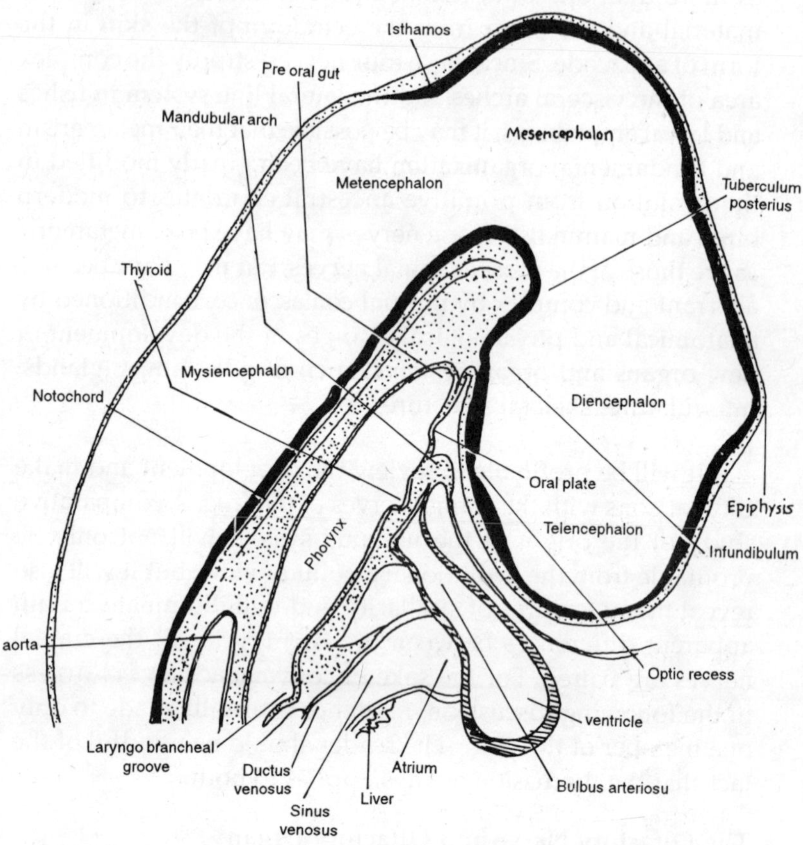

Fig. 9.21.   The 60-hour chick embryo. Sagittal section of the anterior portion.

and are continuous with those of the outer ectoderm: The centrally located, larger epithelial cells will be transformed later into neuroblasts which grow with their axones toward the ventral side of the telencephala (cerebral hemispheres), where they connect with the olfactory area of that part of the forebrain. The smaller and more peripherally located ones multiply and grow backward toward the brain, and form the nerve sheaths by overgrowing the developing nerve fibres. In

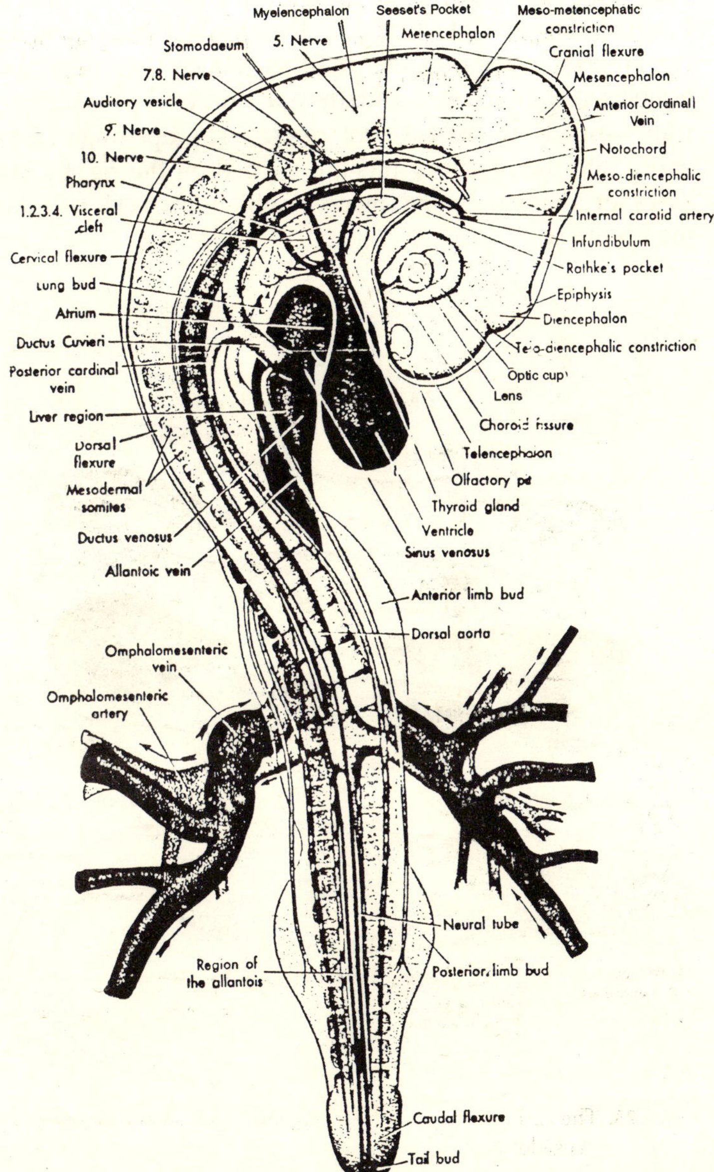

Fig. 9.22.   The 72-hour chick embryo with 35 somites, dorsal view.

their backward growth the nasal sacs utilize the cells located between them and the fore-gut. Eventually the sacs will reach the anterior portion of the pharynx, which they perforate to establish communication with the oral cavity. The two apertures thus formed are the *posterior nares*. The openings (*anterior nares*) into the olfactory pits are located on the right and left sides of the *nasofrontal process*, which projects below the telencephala.

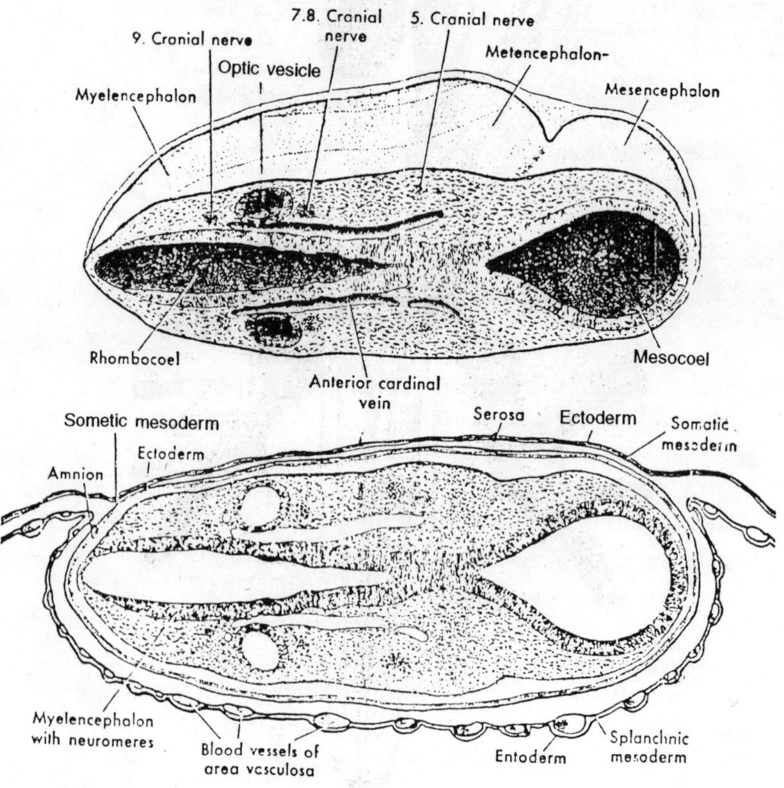

Fig. 9.23. The 72-hour chick embryo. Section through the otic (auditory) vesicle.

The Optic Nerve and the Eye: In preceding chapters the development of the optic nerve and the eye have been followed out as far as the 48-hour chick embryo. At that stage the lens is still a hollow vesicle, opening to the outside by a tiny pore called the *lenticular aperture.* In the 60-hour embryo this pore has disappeared, and the lens with its outer thinner and its innermost thicker layer still contains a flattened lenticular cavity which will persist as far as the tenth day of incubation. The two layers of the optic cup become more closely apposed to each other, but they are still far enough apart in the 72-hour chick embryo to form a shallow remnant of the opticoel. While these changes are in progress, the rim of the optic cup becomes smaller by converging centrally toward the lens, which now becomes located within the diminishing aperture of the optic cup. In this transformation the choroid fissure has become narrower and is continued on the ventral side of the optic stalk as a deep groove. Later on, this groove will harbor the blood vessels leading to the retina and also the visual nerve fibres (axones) which will grow out from the inner layer of cells (retina) of the optic cup. In later development, the groove is entirely eliminated by sinking deeper into the optic stalk and by overgrowth of its sides, thus encasing the optic nerve and the blood vessels in a sheath which was originally the optic stalk. At about the same time the choroid fissure is also eliminated, so that the optic cup becomes really cup-shaped throughout. In this simple manner the optic nerve becomes almost centrally placed on the back of the optic cup, which may now be referred to as the eye-ball.

The outer layer of the optic cup differentiates into a black pigmented layer, whose cells absorb light by means of black granules contained in the cells. The mesenchyme cells furnish a hard protective coat, or *sclera,* and a vascular envelope called the *choroid coat,* to the outside of the eyeball; however, the crystalline liquid called the *vitreous humor,* located between the retina and the lens, is probably formed by the cells of the optic cup itself and is therefore believed to be of ectodermal origin. The outer layer of the transparent cornea is continuous with the epidermis of the skin. The iris, which assists in lending the characteristic colour to the eye, is formed from the thin converging portions of the rim of the optic cup.

The Oculomotor, Trochlearis, and Abducens Nerves: It seems best to treat this group under one heading because they are purely efferent and motor in function, and because they agree in their specific action by innervating the muscles of the eyeball. Of the three, the third or *oculomotor*, nerve appears first and may already be identified in the 60-hour chick. However, it can best be observed in its appearance in sections of the 72-hour embryo, where it emerges from the ventral side of the mesencephalon, and, passing forward as far as the optic cup, ends gradually there near the ventrolateral side of the diencephalon. Since the six muscles of the eyeball are not as yet developed at that time, the nerve has no definite terminal branches, but when they do appear the oculomotor will innervate the *superior, inferior,* and *internal* rectus as well as the *inferior oblique.*

The fourth cranial, or *trochlear*, nerve, also called the *patheticus*, and the sixth cranial, or *abducens*, nerve, develop much later than the oculomotor. Both appear shortly after the fifth day of incubation, the trochlear originating in the depression between the mesencephalon and metencephalon, while the abducens arises near the midventral line of the myelencephalon, just below the seventh nerve. The trochlear never innervates the *superior oblique muscle* of the eye and the abducens nerve supplies the *external rectus.*

The Trigeminalis: When the neural crests are formed in the early stages of the embryo, they extend along the entire neural tube to the tip of the head. There follows a gradual differentiation and adjustment, especially in the anterior part of the embryo, but even in the 33-hour chick these crests can still be observed in the head region as paired condensations of cells extending in the form of paired strands from the tip of the head back to the otic region. These loosely organized strands of neural crest cells, located laterally between the ectoderm and the foregut, make contact with the ectoderm of the skin on their outer borders and organize into massive ganglia. The first and larger one is located just posterior to the first neuromere of the rhombencephalon, and another is located further back, immediately anterior to the otic capsule. The former is the beginning of the *trigeminal ganglion,* also called the *Gasserian ganglion,* and the latter is the ganglion of the seventh nerve, or *geniculate ganglion.* The mass of the Gasserian ganglion is still further increased by the addition of cells which arise as a thickening from the inner portion of the superficial ectoderm in front and above the first visceral cleft. This mass of cells is homologous to the placode of the fifth cranial nerve in the frog embryo.

The trigeminal nerve is a mixed nerve, consisting of motor and sensory fibers, of which the latter arise from the ganglion from bipolar neurones. The ganglion is bipartite in structure, its anterior portion giving rise to the *ophthalmic nerve* which grows forward over the optic cup into the head region. The posterior portion of the ganglion extends downward and branches near the angle of the mouth into the *maxillary* and the *mandibular nerves,* of which the former supplies the maxillary process or upper jaw, and the latter the mandibular arch (first visceral arch) which will form the lower jaw.

The Facialis: The seventh and eighth, or *facial* and *acoustic,* nerves have a common origin and remain closely associated in early chick embryology. Their ganglia are derived from cranial neural crest cells, and from an ectodermal thickening which is continuous with the auditory placode. However, a small part of the geniculate ganglion is derived from an epibranchial placode which is well-separated from the auditory placode. During the fourth day of incubation, the geniculate ganglion, which gives rise to the seventh or facial nerve, is separated from the acousticofacialis complex, as described below in the origin of the acoustic nerve. The remainder of

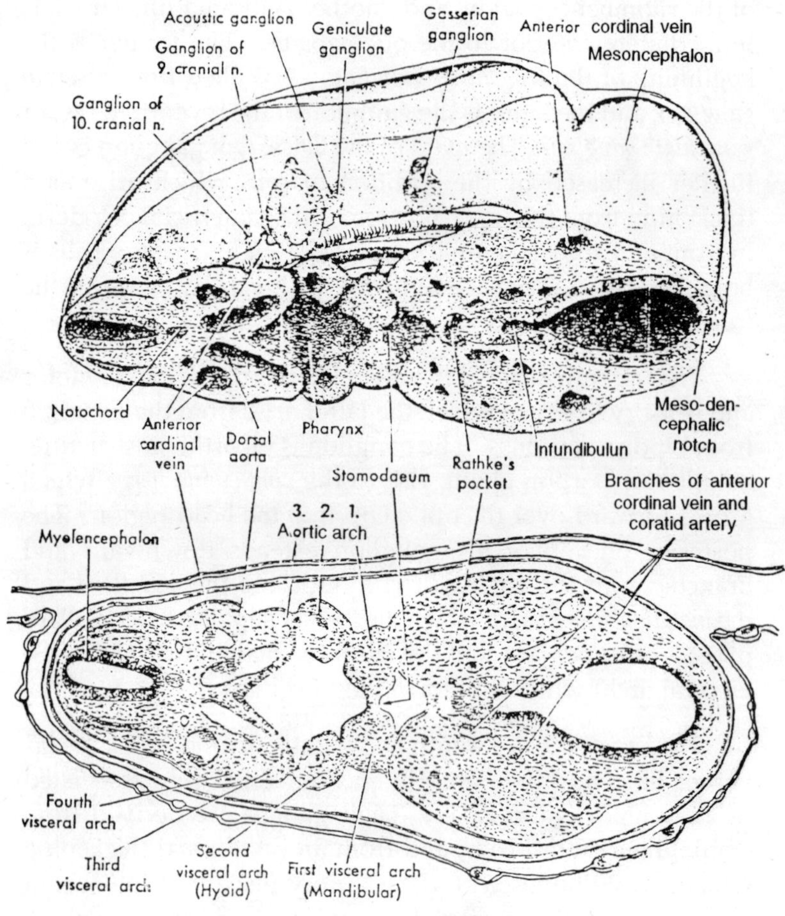

Fig. 9.24. The 72-hour chick embryo. Section through the aortic arches and pharynx.

the complex forms the *auditory ganglion* which becomes located immediately above the second visceral arch. The facial nerve grows downward, dividing above the first, visceral arch into the *pretrematic* and the *posttrematic* branches, of which the former enters the mandibular, and the latter the hyoid arch.

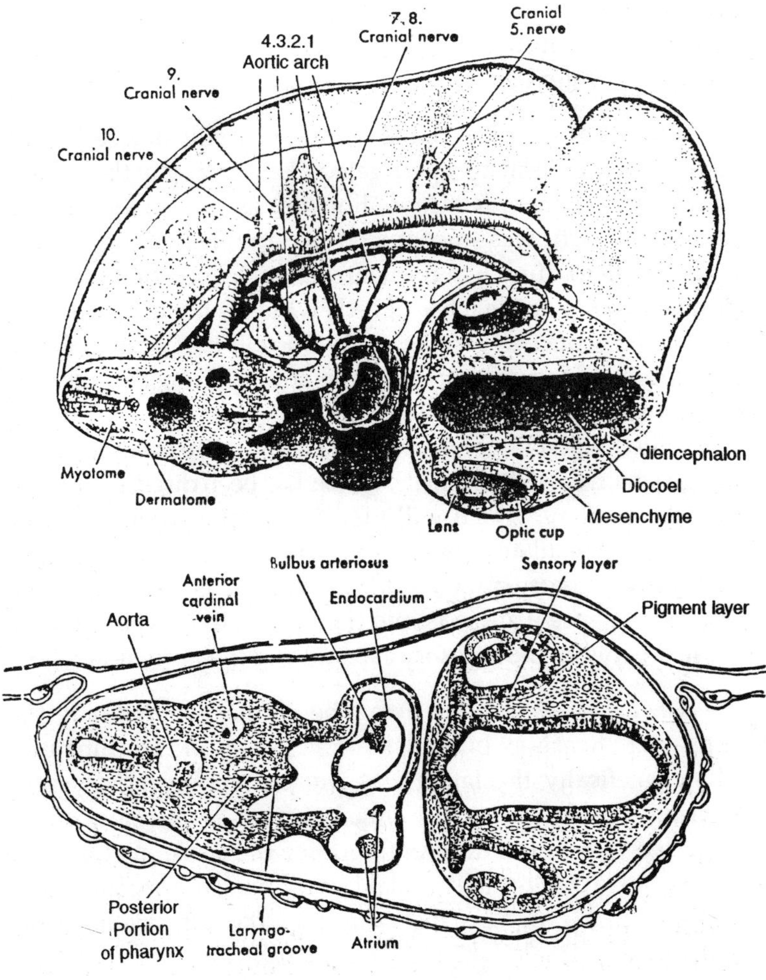

Fig. 9.25. The 72-hour chick embryo. Section through the optic cups.

The Acoustic Nerve and the Ear: The development of the eighth, acoustic or auditory, nerve and the ear are so closely interrelated that it is almost impossible to discuss them apart. In the 72-hour chick embryo the ear has not advanced very far in its differentiation, so that for our present needs not much. However, it is desirable to trace the more general features of the ear from their earliest appearance to their final disposition, and observe how closely this complex organ has been evolved from such primitive beginnings as are present in young vertebrate embryos.

When last mentioned, the otocyst was still attached to the superficial ectoderm by a perforated stalk, ending at the outside in a pore. There was also a short dorsal evagination in the otocyst, which was recognized as the beginning of the endolymphatic duct. On the ventrolateral aspect of the otocyst appears now a dense mass of cells which is the material from which the acoustico facialis ganglion will be formed. Upon separation of the otocyst from the superficial ectoderm, this ganglionic mass moves entirely to its ventral side, and after it has united with cranial neural crest cells, it separates into two ganglia, namely, the *geniculate*, which has been discussed above, and the *auditory*, which will give rise to the auditory nerve. While the geniculate ganglion appears as a definite condensed mass, the auditory appears diffuse and stays attached to the otocyst, whence it spreads into the mesenchyme cells that are located between the auditory capsule and the myelencephalon.

The Labyrinth or Inner Ear: The ear harbors not only the sense of hearing but also the sense of equilibrium. Phylegenetically the latter was present before the sense of hearing was added to the inner ear. It is therefore to be expected that in the differentiation of the sense organs from the primitive otocysts the mechanism for the sense of equilibrium should appear first. This occurs in the formation of the semicircular canals which grow out from the otocyst as flat, pocketlike evaginations. Eventually all three pockets are transformed into semicircular canals by separations of their central areas from the otocyst, only their ends remaining attached. After this transformation the otocyst is referred to as the *utriculus.* Each semicircular canal is supplied with one dilation *(ampulla)* at one end of its attachment to the utriculus. The position of the semicircular canals in all vertebrates from fishes to mammals is uniform; one is located in the horizontal plane, the second in the vertical plane, and the third also in the vertical plane, at right angles to the second.

During the formation of the semicircular canals, the endolymphatic duct changed from its dorsal position on the otocyst to a more central and median one. This change is effected by differential growth of cells, and when it is completed the endolymphatic duct is attached to the utriculus between the ampullae of the two vertical semicircular canals. Another differentiation, called the *cochlear duct,* appears near the utriculus next to the endolymphatic duct, between the ampullae of the two vertical semicircular canals. The entire structure, including the semicircular canals, endolymphatic duct, utriculus, cochlear duct, and cochlea, is now called the *membranous labyrinth.* Later it will become enclosed in a bony envelope, referred to as the *bony labyrinth,* and between the two the liquid *perilymph* fills up the narrow space. The combined membranous and bony labyrinth with its perilymph is usually simply referred to as the *labyrinth* or *inner ear.*

The cavities of the membranous labyrinth are also filled with a liquid which is called *endolymph* because it is contained within the cavities. The latter which have been derived from the original cavity of the otocyst, are lined by a sensory epithelium containing delicate hairlike projections in definite areas, as for example in the utriculus, cochlea, and ampullae of the semicircular canals. The sensory areas become connected with the auditory nerve which originates from the ganglionic mass beneath the otocyst. It divides into two branches, of which the dorsal one supplies the utriculus and the three semicircular canals, and the ventral one the chochlea. This is as far as ear development proceeds in the bird. In the mammal

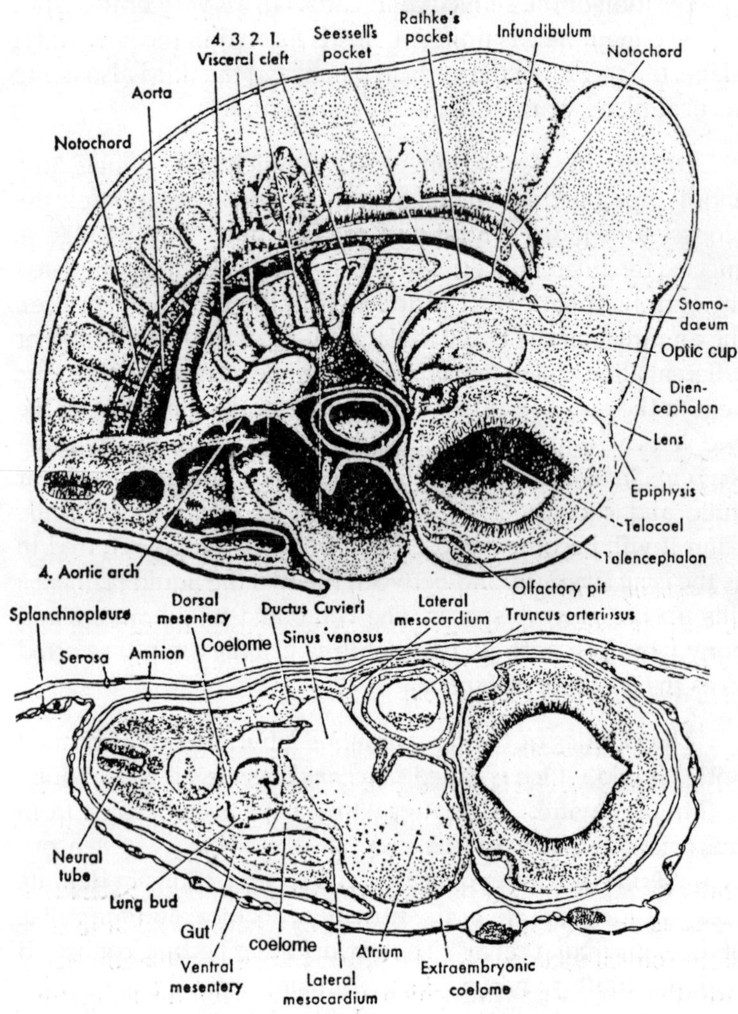

Fig. 9.26. The 72-hour chick embryo. Section through the heart and olfactory pits.

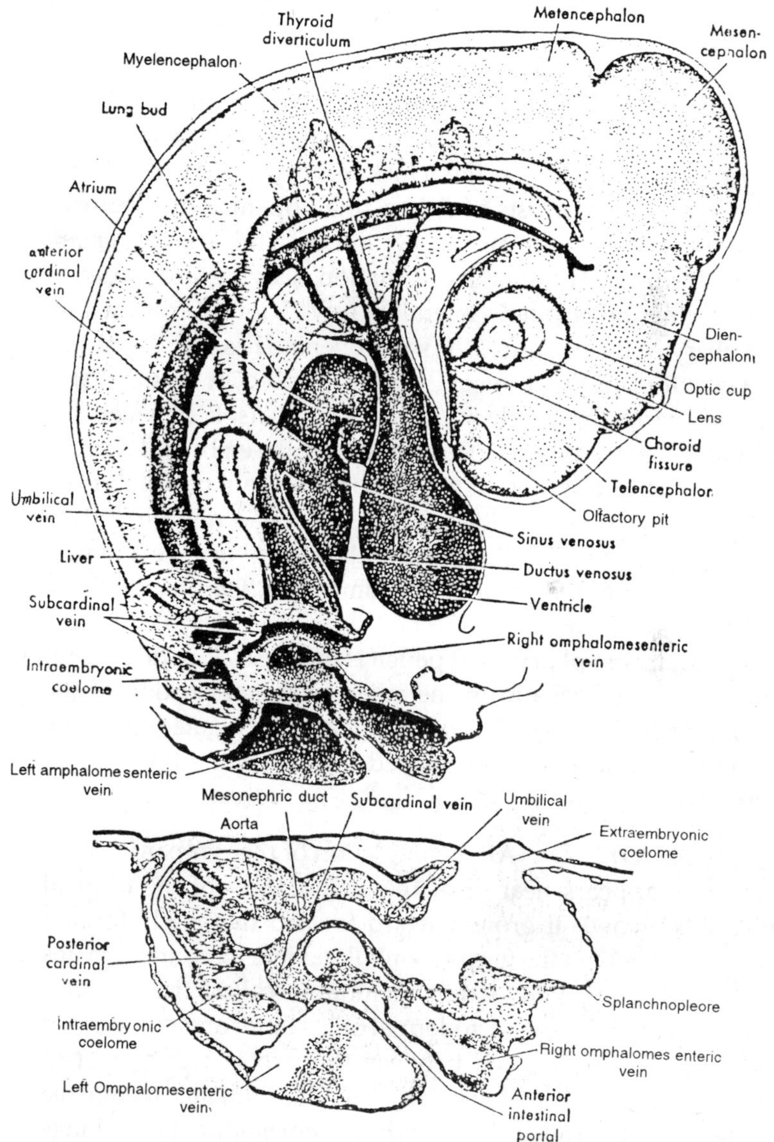

Fig. 9.27. The 72-hour chick embryo. Section through the anterior intestinal portal.

the cochlear duct grows out ventrally from the utriculus, coiling into two to four turns which will develop eventually into the cochlea.

The Tympanic Cavity or Middle Ear: It has been pointed out before that the first visceral cleft (hyomandibular cleft) is never completely formed in the chick embryo. At its fullest development it consists of only a small opening at the upper region of the first visceral furrow, and even this small aperture disappears shortly after its formation, leaving a depression resembling the original visceral furrow. Viewed from within the pharynx, the closure of the small opening has not diminished the depth of the pharyngeal pouch. Indeed, it seems that it has increased in depth and has pushed itself between the labyrinth and the outer ectoderm or skin of the embryo. Presently there occurs a downward growth of a delicate partition from the dorsal wall of the pharynx, thus adding a minute portion of the pharynx to the distal end of the first pharyngeal pouch. This newly formed chamber becomes the *tympanic cavity* or the *middle ear*, and its connection with the pharynx through the tubular first pharyngeal pouch is the *Eustachian tube*. Later on, a group of cells from the adjacent upper portion of the hyoid arch (second visceral arch) are segregated into the tympanic cavity and transformed into a minute bone called the *columella auris*.

The External Ear: After completion of the tympanic cavity a groove appears near or at the old location of the original visceral furrow. It grows inward toward the newly formed tympanic cavity in the form of a tubular depression, and remains separated from it by a thin membrane called the *tympanic membrane*. The latter is ectodermal on its outer and entodermal on its inner surface, and between the two is a thin layer of mesoderm cells. The tubular canal from the tympanic membrane to the outside is called the *external auditory meatus;* it constitutes with its opening the *external ear*. The columella auris, mentioned above, is attached by one end to the inner surface of the

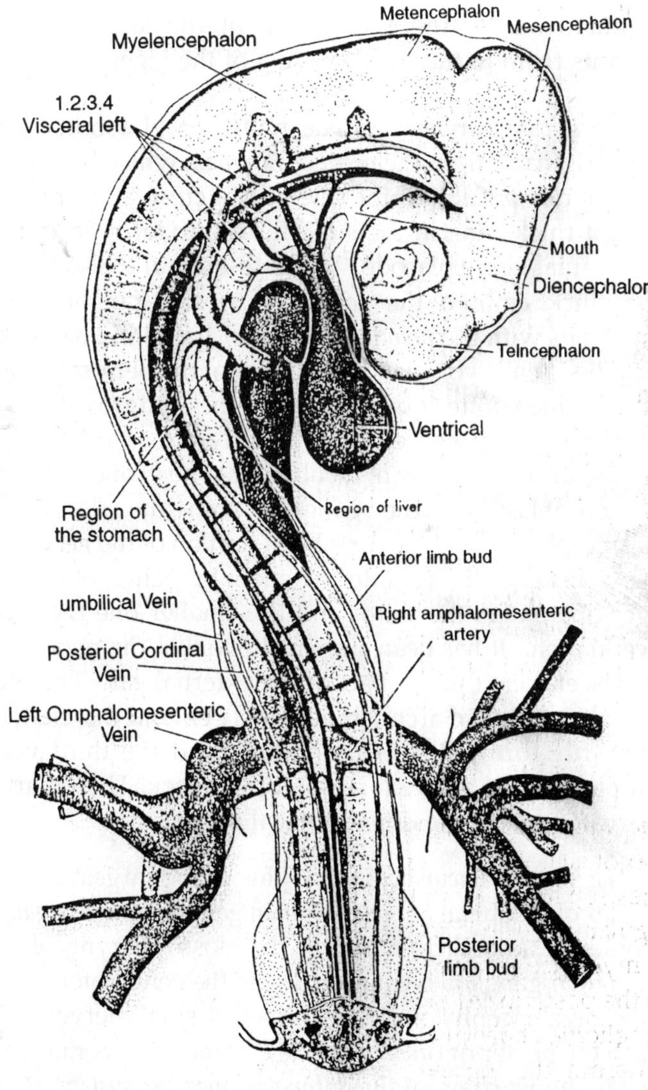

Fig. 9.28. The 72-hour chick embryo. Section through the allantois.

tympanic membrane, and with the other if fits into the *fenestra ovalis* of the labyrinth. Sound waves striking the tympanic membrane are relayed by the columella auris through the fenestra ovalis to the endolymph of the labyrinth, which transmits them to the cochlear duct of the lagena.

The Glossopharyngeus: The ninth, or glossopharyngeal, nerve originates from the *petrosal ganglion,* which is formed by the posterior portion of the otic neural crest segment and an adjacent thickening in the superficial ectoderm, forming the third cranial nerve placode. In the 48-hour and also in the 60-hour chick embryo the *postotic (petrosal)* ganglion is still continuous with the *preotic (geniculate)* ganglion: however, in the 72-hour embryo the independence of the petrosal ganglion has become quite obvious, and the outgrowth of the glossopharyngeal nerve occurs shortly afterwards, between the third and fourth day of incubation. The axones grow from the neuroblasts of the ganglion into the floor of the myelencephalon, and the peripheral portion of the nerve grows downward and divides into two main branches, of which the larger one enters the third, and the smaller one the second, visceral arch. It has been shown above that the region of the first visceral cleft forms part of the external ear. The second visceral arch (hyoid arch), immediately behind this cleft, must necessarily form parts of the mouth, and the third visceral arch parts of the adjacent region of the neck. These parts are innervated by the glossopharyngeal nerve.

The Vagus (Pneumogastric): This is the largest and most complex of all cranial nerves. Its main ganglion *(jugular ganglion)* becomes separated from that of the glossopharyngeal nerve. The jugular ganglion is connected with the ventrolateral portion of the myelencephalon by a number of small nerve roots, of which the posterior ones organize a branch that connects with the ganglionic chain of the *autonomic nervous system.* A thick branch from the jugular ganglion connects with the more ventrally located *nodosal ganglion,* from which a branch for the fourth and also one for the fifth visceral arch originates, and, after having supplied these structures, continues posteriorly toward the viscera of the body cavity as the *pneumogastric nerve,* supplying the heart, lungs and stomach. Similar to the nodosal ganglion, the jugular ganglion is formed partly by neural crest elements and partly by cells from the thickened superficial ectoderm.

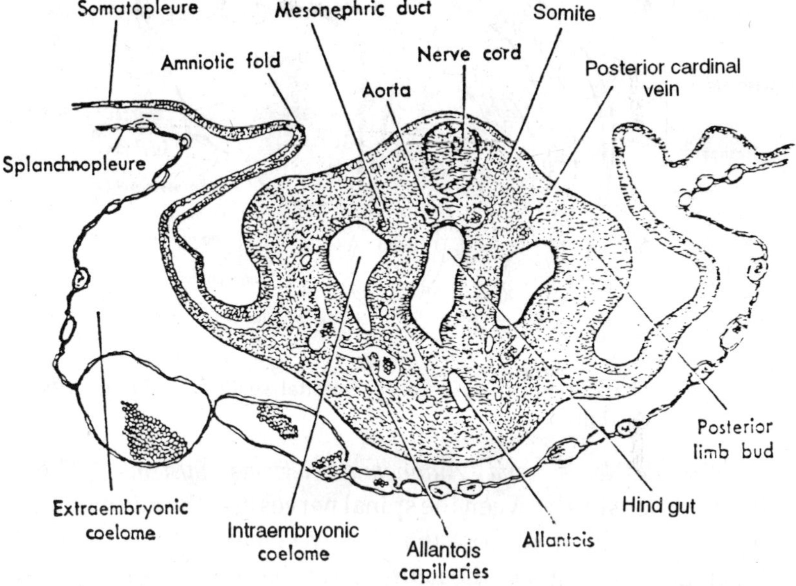

Fig. 9.29. The 72-hour chick embryo. Section through the allantois.

The Accessorius and the Hypoglossus: There is no doubt that the eleventh, or *spinal accessory,* and the twelfth, or *hypoglossal,* nerve were once upon a time spinal nerves and that they have been added to the cranial nerves by a process of cephalization. They are almost exclusively motor (efferent) in function and therefore represent chiefly the ventral roots of spinal nerves with which they are actually continuous. Both of these nerves appear during the fourth day of incubation and are characterized by the absence of ganglia and the lack of any contribution from the superficial ectoderm. They innervate chiefly regions of the neck.

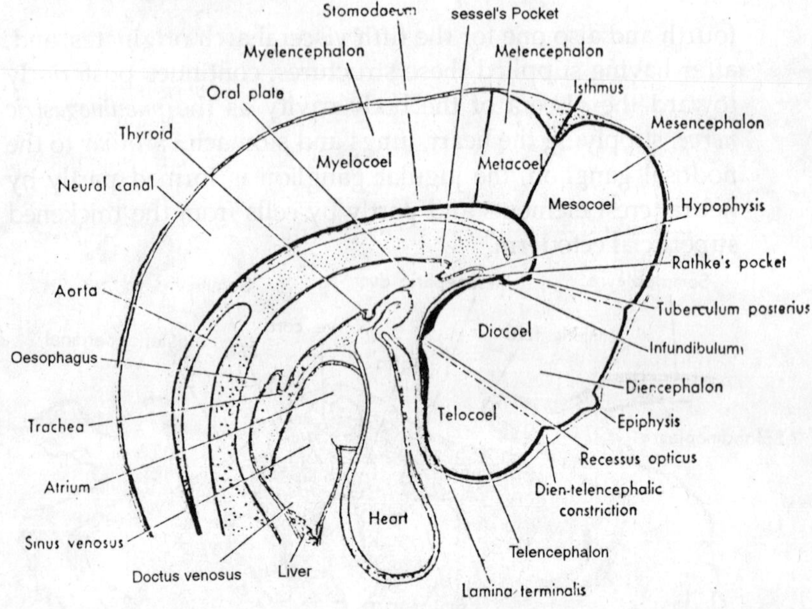

Fig. 9.30. The 72-hour chick embryo. Sagittal section of the anterior portion.

Spinal Nerves and Autonomic Nervous System: The interrelationship between the spinal nerves and the autonomic nervous system is so close that the two have to be discussed together. The origin of the neural crests has been described in Chaps. XIII and XIV, so that we shall consider here only the formation of the spinal nerves and their relation to the autonomic system. When in the following discussion the singular form is used in the description of nerves and ganglia, it should be remembered that spinal nerves, like cranial nerves, are also paired and that the discussion is applicable to both members of the pair. Morphologically, each spinal never is composed of two strands known as the dorsal and ventral roots, of which the former contains the spinal ganglion. Furthermore, the nerves originating from the spinal cord are connected with the autonomic nervous system by a communicating strand — the *ramus communicans.* Physiologically, the dorsal root of each spinal nerve is afferent and sensory, and the ventral root efferent

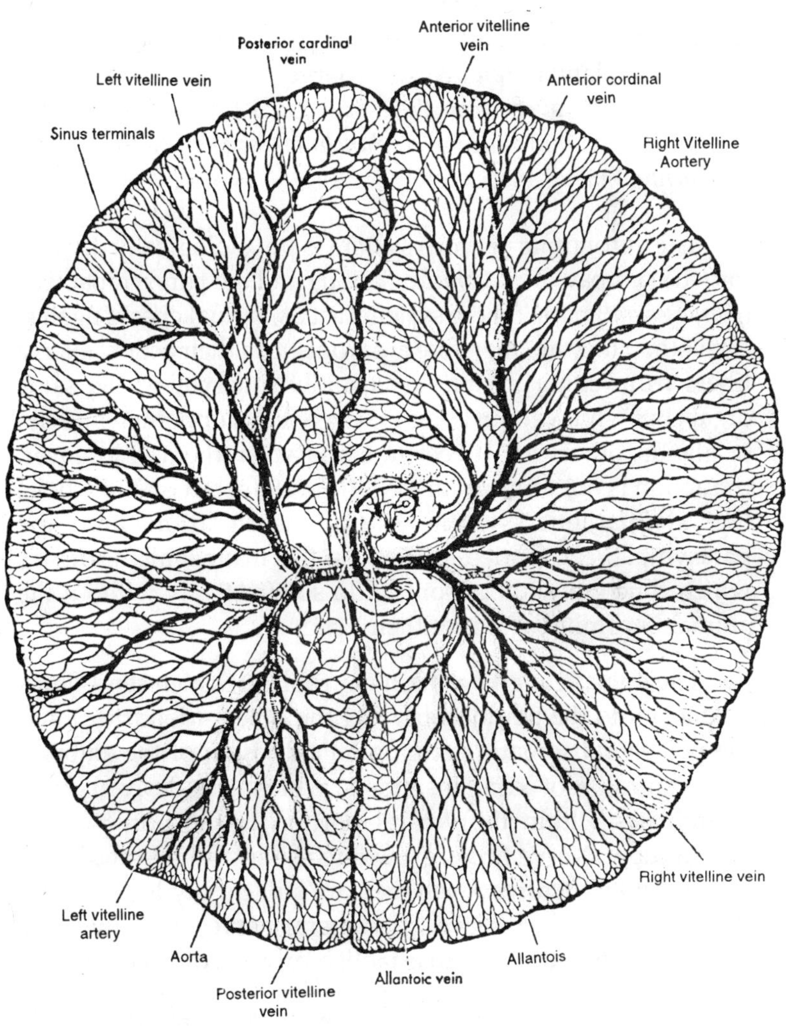

Fig. 9.31. Blood circulation in the 96-hour chick embryo, dorsal view.

and motor. A greater complication in this respect aries with the nerves connected with the autonomic nervous system. The nerve fibres which innervate the somatic peripheral organs and systems are referred to as being *somatic sensory* and *somatic motor* fibers.   On the other hand, the organs and systems connected with the autonomic nervous system are derived from the splanchnopleure, and nerve fibres of such organs going to or coming from the nerve cord through the spinal nerves are called *visceral (splanchnic) sensory* and *visceral (splanchnic) motor* respectively.

The autonomic nervous system is physiologically divided into the sympathetic *(thoracico-lumbar)* and the parasympathetic *(cranio-sacral)* parts. These systems act as mutual checks against each other; they are therefore antagonistic in their action. Nerve fibers from either system are usually innervating the same organ, but while certain fibres may accelerate or tighten muscles or glands, others may decelerate or slacken them. Moreover, the action is not uniform in any one system, but on the whole, the sympathetic system is for acceleration and the parasympathetic for deceleration. In this manner all the organs which function without control do so effectively without the animal becoming aware of their action.

The spinal ganglia contain neuroblasts which send fibres (axones) into the neural tube, and the others (dendrites) into the peripheral somatic or autonomic division of the embryo. The ventral root contains fibres which are derived from neuroblasts within the ventral portion of the horns of the neural tube. Nerve fibres grow into the myotomes and other organs and structures when they are still closely located near the neural tube. However, peripheral nerves actually grow along definite pathways to establish proper connections.

The autonomic nervous system with its paired sympathetic ganglia originates during the fourth day by strands of cells which are derived from the neural tube and the neural crest material. These strands extend in metameric order from the neural crests to the sides of the aorta, where they establish terminal thickenings that will later form the sympathetic ganglia. The latter become united by connections between the ganglia to form the characteristic sympathetic chains of the adult. Anteriorly the chains are connected with the sympathetic system of the cervical region, and the paired *pneumogastric* or *vagus* nerves.

The Alimentary Canal and its Derivatives

Growth of the Enteron: In embryos of less than one somite, the alimentary canal is an open gut, with the cellular entodermal roof above and the non-cellular yolk floor below. When the head fold appears, the entoderm follows by pushing into it, to differentiate the fore-gut. Similarly, the hind-gut is established by the growth of the entoderm into the tail bud. The elevated head and tail are undercut by the subcephalic and the *subcaudal pockets* respectively, and the two are continuous with the lateral body folds, which constrict the embryo on both sides from the underlying blastodisc. There is a progressive undercutting on all sides, shaping the contour of the embryo and limiting the open gut to two small areas, one over the yolk stalk and the other over the allantoic stalk. The body folds which accomplish this constriction converge on the ventral side of the embryo, surroundings the yolk stalk and allantoic stalk on all sides. In this manner a new stalk is formed, referred to as the *unbilical cord.*

As the embryo grows in size, the yolk sac diminishes accordingly, so that shortly before hatching it has become depleted and is small enough to be withdrawn into the alimentary canal. On the other hand, the allantois remains large and completely functional almost to the day of hatching. Shortly before emergence the chick begins to use its lungs for respiration, thus relieving the allantois of this function. At the same time the vascular system establishes its pulmonary circuit, resulting in a progressive diminution of respiratory function in the allantois. The latter now begins to dry up, its attachment to the body *(umbilicus)* is severed, and the chick is ready to pick its way through the calcareous shell.

The Mesenteries: As the embryo becomes constricted from the underlying yolk, its alimentary canal becomes more or less tubular, except in the restricted areas of the umbilicus where it still remains open. Moreover, its various portions become differentiated and assume their permanent position in the body cavity. During these changes the gut moves away from its original location immediately beneath the notochord and becomes suspended by the *dorsal mesentery*, composed of splanchnic mesoderm, which is a reflected portion of the somatic mesoderm. Since the splanchnic mesoderm extends also over the yolk sac and the allantois, a ventral mesentery must necessarily be formed when the yolk stalk becomes progressively smaller and constricted, thus leaving the closed gut attached to the ventral body wall by a second mesentery. Eventually this *ventral mesentery* disappears, with the resulting union of the two lateral coelomic cavities. Only in the region of the liver, the hind-gut, and the oesophagus does the ventral mesentery remain. As a matter of fact, the liver diverticulum growing out from the alimentary canal in the 48-hour embryo is caught between the two approaching sheets of the splanchnic mesoderm and is permanently held in place by them.

With the exception of the stomodaeal and the proctodael portions, the entire alimentary canal is differentiated from the entoderm, which, in the adult, still forms its lining. Moreover, all glands and organs which are formed by outgrowths from the undifferentiated enteron retain their entodermal heritage in a similar manner. To the original entodermal layer are added mesenchyme cells which form the muscular layers, and the outer covering of the canal with all its derivatives is furnished by the splanchnic mesoderm.

The Fore-gut: The entire alimentary canal is conveniently divided into the fore-gut, mid-gut, and hind-gut, of which the fore-gut extends from the mouth to the pancreas. The mouth forms in the 60-hour embryo by the rupture of the oral plate, which is located between the two mauidibular arches. During this process the superficial ectoderm is drawn inward so that the oral cavity becomes lined by it anterior to the pharynx. The transition between the ectoderm and entoderm is soon lost and has become imperceptible in the 72-hour chick.

The Pharynx and Visceral Arches: The mouth is followed by the pharynx, which is chiefly characterized by the paired pharyngeal pouches.   These have been discussed quite thoroughly in the preceding chapters, and very little need be added here to complete the account. Of the four pharyngeal pouches developed in the embryo, only three pairs form apertures to the outside. The fourth, or last one, never opens but remains a pouch from whose posterodorsal border the *postbranchial bodies* are developed. These bodies contain tissue which resembles that given off for the organization of the *thymus gland* by the third and fourth pairs of pharyngeal pouches. They also contain cells similar to the *epithelioid bodies* given off by the third pharyngeal pouches. However, the postbranchial bodies never enter into the formation of the thymus gland.

The visceral arches separating the pharyngeal pouches are short, stubby columns which spread out fanwise from the level of the truncus arteriosus to the paired dorsal aortac. They are covered on the outside by ectoderm and toward the pharynx by entoderm, and since the pharyngeal pouches are considerably deeper than the visceral furrows, the contribution of the entoderm is much greater than that of the ectoderm. The core of each visceral arch consists of mesenchyme, which differentiates into the branchial blood vessel (aortic arch) and skeletal support. The branchial skeleton is composed at first of cartilage, but later this is replaced by bone in certain visceral arches, resulting in the partial ossification of the *columella auris* of the middle ear, and some of the bony elements for the mandible or jaw. Furthermore, each arch is supplied by one or more nerves, as outlined above in the discussion of the cranial nerves.

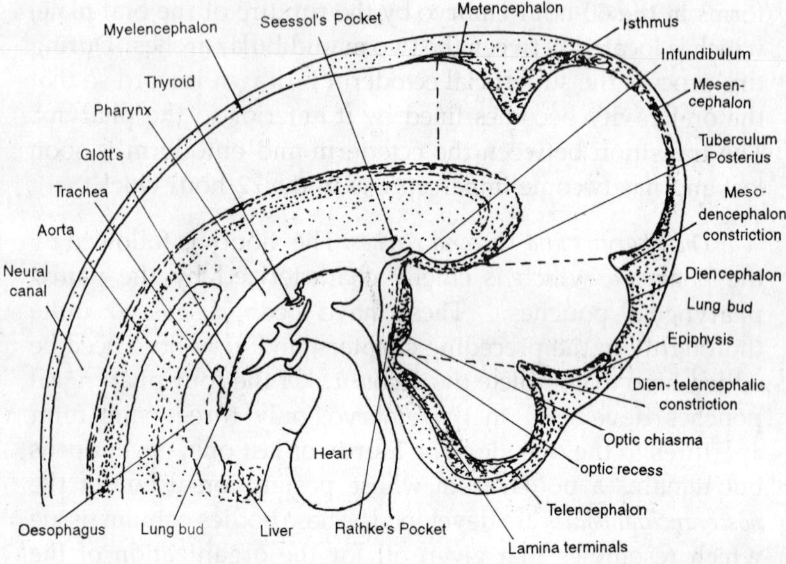

Fig. 932.　The 96-hour chick embryo. Sagittal section of anterior portion.

The first pair of visceral arches, or the mandibular arches, project downward on both sides, with the mouth opening between them. During the fifth day of incubation they unite ventrally to form the lower jaw. The upper jaw is formed later by the approach of the two *maxillary processes* which are derived from the uppermost portion of the mandibular arches, and the *frontal process*, which is located centrally and projects downward toward the mouth.

The Thyroid Gland: The thyroid gland appears as a shallow depression in the 48-hour embryo between the second pair of pharyngeal pouches. As its grows down-ward it becomes constricted at its origin in the floor of the pharynx. In the 72-hour chick it has become vesicular, and shortly afterwards, in the 4-day embryo, it becomes entirely detached from the alimentary canal. After elimination of its cavity it begins to differentiate into its characteristic glandular form. Early during the fifth day of incubation, the same region of the pharynx which gave rise to the thyroid gland develops the *tongue*.

The origin of the thyroid gland in the chick is to strikingly like that of the frog that there seems to be a close relationship between the two. Indeed, this relationship can be carried back to amphioxius, in which the endostyle (hypopharyngeal groove) is apparently homologous to the thyroid gland of vertebrates. The ammocoetes larva of the lampreys bridges this gap since it has an endostyle in its larval form which becomes constricted off and functions as a thyroid gland in the adult.

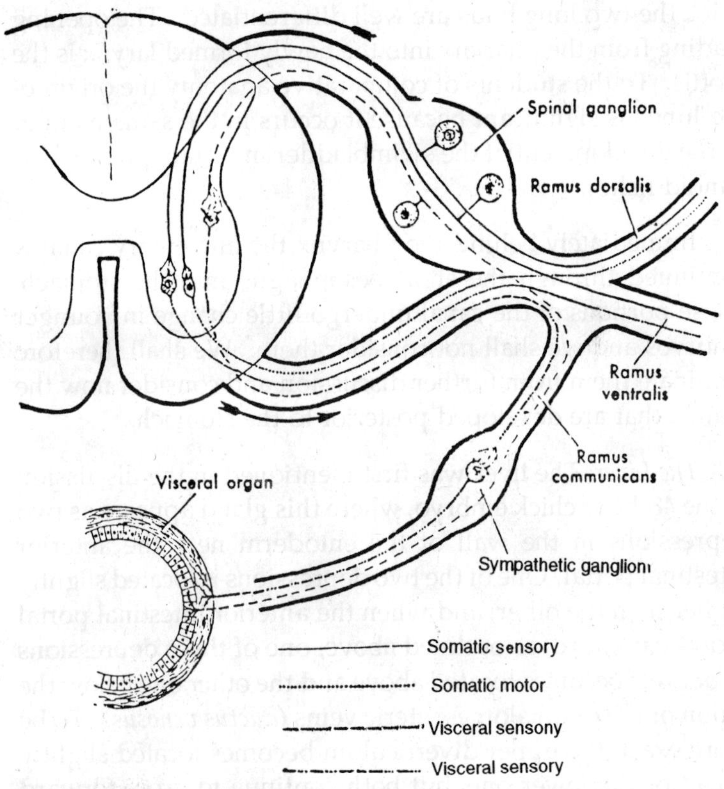

Fig. 9.33. A spinal nerve and its components.

The Pulmonary Apparatus: In the posterior part of the pharyntx, where it becomes transitional with the oesophagus, another depression is formed in the center of the floor, called the *laryngotracheal groove.* At first it is very narrow, but it soon deepens, becoming saclike, and as it pushes away from the pharynx it is converted into an elongate diverticulum, divided into two lobes at its distal end. This diverticulum represents the beginning of the pulmonary apparatus, consisting in this early condition of an embryonic *larynx, trachea,* and *lung buds.* The appearance of the laryngotracheal groove may be observed shortly after the second day of incubation, and in the 72-hour chick the two lung buds are well differentiated. The opening leading from the pharynx into the newly formed larynx is the glottis. To the students of comparative anatomy the origin of the lungs is significant because it occurs in the same manner as the development of the swimbladder in certain present day ganoid fishes.

Immediately behind the pharynx the alimentary canal is continued through the short oesophagus into the stomach. These portions of the canal undergo little change in younger embryos and we shall not consider them. We shall therefore eliminate them from further discussion and consider now the glands that are developed posterior to the stomach.

The Liver: The liver was first mentioned in the discussion of the 48-hour chick embryo, where this gland appears as two depressions in the wall of the entoderm near the anterior intestinal portal. One of the two depressions is located slightly higher than the other, and when the anterior intestinal portal grows backward, as outlined above, one of these depressions or pockets becomes located above and the other one below the union of the omphalomesenteric veins *(ductus venosus).* To be more exact, the upper diverticulum becomes located slightly ahead of the lower one, but both continue to grow forward into the ventral mesentery so that in the 60-hour embryo the liver has reached the level of the sinus venosus. During this time the two primary diverticula have branched and rebranched

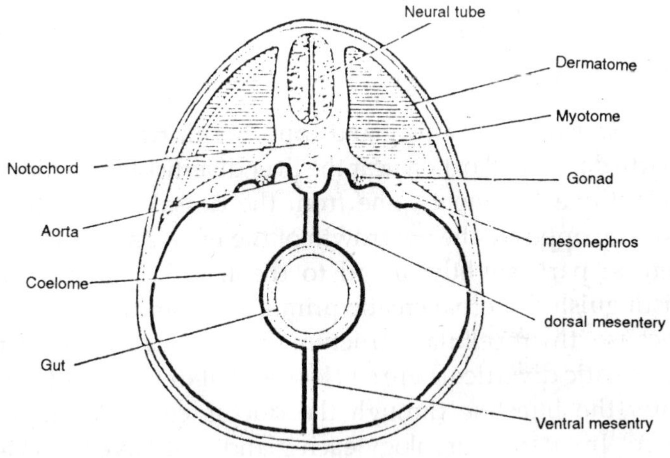

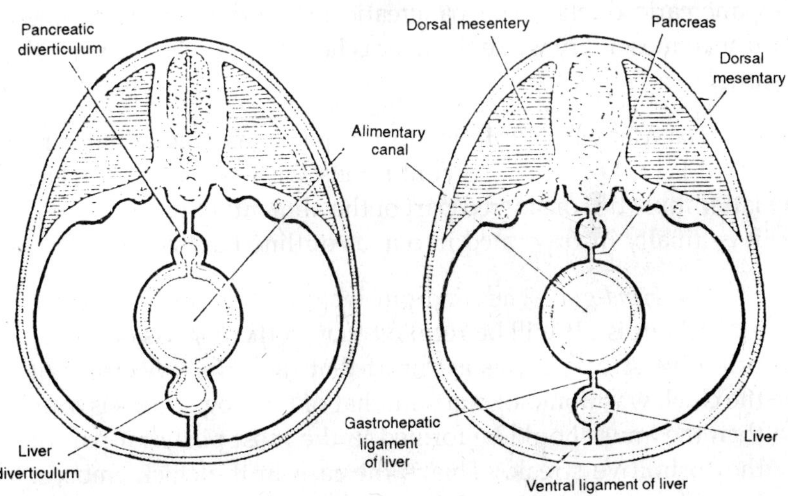

Fig. 9.34. The dorsal and ventral mesenteries and their relation to the gut, liver and pan creas.

into a profuse network of liver tissue surrounding the entire ductus venosus. The bile duct is formed by the fusion of the two liver diverticula which takes place, when the embryo in its forward growth pushes the intestinal portal further back. The resulting single duct (*ductus choledochus*) will remain to act as bile duct in the adult.

The Pancreas: This gland appears during the third and fourth day of incubation near the liver diverticulum. It develops from three rudiments, one from the dorsal part of the gut, directly opposite the outgrowth of the liver, and two from the ventral part, slightly lateral to the liver. It is difficult to distinguish these pancreatic primordia from that of the liver because their cellular structure is identical. The ventral pancreatic diverticula are so close to that of the liver that they enter the intestine through the ductus choledochus of the liver. In later embryology each gland will have its separate duct; as a matter of fact, the adult chick has two or three pancreatic ducts. The pancreatic primordia grow into the dorsal mesentery while the liver is held in place by the ventral one.

The Mid-gut : Posterior to the pancreas the alimentary canal is still open and is continuous with the yolk sac. The latter must be considered part of the alimentary canal because it is finally incorporated into it, as outlined above.

The Hind-gut: The hind-gut includes the rectum, cloaca, and allantois. It will be recalled that in the discussion of the primitive streak, it was assumed that the primitive streak of the chick was homologous with that of the frog. If this is true, then the anus should be formed in the most posterior part of the primitive streak. This is the case in the chick embryo. Immediately posterior to the primitive streak in the 33-hour chick embryo, there is a spot where the ectoderm and entoderm come together without any intervening mesoderm. This is the anal plate. When, in later development, the tail bud appears at the posterior end of the vanishing primitive streak, the anal plate is at first overgrown by it and then pushed to its ventral side. In the 72-hour embryo the anal plate has become very thin and will be ruptured shortly afterward.

The growth of the tail bud take place at a rapid rate, and as it just backward and downward it carries the hind-gut along within itself, so that the latter extends a good distance beyond the anal plate. There is apparently no significance attached to this unusual extension of the gut, because it disappears in later embryos without leaving a trace. Phylogenetically it should be homologous to the same structure in the frog embryo, in which it was considered to be a remnant of the neurenteric canal. It has also been suggested that the tail gut as well as the preoral gut may represent an ancestral condition of terminal anus and terminal mouth respectively, and that the present anus and mouth are neogenic subterminal apertures. One significant fact should be added here to the discussion of the allantois. It is located immediately anterior to the anal plate. This part of the chick embryo is destined to form the cloaca, and when the allantois is finally discarded its intraembryonic position will still remain as a ligamentous strand, while in the mammal it will functions as the urinary bladder.

The Vascular System

The entire vascular system of the embryonic chick may be divided into an *intraembryonic* and an *extraembryonic* system, of which the latter may again be classified into the *vitelline* and *the allantoic* circulation. However, such distinctions are arbitrary and conventional, because these subdivisions are merely integrated parts of one great system. We shall consider the intraembryonic circulation first, but in doing so we shall have occasion now and then to enter the domain of the vitelline as well as the allantoic circulation.

The Heart: The development of the heart has been followed as far as the 48-hour chick, and through it has increased in size and also changed in form in the 72-hour embryo, its fundamental plan and structure have been very little modified. The most obvious external transformations of the heart, viewed from its posterior and its progressive changes from its simple, curved tubular form in the 33-hour embryo to the four-day chick can be followed there.

In the four-day embryo the sinus venosus leads anteroventrally into the atrium, which is separated from it by a faint constriction. Posteriorly the sinus venosus leads into the ductus venosus, which has been formed by the union of the bases of the two omphalomesenteric veins. The atrium has a slight central, longitudinal depression, indicating its future division into a smaller right and a larger left auricle. The atrium leads into the muscular ventricle through the *atrioventricular* aperture, and the ventricle projects forward and upward, tapering off into the *bulbus* and *truncus arteriosus.*

The valves of the heart are still absent in these early embryos. The blood is therefore still propelled by peristaltic waves which prevent the blood from regurgitation.

The Aortic Arches: In younger embryos, as for example in those of thirty to thirty-seven hours of incubation, it is possible to identify two ventral aortae, or aortic roots, beneath the pharynx which are continued dorsally as the paired dorsal aortae or dorsal aortic roots. Even in the 48-hour embryo it is not difficult to identify two paired ventral vessels leading away from the truncus arteriosus. In the 60- and 72-hour chicks these two branches have been reduced to such minima that it is difficult to identify them as separate units. It appears that the truncus arteriosus splits up immediately into the paired aortic arches.

To the first and second pair of aortic arches which were seen to be present in the 48-hour chick embryo, the 60-hour chick has added the third, and the 72-hour embryo the fourth pair. Indeed, the first aortic arches are in the process of atrophy by the time the fourth pair has been completely formed; and while the fifth and sixth pair are added on the fourth day of incubation, the second pair also degenerates. At no time during the development of the early are there ever more than three pairs of functional aortic arches present.

In the chick embryo the ultimate fate of the aortic arches differs slightly from that of those in the frog larva. In both animals the most anterior two pairs are eliminated and the dorsal connections between the third and fourth arches disappear also, leaving the two ventral aortae as the external, and the third arches plus the anterior portions of the two dorsal aortae as the internal, carotid arteries. The fourth pair is represented by its right member in the formation of the arch of the systemic aorta (*dorsal* or *abdominal aorta*), while its left counterpart persists in this proximal portion as the *Left subclavian artery*. The left subclavian artery in the birds has a second root from the carotid artery, and the original root of the fourth arch is eventually lost; it is therefore absent in the adult. The fifth and sixth are so closely united that the fifth seems to be a mere loop on the sixth. They have common origins ventrally and enter the dorsal aortic roots also as united vessels. The connection of the united fifth and sixth arches into the posteriorly directed left root of the aorta is also obliterated. However, the root of the combined fifth and sixth arches remains and will be changed in function, as explained in the next paragraph.

Toward the end of the first week of incubation, a septum appears in the truncus arteriosus which is continued into the bulbus and finally into the ventricle, thus dividing the latter into a right and a left ventricle, and the bulbus and truncus into two vessels, one leading away from the right and the other from the left ventricle. The left ventricle is from then on connected only with the systemic aorta, which was the fourth right aortic arch. The right ventricle, which is now connected to the other vessels, is connected to the sixth aortic arches. By that time the abortive fifth pair of aortic arches has disappeared entirely, and the sixth pair has developed a posteriorly protruding vessel on each arch, which connects with the capillaries that have developed in the lungs. The vessels of the sixth aortic arches will therefore become the *pulmonary arteries* of the adult. Since the lungs are not used for respiration until a day or two before hatching, the function of the pulmonary arteries is rather limited in the embryo, being confined merely

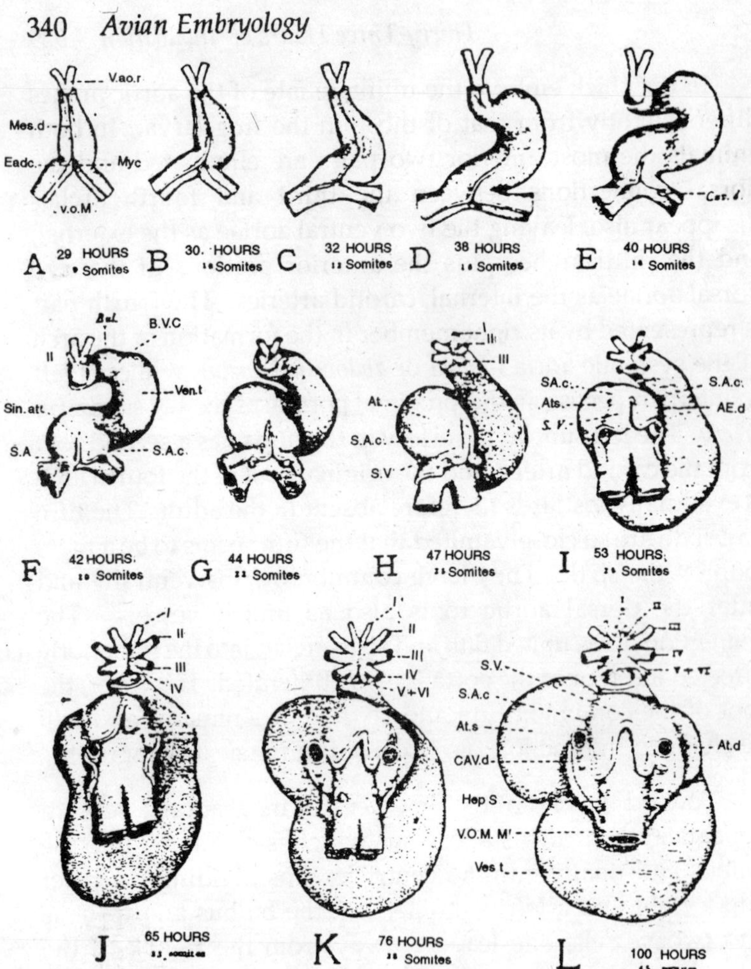

Fig. 9.35. Dorsal views of a series of hearts. I-VI, aortic arches I to VI; At., atrium (d. right), (s. left); Bul., bulbus cordis; B-V. c., bulboventricular constriction; *Cuv.d.*, duct of Cuvier; *Endc.*, endocardium; Hep.s., stubs of some of the larger hepatic sinusoids; *Mes.d.*, dorsal mesocardium; Myc., cut edge of epi-myocardium; S-A.c., sino-atrial constriction; *Sin-at.*, sino-atrial region (before its definite division); *S.V.* sinus venosus; *C.a.o.r.*, ventral aortic roots; *Vent.*, ventricle; *V.O.M.*, omphalomesenteric veins; *V.O.M.M.*, fused omphalomesenteric veins.

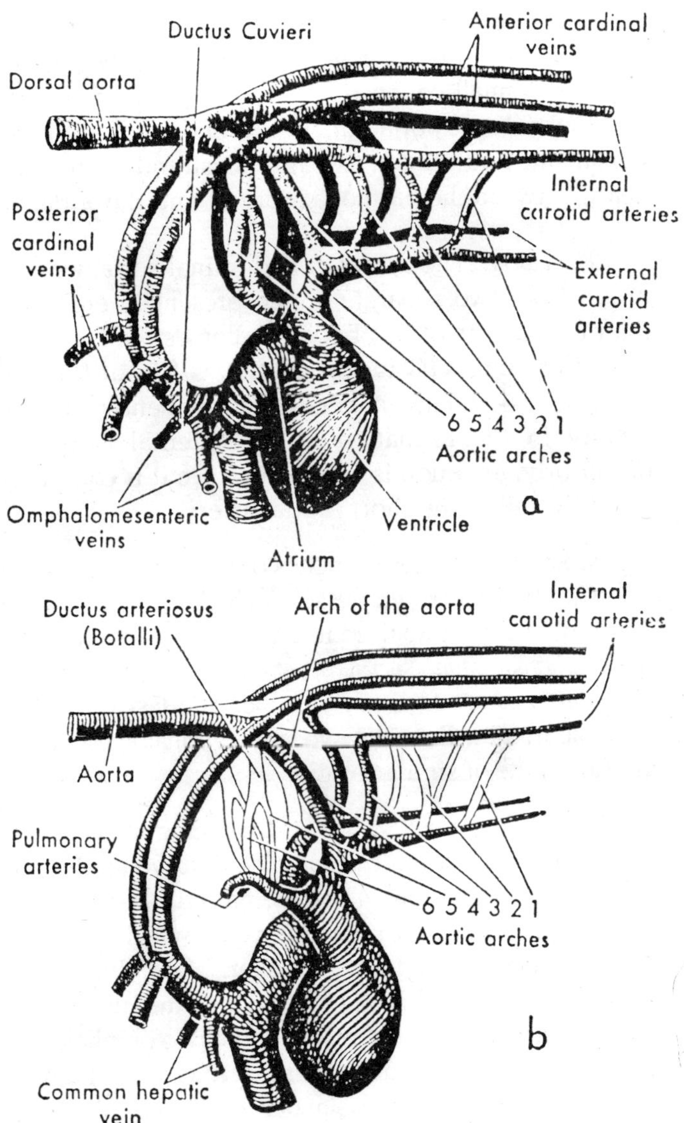

Fig. 9.36. The aortic arches of the chick embryo and their later transformation.

to the supply of blood for the growth of the lungs. During all that time the blood is shunted from the pulmonary arteries to the systemic aorta through the functional sixth right arch. When, at the time of hatching, the chick begins to use its lungs, the small connection between the pulmonary arteries and the systemic aorta (*ductus Botalli* or *ductus arteriosus*) will degenerate and remain as a fibrous strand slightly above the heart, between the pulmonary artery and the aorta.

The Aorta: The dorsal or abdominal aorta, which took its origin from two separate aortae present in embryos of less than forty hours of incubation, becomes the largest artery in the adult body. Its fusion continues forward as far as the aortic arches, where it persists permanently in its original duality in the formation of the internal carotid arteries. Posteriorly, its union is complete, since it is continued to the tip of the tail as the short caudal artery.

Most of the larger branches of the aorta are derived from intersomitic segmented vessels, which are still present in the adult in the form of paired intercostal arteries. In this manner the *subclavian, renal, spermatic, ovarian, posterior mesenteric*, and *sciatic arteries* are formed, of which the last gives rise to the allantoic arteries. Even the two large omphalomesenteric arteries are said to be of similar origin.

The Cardinal Veins. The cardinal veins are as prominent in the venous circulation of the chick embryo as the aorta is in the arterial circulation. They are paired vessels consisting of two anterior and two posterior veins which enter the sinus venosus through the ducts of Cuvier. These ducts are situated and held in place in the anterior region of the *lateral mesocardium,* which was formed by the fusion of the two omphalomesenteric veins and the somatic mesoderm. It may be of interest here to digress for a moment and point out that the lateral mesocardium, in conjunction with the dorsal mesocardium and the ventral mesentery, constitutes the septum transversum. The latter is located immediately behind the heart and gives rise to the muscular *diaphragm* in the mammals, which divides the general body cavity into the thoracic and the abdominal cavity.

The anterior cardinal veins, or internal jugular veins, empty through the ducts of Cuvier into the sinus venosus. When the atrium becomes divided into a smaller left and a larger right auricle, that part of the sinus venosus which is connected with the two ductus Cuvieri becomes incorporated into the right auricle. Thus it happens that the two jugular veins flow into this heart chamber. Later on the two ducts of Cuvier become the anterior (superior) vena cava. The posterior cardinal veins require out special attention with regard to their modification, though these changes do not occur in early embryos of the ages that are usually studied in the college laboratory. However, they are of great interest to the comparative embryologist and anatomist, because they are instrumental in the formation of the posterior (inferior) vena cava, which is the largest vein in the adult body of birds as well as of mammals.

Just prior to the third day of incubation the posterior cardinal veins are paralleled by the subcardinals, which develop on either side of the embryo beneath the larger posterior cardinal veins, and slightly to one side of the mesonephroi. The subcardinals are really the anterior paired continuations of the single caudal vein, which bifurcates and passes along the kidneys to enter the posterior cardinals just prior to their entry into the ducts of Cuvier. At the time of their appearance the posterior cardinal veins develop a profuse vascular network, growing into the tissue of the mesonephroi, while the subcardinals send out fine branches to make connections with this capillary system. In this manner a renalportal system is formed by means of which the blood drained from the body by the posterior cardinal veins is filtered through the kidneys and is received by the subcardinals, which send it forward into the anterior end of the posterior cardinal veins, whence it reaches the heart.

It is well to remember here that the posterior cardinal veins lay the foundation of the venous system of the adult by sending into the body wall their segmented intersomitic branches, similar to those of the aorta. The large veins that parallel the arteries of the adult are merely modified metameric vessels of the original posterior cardinal venous system.

The Posterior Vena Cava. Toward the fourth day of incubation a short vessel is organized from small blood cavities (*sinusoids*) in the dorsal part of the right lobe of the liver. To this is added another short piece derived from small cavities that developed in the thickened dorsal mesentery near the anterior portion of the liver. These two elements form the anterior part of the posterior vena cava, which connects with the *ductus venosus* located within the liver.

In later development the vena cava grows from the right side of the liver backward toward the right mesonephros, where it connects with the anterior portion of the right subcardinal vein. After approximately one week of incubation the anterior portions of the posterior cardinal veins degenerate, so that their blood enters the ductus Cuvieri directly through the subcardinal veins. This venous modification persists during the existence of the mesonephros, but when the latter atrophies and the metanephroi, or permanent kidneys, make their appearance, the posterior remnants of the cardinal veins unite and fuse with the subcardinals on both sides. The posterior vena cava is finally made up from its anterior portion in the liver and ductus venosus, the right subcardinal vein, some renal veins which connected the posterior cardinals and subcardinals, the posterior united cardinals, and the caudal vein.

The Vitelline Circulation. Allusions to the fate of the vitelline circulation and also in the discussion of the alimentary canal in the present chapter. This vitelline venous circulation becomes confluent in the ductus venosus, which enters and is continuous with the sinus venosus. The ductus venosus represents the united omphalomesenteric veins. They drain the blood from the right and left sides of the area vasculosa through lateral vitelline veins which connect with the sinus terminalis. After the third day of incubation a posterior and an anterior vitelline vein are added to the left omphalomesenteric vein, so that this vessel carries more blood than its mate on the right side. The anterior vitelline vein is organized from two anterior vitelline vessels which are still separated from each other in the 48-hour chick. They have become partly fused in the 60-hour embryo and continue to do so throughout their entire length, except near the body of the embryo where the right anterior vitelline vein remains as a small spur.

The two omphalomesenteric veins undergo radical changes by partial fusion, by transverse connections dorsal and ventral to the fore-gut, and by elimination of short parts, thus utilizing in restricted regions only sections of one or the other of the two veins. In this manner those two veins become united into one as far as the yolk stalk. At the same time the single vein becomes twisted around the fore-gut. The omphalomesenteric vein is later incorporated into the hepatic-portal vein, which is formed independently in the dorsal mesentery, becoming confluent with the omphalomesenteric vein near the pancreas. In the seven-day embryo the ductus venosus becomes obstructed by the ingrowth of liver tissue and ceases to be a direct passage for the blood from the omphalomesenteric vein to the sinus venosus. From then on the blood from the yolk sac enters the liver through the posterior remnant of the omphalomesenteric vein, which is now called the hepatic-portal vein. It ramifies through the liver, and leaves by two vessels referred to as the hepatic veins.

The two omphalomesenteric arteries do not change appreciably in later stages of development. Similar to the omphalomesenteric veins, they unit into one artery which leaves the dorsal aorta between the posterior limbs, near the twenty-third somite, and spreads out over the yolk mass. It is still present in the adult as the anterior mesenteric artery. The coeliac and the posterior mesenteric artery develop as independent outgrowths from the dorsal aorta, slightly more posteriorly.

The allantoic Circulation. During the third day of incubation just prior to the appearance of the allantois itself, the lateral body walls of the embryo develop the umbilical or allantoic veins. When the allantois begins to differentiate from the cloacal region of the alimentary canal, it develops a protuse capillary system which becomes connected with the umbilical veins and also with the two umbilical arteries. The latter are branches of the iliac arteries which supply the developing posterior limb buds. The two allantoic veins extend forward and enter the ducts of Cuvier near the sinus venosus. During the fourth day of incubation the right umbilical vein begins to atrophy. The left one, which becomes correspondingly larger, develops at its distal end a branch which passes the blood through the liver directly into the ductus venosus, and from there into the heart. Its primary connection to the ductus Cuvieri is then entirely lost, so that from then on all the blood returning from the allantois to the heart has to pass through the liver. At the time of hatching, when the allantois is lost, the umbilical arteries become greatly reduced in size, supplying a small area near the cloacal region of the alimentary canal from which the allantois extended into the seroamniotic cavity in the embryo. The umbilical vein also remains functional in the adult chick, where it assumes a minor function remaining as a small vessel leading from the body wall to the left hepatic vein.

The Lymphatic System. The lymphatic system is another phase of the circulatory system which develops comparatively late in the chick embryo. It appears toward the end of the first week of incubation at two centers, one in the pelvic region and, slightly later, a much larger one in the anterior part of the body at the base of the neck. The two become connected by the organization of lymphatic spaces within the mesenchyme, and by their confluence form the lymph channels.

Lymph is a body fluid composed of plasma and white corpuscles, or lymphocytes. The entire system is rather loosely organized so that lymphocytes are able to pass through the

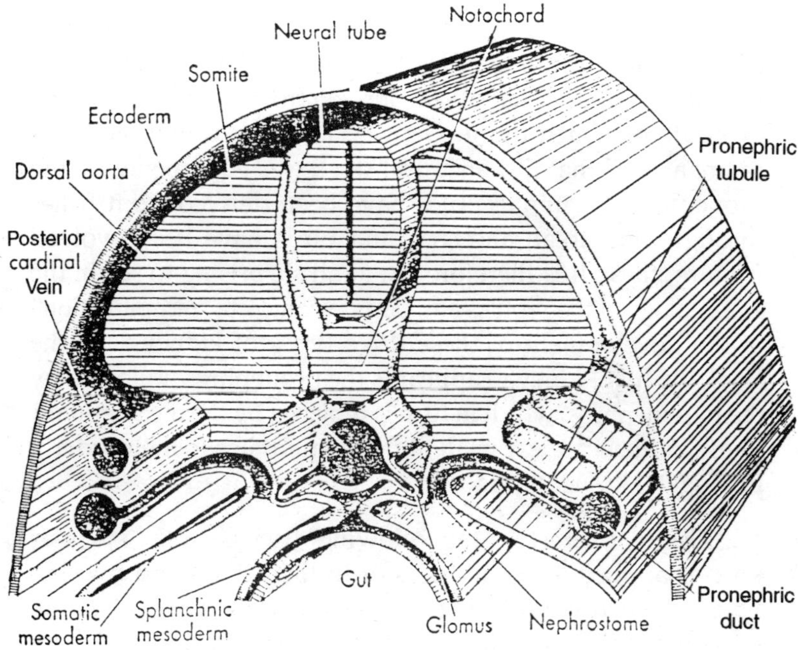

Fig. 9.37.   Stereogrammatic Diagram of the pronephros. Posterior cardinal vein is absent on the right side.

walls of the lymph vessels and return to them in other places of the body. Likewise, the plasma may pass through the walls and be returned to the system again. The walls of the lymphatics are composed of a thin, mostly simple squamous endothelium, except in localized plexuals or lymph sacs and in the heavier contractile tubes called lymph hearts.

All the lymphatic vessels finally empty into the venous system through the *thoracic ducts* as the junction of the subclavian and jugular veins.

The Excretory System

Three different types of kidneys are found in the vertebrates: namely, the pronephros, the mesonephros, and the metanephros. Of these the pronephros is present in adult cyclostomes, the mesonephros in adult fishes and amphibians, and the metanephros in adult reptiles, birds and mammals. It is highly probable that fishes have been derived from a type of primitive chordate of which the modern cyclostomes are an aberrant remnant, and that this ancestral stock was equipped with the pronephros. Again, reptiles have been derived from fishes through amphibian ancestry, and birds have been evolved from reptiles. It is therefore not at all surprising to see in the embryology of the reptiles and birds a complete "recapitulation" of the development of these three types of kidneys. In the chick the first kidney to appear is the pronephros, followed by the mesonephros, which in turn is replaced by the metanephros. These kidneys appear in all higher vertebrates in their orderly phylogenetic succession.

The Pronephros. In the following discussion the singular form is applied to the kidneys or to their ducts and tubules. This is done to avoid ambiguity and clumsiness of expression. It should always be borne in mind that the kidney system consists of paired organs.

The pronephros begins to appear in chick embryo of thirty-three to thirty-four hours of incubation. It develops from the intermediate mesoderm, which is really the kidney-building or nephrogenous tissue. It is therefore equivalent to the uppermost portion of the lateral mesoderm, described in the frog embryo. It will be recalled that in the amphibian embryo this part of the lateral mesoderm becomes isolated to form the pronephros.

In the chick the pronephros never develops hollow pronephric tubules. They appear as solid strands of cells in the intermediate mesoderm, one end of which remains attached to the lateral mesoderm, the other growing outward toward the superficial ectoderm. When they have reached a point halfway between the latter and their origin from the lateral mesoderm, they turn backward at an angle of ninety degrees, so that all of these strands become joined and confluent with each other from the fifth to the sixteenth somite. Beyond the latter the nephrotome forms a hollow duct (*pronephric duct or segmental duct*) which continues as far as the thirty-third somite, where it enters the alimentary canal.

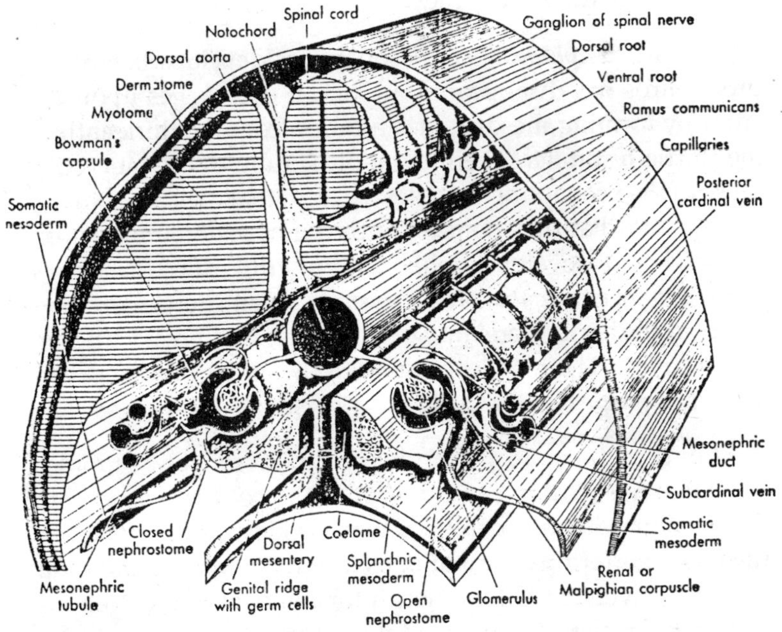

Fig. 9.38. Stereogrammatic diagram of the mesonephros.

It is obvious that in the chick embryo the pronephros cannot be functional, because its "tubules" are not hollow. Neither do they have nephrostomes, though attempts to form such openings are indicated by a notched appearance of the strands near their origin bordering the coelome. It appears, therefore, that the pronephros represents an earlier phylogenetic condition of the bird embryo. That the pronephric duct is hollow beyond the sixteenth somite is probably due to the fact that the succeeding kidney (mesonephros) has hollow tubules and uses this duct to transmit its excretory products, and instead of forming another duct, uses the pronephric duct for this purpose. It is for this reason that the pronephric duct becomes automatically the *mesonephric* or *Wolffian duct* with the arrival of the next kidney.

The Mesonephros. The mesonephros develops before the pronephros has disappeared; in fact, it encroaches upon the latter by forming mesonephric tubules from the thirteenth to the thirtieth somite. The mesonephric tubules are much like those of the pronephric "tubules", except that they are actually hollow and that they connect with the mesonephric duct on one end and develop cuplike outgrowths called *Bowman's capsules,* or *renal capsules,* on the other. A small artery, branching off from the aorta, leads into each capsule and breaks up into a *glomerulus.* The renal capsule with the glomerulus is called a *renal corpuscle* or a *Malpighian corpuscle.*

Some of the more anteriorly located mesonephric tubules still have nephrostomes leading into the coelome, similar to the pronephric tubules described in the from embryo. However, the more posteriorly located ones lose the nephrostomes and rely exclusively on the glomeruli for removal of nitrogenous wastes and water. At its highest development the mesonephros is confined chiefly to the region limited by the twentieth and thirtieth somites. In this area the simple metameric organization of mesonephric tubules becomes obliterated by the formation of several more tubules in each segment, and these again bud off primary, secondary, and tertiary tubules, which evaginate into the Wolffian duct to act as *collecting tubules* for the urine. Thus the mesonephros, or *Wolffian body,* has increased considerably in size and bulk, and has assumed a definite, elongate shape. In a little over one week of development the mesonephros has been completed, and it remains functional almost to the end of incubation.

Nitrogenous wastes are removed from the blood by extracting them in the glomeruli. Apparently not all wastes are removed in this manner. After leaving the glomeruli, the blood is again broken up into capillaries surrounding the so-called *secretory portion* of the mesonephric tubule, where more waste and water are removed. From here the excretions are carried into the collecting tubules and sent into the Wolffian (mesonephric) duct. The blood is thus filtered through the kidney and enters the subcardial vein, whence it is transported to the heart.

The blood supply from the aorta to the glomeruli is gradually decreased with the development of the renal-portal system. In its later stages the blood to the mesonephros is chiefly supplied by the posterior cardinal vein, filtering through the mesonephros and entering the subcardinal vein.

The Metanephros. This kidney also develops while the mesonephros is still functional, but it is fully developed at the time of hatching, so that the young chick is equipped with a functional metanephros. This kidney develops from a very limited region, localized to the nephrotomes between the thirty-first and thirty-third somites. The tubules which constitute the metanephros are somewhat like those of the preceding kidney. The Wolffian duct is no longer utilized for the transportation of urine. Instead, a new tube, the ureter, grows out at the base of the Wolffian duct near the cloaca, and as it grows forward it sends out pocketlike evaginations (*collecting tubules*) which unite with the *excretory tubules* formed by the nephrotomes. Other elements associated with the ureters in its forward growth form the renal corpuscles and secretory tubules. The ureters of both sides of the embryo lead independently into the cloaca. The mesonephric ducts, or Wolffian ducts, which are also separated from each other, lead into the cloaca slightly anterior to the ureters.

The Reproductive System

In their earliest stages of development the gonads are

paired organs in the chick embryo. This condition prevails throughout the life of the male. In the female, however, the right ovary degenerates completely, leaving only the left as the functional sex gland. As in the discussion of the excretory system, the singular form will be used in the description of the development of the gonads.

The Genital Ridge and the Germ Cells. The first indication of the presence of sexual development can be observed in the late 4-day chick. There is a slight swelling of the peritoneal epithelium, located between the mesonephros and the dorsal mesentery, for about seven somites immediately behind the omphalomesenteric arteries. This swelling, known as the *genital ridge*, is covered by epithelial cells (*germinal epithelium*), and contains mesenchyme cells, and the primordial germ cells which are conspicuous by their larger size. The latter are found in the germ wall of the blastodisc at the late primitive streak stage. They remain there until the blood circulation has been established, which they enter the blood stream to be transported all over the body of the embryo. Many of them are lost, but there are some that will finally arrive near the dorsal mesentery where they have the blood stream by amoeboid movement to take their place in the genital ridge (Figure 146).

The Testis. With the growth and protrusion of the mesonephros into the coelome, the genital ridge becomes covered by peritoneal epithelium and located on that side of the mesonephros which faces the mesentery. In fact the genital ridge becomes now incorporated into the mesonephric protrusion and the two are jointly referred to as the *urinogenital ridge*. This intimate union with the kidney is further increased by poliferation of the epithelial cells closest to the mesonephros. In this region the primitive testis forms strands (*rete cords*) which grow toward the capsules of the renal corpuscles and eventually connect with the mesonephric tubules.

At the same time the mesenchyme cells have differentiated into connective tissue (stroma) which permeates the developing

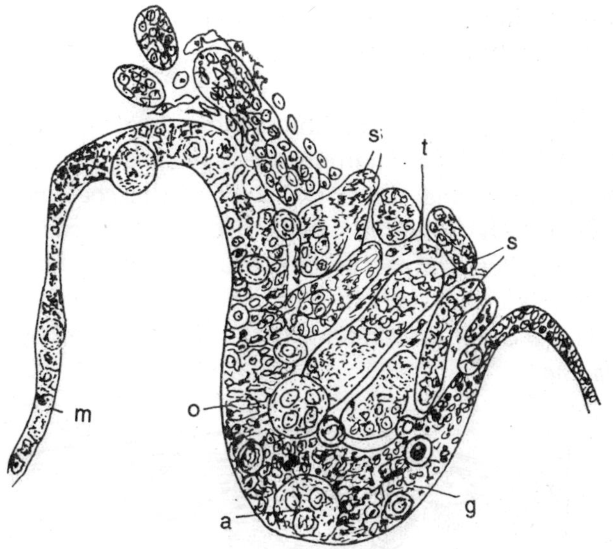

Fig. 9.39. Section through the gonad of a chick, the middle of the fifth day, showing the sexual cords growing inward from the germinal epithelium, g, germinal epithelium; m, epithelium of the necessary (peritoneum); o, primordia germ cells; s, sexual cords; t, connective tissue stroma.

testis. The stroma becomes invaded by another growth of cellular strands *(sexual cords)* which grow inward from the periphery of the young testis and are composed of germinal epithelium containing numerous germ cells. In later development the sexual cords differentiate into the *seminiferous tubules* and establish connection with the rete cords, which will also become tubular and form the *rete testis.* Eventually the mesonephros will lose its excretory function in the region of the gonad and degenerate except in its mesonephric tubules, which have made connection with the rete cords. These tubules retain their connection with the testis and are known as the *vasa efferentia.* Together with the mesonephric duct (Wolffian duct) in this region, which will be converted into the *vas deferens,* they constitute the *epididymis.* Some mesonephric tubules remain without connection and do not act as vasa efferentia. These constitute the *paradidymis.*

In its final transformation to the compact glandular testis, the genital ridge shortens and pushes away from the mesonephros and the dorsal wall of the coelome. In this manner the testis become suspended by the *mesorchium,* which is a mesentery formed by the splanchnic mesoderm. Within this mesodermal sheet the entire gonad becomes surrounded by a layer of compact connective tissue cells forming the *tunica albuginea.*

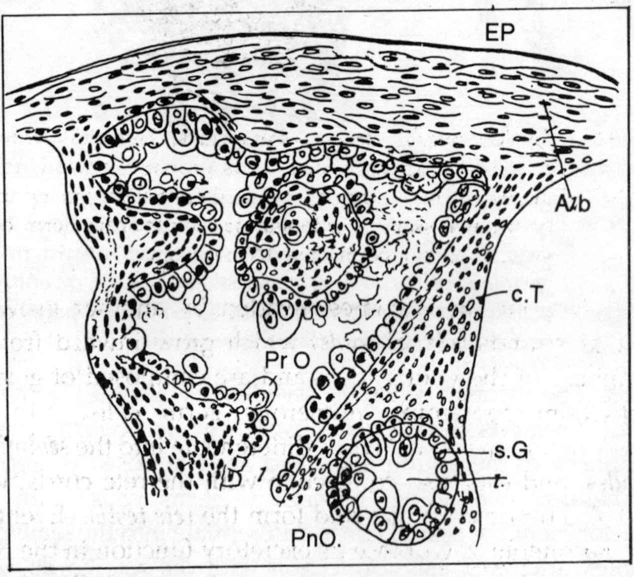

Fig. 9.40. Cross section through the periphery of the testis of a chick that has just been hatched. The sexual cords are differentiating into seminiferous tubules in which the spermatogonia and supporting or Sertoili cells from the lining. The tunica albuginea is bordered by the germinal epithelium on the outside and is continuous with the sexual cords on the inside. Alb., albuginea, c.T., connective tissue of the stroma, or septulae testis. *Ep.,* remains of the germinal epithelium now forming the outer most of serious covering of the testis. l., lumen of the sexual cords. *pr.o.,* spermatogonia s.C., sexual cord, lined by supporting cells and spermatogonia.

The Ovary. The ovary develops in the same manner as the testis in its earlier stages, except that the one on the left side becomes increasingly larger and the right one eventually degenerates entirely, thus leaving only the left one as the functional ovary in the adult. The developing ovary also forms rete cords and sexual cords containing germ cells. They even establish connection with each other, but they disappear almost completely shortly after they have been formed. Moreover, these cellular strands are so rudimentary that there is no connection with the mesonephros. The mature ovary has therefore no ducts or tubules that lead away from it.

After one week or a few days more, the young ovary enters a second period of cellular proliferation in its peripheral region, producing again strandlike projections of germinal epithelium and germ cells into the stroma. Since these strands begin in the cortical layer they are called *cortical cords.* Small nests of cells separate from these cords to form the ovarian follicles, each containing one oocyte with an enveloping follicular layer of smaller cells.

The oviduct, or Mullerian duct, appears as a band of cells on the outer side of the Wolffian duct. It remains open at its anterior end *(ostium tubae* or *infundibulum)* and is connected with the cloaca at its posterior termination. The ostium tubae represents probably the nephrostome of a mesomephric tubule. The Mullerian duct appears in both sexes. In the males it disappears shortly after its formation. Furthermore, the right Mullerian duct disappears also, probably because the right ovary atrophies.

The ovary is suspended by the mesovarium, a mesentery which is homologous to the mesorchium in the male. The degenerating mesonephros leaves two remnants to its organization near the ovarian region: namely, the *epoopheron,* homologous to the epididymis, and the *paroopheron,* which is homologous to the paradidymis in the male.

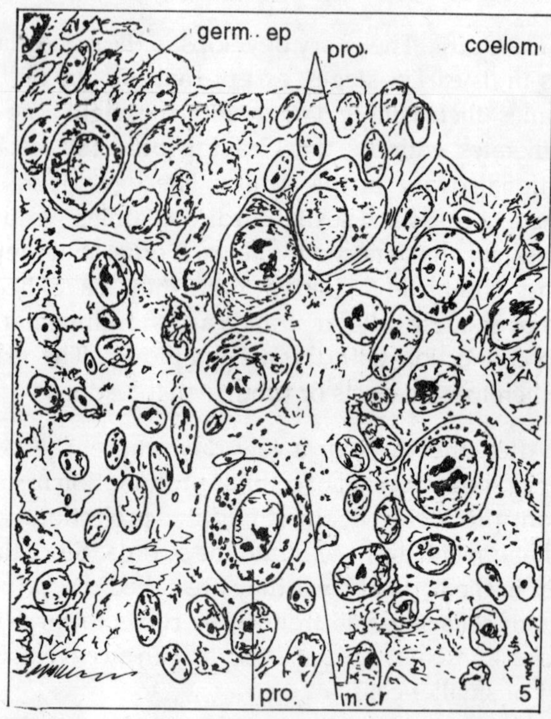

Fig. 9.41. Portion of a transverse section through an ovary of a 6-$\frac{1}{2}$ day chick embryo (after Swift). germ.ep., germinal epithelium, m.c., sexual cord. pr.o., primordial germ cells.

10

Extra Embryonic Membranes

Evolution and Homology of Fetal Membranes

The fetal or embryonic membranes of the avian and reptilian embryos are not only interesting from the stand-point of comparative embryology, but they also illustrate in their changes of function and special adaptations to the environment some of the fundamental laws of organic evolution. The fetal membranes as found in the chick embryo are also present in the embryos of mammals, though in the latter they have undergone radical changes, especially in the embryos of placental mammals. It is therefore very desirable that the development of the embryonic membranes of the chick be thoroughly understood, so that their homologues in the embryos of mammals may be identified.

The chick embryo has four different embryonic or fetal membranes: namely, the *yolk sac,* the *allantois,* the *amnion,* and the *chorion* or *serosa.* Of these four, the first two are present in rudimentary form in the frog and other amphibian embryos; however, the last two, namely, the amnion and the serosa, are neogenic structures which were developed first in the reptilian embryo. The birds, having been derived from reptiles, retained these newly-acquired embryonic features in their development. In fact, as indicated above, they have also been handed down to the mammals in a modified condition. At hatching the greater part of the fetal membranes are discarded. The amnion and the serosa are completely eliminated, and the allantois is also discarded to its greater extent. However, the yolk sac is incorporated into the small intestine.

The Constriction of the Embryo from the Yolk Sac

The closely apposed ectoderm and somatic mesoderm (somatopleure), as well as the combined entoderm and splanchnic mesoderm (splanchnopleure), extend from the embryo peripherally into the extraembryonic area. The origin of these folds, known as the *limiting body folds*, begins with the formation of the crescent-shaped headfold in embryos of less than twenty hours of incubation. It extends laterally from the subcephalic pocket, turning backward on both sides near the heart of the embryo toward the posterior region in the form of the lateral body folds. Later, when the tail bud is established, the lateral body folds become continuous with the fold around and posterior to it, thus limiting the embryo from the underlying yolk mass by a marginal groove extending from the head fold through the lateral body folds to the tail fold, undercutting and separating the embryo from the yolk mass as it becomes older.

The Yolk Sac and Its Relation to the Gut

In chicks of sixteen hours of incubation, i.e., those of the primitive streak stage and slightly older, the gut, or enteron, is represented as a flat, circular cavity beneath the primitive streak. Its roof is formed by splanchnopleure and extends peripherally into the extraembryonic area, while its floor is composed of the non-cellular yolk. In the 24-hour embryo the gut has still the same relationship, with the exception of the small enteric pocket which has developed into the developing, forward-jutting head. This pocket, known as the fore-gut, is the first differentiation of the alimentary canal that is completely surrounded by mesoderm or mesenchyme.

In the 48-hour embryo a similar process takes place in the formation of the tail bud, so that a saclike cellular hind-gut is produced. The undifferentiated part of the enteron located between the fore-gut and hind-gut is called the mid-gut. As the embryo grows older and the amniotic head fold, tail fold, and lateral folds, undercut the embryo, thus increasing the length of the fore-gut and hind-gut, the mid-gut becomes

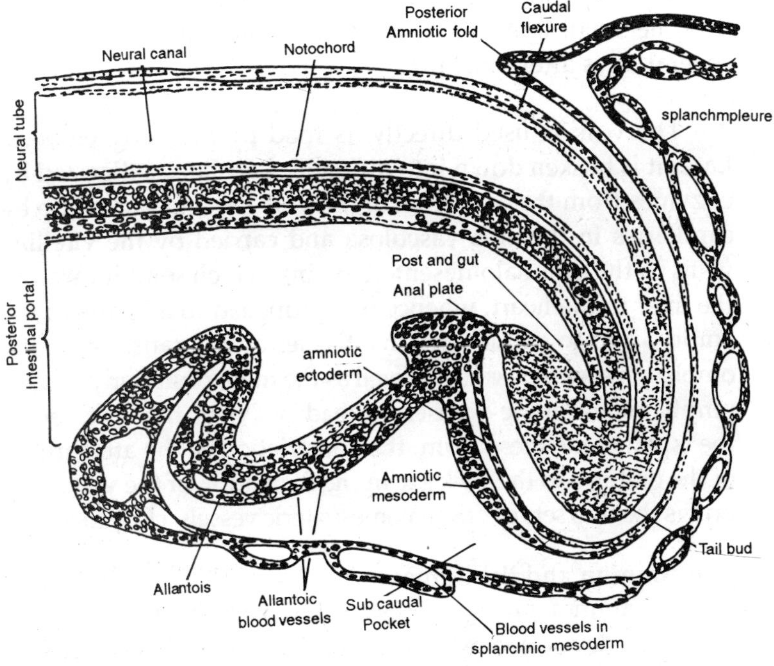

Fig. 10.1. The 72 hour chick embryo. Sagittal section through posterior region to show the beginning of the allantois.

correspondingly smaller. Eventually the latter is restricted to a small portion of the small intestine posterior to the pancreas, where it leads into the reduced yolk sac by means of the yolk stalk. The embryo consumes the yolk as it grows, and the yolk sac therefore becomes correspondingly smaller, so that two or three days before hatching it has become so diminutive that it is drawn within the tissues of the embryo and shows itself as a protuberance from the small intestine. The splanchnopleure, forming the roof of the enteron in the earliest chick embryos, and forming the complete fore-gut and hind-gut, is the same cellular layer that covers the yolk sac. As the latter diminishes in size, the splanchnopleure follows the yolk sac into the embryo, where it is utilized in the lining of the mid-gut. It appears, therefore, that the yolk sac is really part of the gut, as it was in the frog embryo. It is finally incorporated into the small intestine of the chick, where it is still present several days after hatching.

The yolk is used directly as food by the early embryo. Later it is broken down into diffusible substances digested by enzymes from the entoderm of the yolk sac and picked up by capillaries in the area vasculosa and carried by the vitelline veins to the omphalomesenteric veins, which send it through the liver to the heart, whence it is propelled to all parts of the embryo and to extraembryonic tissues and organs. Since the omphalomesenteric veins, as well as the more posteriorly located omphalomesenteric arteries, spread out over the yolk sac in the splanchnic mesoderm, the constriction of the attachment of the embryo to the yolk sac in the formation of the yolk stalk brings the two sets of omphalomesenteric vessels close together.

The Amnion and the Serosa

The amnion and the serosa are two fetal membranes that have been developed in the evolution of the reptiles and passed on to the birds and mammals. The amnion is a membrane enveloping the embryo, separating it from its immediate contact with the environment. It acts, therefore, much like a bag in which the embryo is contained. Between the embryo and the enveloping amnion is a space (*amniotic cavity*) containing a saline liquid (*amniotic fluid*). Since the reptilian and avian eggs are deposited on dry land, it seems obvious that the amnion and the amniotic fluid have been developed to keep the embryo moist and protect it from desiccation. This original function of the amnion has changed in the mammals, where the amnion is chiefly a protective organ, equalizing and distributing pressure brought against the embryo by phisical forces from the environment, and, furthermore, where it serves as a prevention of adhesion by favouring movement of the embryo.

The phylogenetic origin of the amnion is still problematical, though the following theory is sometimes advanced for its origin. The reptilian eggs are more abundantly provided with yolk than are those of the amphibians. It is assumed, therefore, that due to its weight the reptilian embryo sank into the soft yolk and that the blastodisc around the embryo was thus elevated over it in the form of a crater. In the course of evolution the more of less circular margin of the raised blastodisc grew over the embryo, hiding it from view by covering it with a membrane (amnion).

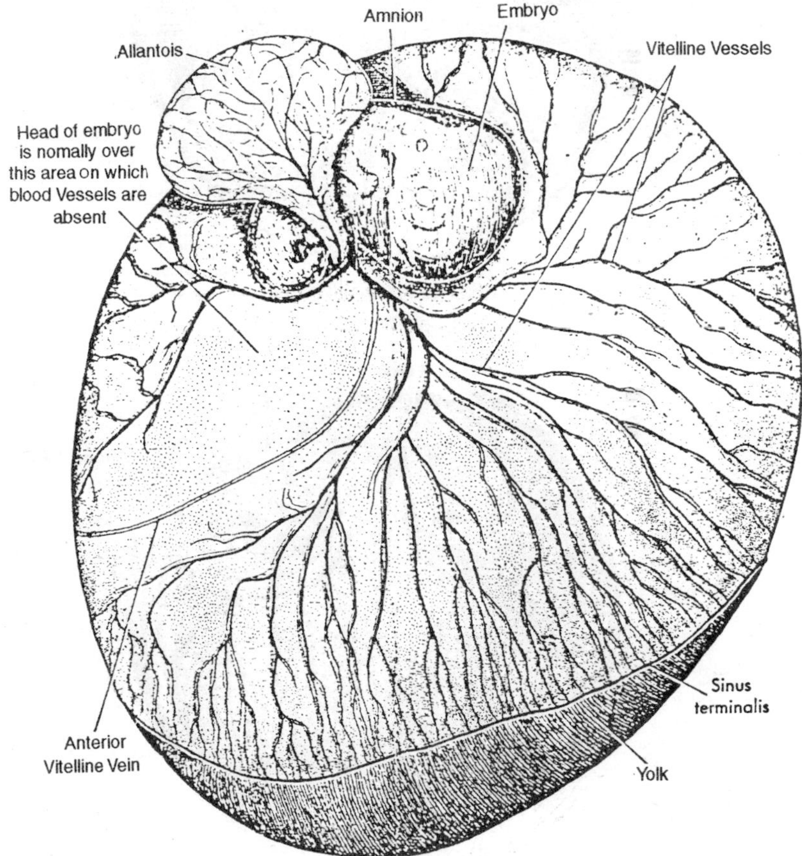

Fig. 10.2. Chick embryo of six days of incubation removed from shell, albumen and serosa eliminated. The embryo has been pushed aside and the allantois has been elevated.

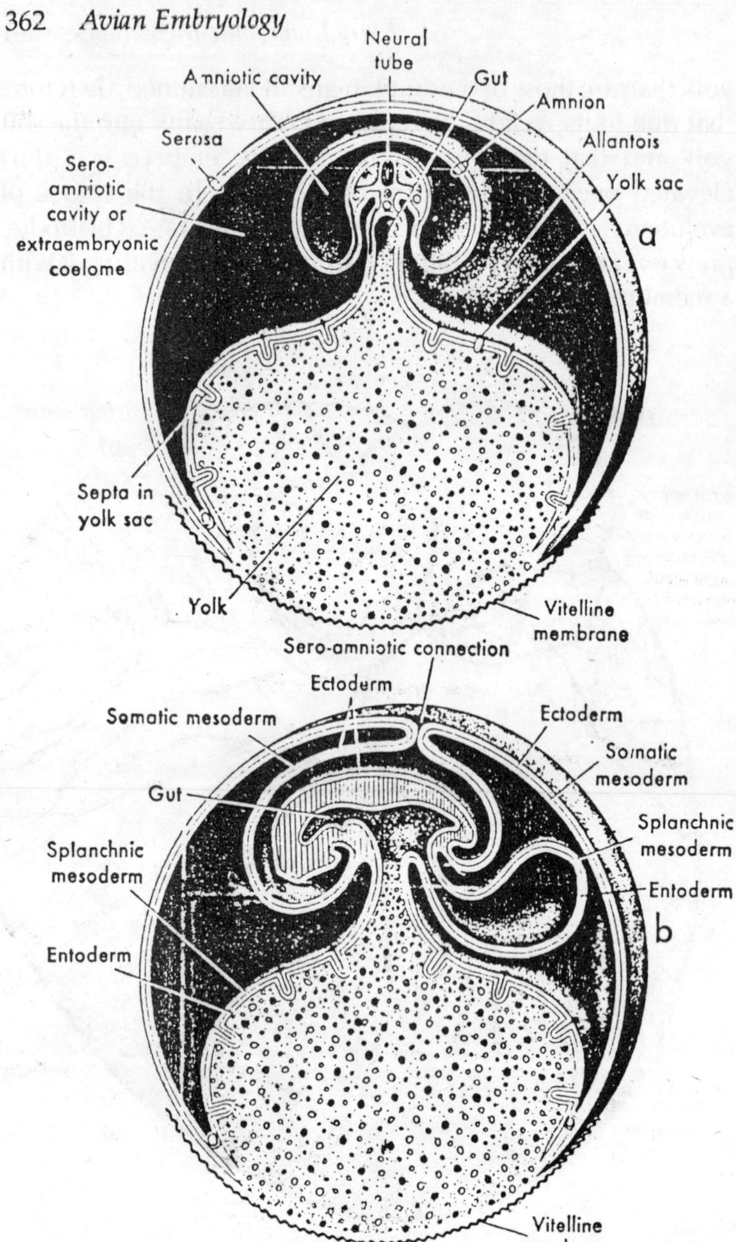

Fig. 10.4. The fetal or embryonic membranes of the chick embryo at an early stage of incubation.

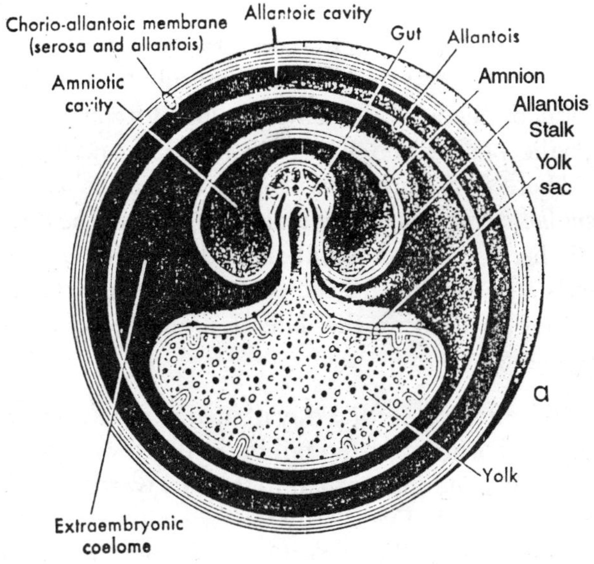

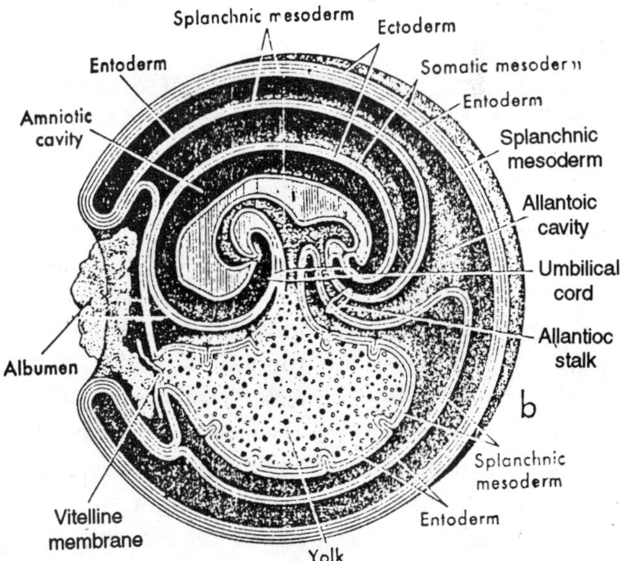

Fig. 10.4. The fetal or embryonic membranes of the chick embryo at a late stage of incubation.

That this theory is not so fantastic as it appears at first sight is indicated by the actual development of the amnion in the chick embryo. At thirty hours of incubation the large and well-defined head of the embryo juts forward over the subcephalic pocket and has sunk into the underlying yolk mass. There is a simultaneous elevation of a blastodermic fold (*amniotic head fold*) immediately in front of the head. The amniotic head fold is crescent-shaped in its initial stages, its free ends pointing toward the posterior region of the embryo where they will form the lateral amniotic folds in later development.

Shortly after the second day of incubation the *amniotic tail fold* appears posterior to the tail bud. It grows forward over the embryo in a manner similar to the backward growth of the amniotic head fold. Eventually the two will meet and fuse in the region just slightly posterior to the omphalomesenteric arteries during the fourth day of incubation. The location of the final union of the amniotic folds is marked by a scar called the *seroamniotic connection*.

The amniotic folds are formed from somatopleure, which is composed of ectoderm and somatic mesoderm. Since a fold is composed of an inner and an outer layer, two separate sheets must be produced by the union of the amniotic folds, an inner and an outer one, the former being the amnion and the latter the serosa. We may compare the amniotic folds of the embryo with an oval crater of a volcano that has a well-developed rim on one side and a very small one on the other. The former represents the amniotic head fold and the latter the tail fold, both being connected by the remainder of the rim of the crater, representing the *lateral amniotic folds*. If one assumes further that the wall of the crater is hollow and that it is therefore composed of an inner and an outer surface, one can see immediately that if the rim of the crater should grow and fuse centrally, an outer and an inner sheet would be produced, both covering the central depression of the volcano.

The amnion, representing only the innermost half of the amniotic folds, is a transparent membrane composed of somatopleure, with somatic mesoderm on the outside and ectoderm on the inside. The latter is continuous with the skin ectoderm of the embryo near the yolk stalk and the allantoic stalk (to be discussed later). The serosa, representing the outer half of the amniotic folds, is also composed of somatopleure, having the somatic mesoderm on the inside and ectoderm on the outside. Between the amnion and the serosa is the seroamniotic cavity (*extraembryonic coelome*), which is in direct communication with the coelome within the embryo (*intraembryonic coelome*). The serosa grows around the yolk sac, enveloping it completely at the end of the second week. It expands and adheres to the entire calcareous shell within the shell membrane, so that the embryo with all its fetal membranes is encompassed by it. Its intimate union with the allantois will be discussed in the next paragraph.

The Allantois

To the comparative embryologist the allantois is of special interest because it develops in conjunction with the serosa into the *chorion* of the mammalian embryo. Indeed, its change in structure and function in the development of mammals can be traced step by step from its simple beginnings in the reptilian and the avian embryos. Moreover, unlike the amnion and the serosa, the allantois is not entirely discarded in the mammal at the time of birth. Its proximal or intraembryonic portion is retained to function in the adult as the urinary bladder.

The allantois is present in the 72-hour chick embryo as a diverticulum in the floor of the posterior portion of the hind-gut. It begins as a saclike diverticulum posterior to the yolk stalk, where it grows out rapidly and invades the extraembryonic coelome as early as the fourth day of incubation. Its cavity is filled with a saline liquid, imparting to it a baglike shape in embryos of five to six days of incubation. Its distal portion, called the allantoic vesicle, rapidly replaces the extraembryonic coelome *(Seroamniotic space)*, wedging itself between the amnion and the serosa. The proximal or intraembryonic portion of the allantois consists of a narrow stalk called the allantoic stalk, which, in mammals, is utilized in the formation of the urinary bladder.

The allantois is instrumental in the storage of excretory substances which are given off in development. Being an outgrowth of the gut, the allantois must necessarily be composed of splanchnopleure, the same layer which forms the yolk sac. In fact, the splanchnic mesoderm of the allantois also parallels that of the yolk sac in the formation of a complex capillary blood system. It connects with the intraembryonic circulation by means of the paired allantoic arteries and veins. Upon attaining its full size the allantois replaces the entire extraembryonic coelome and extends itself around the entire embryo, apposing itself closely to the somatic mesodermal layer of the serosa. The latter fuses with the splanchnic mesoderm of the allantois, forming the chorio-allantoic blood vessels in this area, and extending the capillary circulation of the allantois in this manner over the entire inner portion of the calcareous, porous shell. The exchange of respiratory gases is easily accomplished through the egg shell, and though the allantois is partly excretory in function and also aids in the absorption of a large portion of the albumen, its main function must be considered respiratory. In later stages of development the allantoic circulation also absorbs calcium from the egg shell, which is used in the embryo for the formation of bone. At the same time this renders the egg shell thin and fragile, thus facilitating its rupture by the chick at the time of hatching.

11

Nutrition and Utilization of Albumen and Yolk

The fertilized egg must contain all the substances necessary for development if a viable chick is to result. This means that the hen herself must be fed on an adequate diet if the composition of the egg is to be satisfactory. It is not intended to consider here the effects of specific deficiencies in any great detail: for discussions of these topics the reader is referred to *Beer* (1969) and *Romanoff* (1972).

Nutrition of the Developing Bird

The difficulties of defining the nutritional requirements of the developing bird are immense. Various techniques have been employed — whole embryo cultures *in vitro, perfusion techniques,* etc.: notwithstanding the considerable advances made, our knowledge is still meagre, though some progress has been made in defining those substances that are essential for normal development.

Essential amino acids

Klein et al. (1962), Grau *et al.* (1965, *Austic et al.* (1966), *Grau* (1968a, b) and *Austic & Grau* (1971) have established that leucine, lysine, methionine and phenylalanine are essential while reginine, tyrosine, valine and tryptophan may be. Neither aspartic acid nor proline are essential amino acids.

It is possible that certain complete proteins are also needed for development.

Carbohydrate

It has been established that glucose is essential for development. It is almost certainly used as an energy source, though its reaction with egg white proteins may also be important in making these available to the embryo for subsequent utilization.

Essential fatty acids

Little definite information seems to be available though the data of Menge *et al.* (1964) indicate that linoleic acid is essential for normal development.

Minerals, trace elements and vitamins

The egg contents are relatively deficient in calcium; that which is present is virtually exclusive to the yolk and is in the bound form. It is possible that the egg white proteins move across the vitelline membrane for this purpose. It has been shown that calcium salts are deleterious and even teratogenic to early development. The calcium required for skeletal calcification is provided by the shell during the later stages of incubation.

The functions of the various trace elements have not been fully elucidated but the following involvements seem likely: manganese acts as an enzyme co-factor and is concerned in skeletal and feather development; zinc is a co-factor, is needed for the maturation of the hepatocytes and is concerned in skeletal and feather development copper is essential in the early development of the chick and is required for the synthesis of connective tissues, particularly tissues of the blood vessels; iodine is needed for normal thyroid development and function iron is essential for haem synthesis and may be required for certain feather pigments.

Vitamin A is very important both in the early stages of development during the establishment of the circulatory system in the area vasculosa and later during bone mineralization. Vitamin D_3 is essential for normal calcium metabolism. Vitamin

E is required for the development of a proper circulation while vitamin K is required for the blood coagulation mechanism.

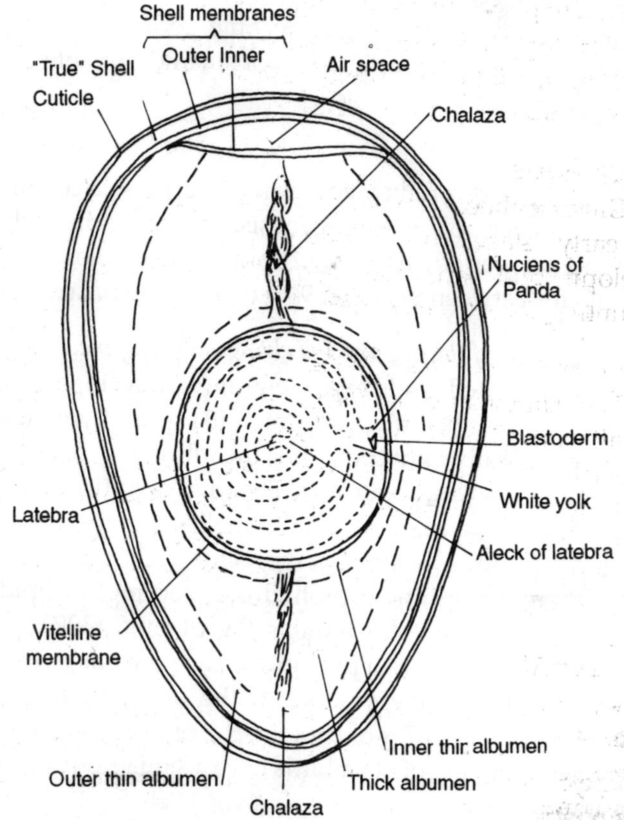

Fig. 11.1. Generalized structure of the fertilized avian egg.

Of the water-soluble vitamins only the B complex is necessary for development. The bird, with only a few exceptions (e.g., the bulbul), is able to synthesize ascorbic acid. Thiamin (vitamin B_1)is an important co-factor in carbohydrate metabolism, including the pentose phosphate pathway. Riboflavin (vitamin B_2) is required by the embryo for the myelination of the nerves and for normal development. Nicotinic acid is required for normal muscle development.

Pyridoxine (vitamin B_6) acts as an essential co-factor and is required for proper development. Both biotin and folic acid are necessary for skeletal development and the latter is also involved in purine synthesis. Embryos deficient in pantothenic acid develop haemorrhages and become oedemic. Vitamin B_{12} is required for normal lipid metabolism and possibly in the uptake of iodine by the thyroid

Energy sources

Glucose almost certainly acts as the energy source during the early stages of embryogenesis. Over the whole developmental period, however, lipid is the major source, accounting for between 84% and 98% of the material oxidized.

Gross composition of the egg and the changes during incubation

The unincubated hen's egg, weighing an average of 60 g overall, comprises shell (6 g), albumen (34 g) and yolk (20g). The shell is virtually anhydrous but the albumen contains 30 g of water and the yolk about 10 g. The total quantity of solids in the egg is 20 g, 6 g are found in the shell, 4 g in the albumen and 10 g in the yolk. Over 90% of the solids in the albumen are proteins, there being little carbohydrate (300 mg) or lipid (5 mg). In contrast, the yolk contains 6 g of lipid (30% of the solids) but similar amounts of protein and carbohydrate. Overall, therefore, the developing embryo of the domestic fowl has access to 40 g of water, 7 g of protein, 6 g of lipid and 425 mg carbohydrate. The eggs of other birds have a similar proportional composition.

Fig. 11.2. and 11.3 Show the gross changes in composition of albumen and yolk during development. Not unexpectedly, considering the size of the embryo, little changes can be detected during the early part of incubation. After 5 days changes can be detected in the yolk but it is not until the thirteenth day that changes occur in the albumen. These latter changes are related almost entirely to the movement of the albumen through the seroamniotic connection and through the yolk sac umbilicus.

By comparing the net changes in the non-embryonic and embryonic portions we can see that most of the observed changes can be explained in terms of the redistribution of substances from extra-embryonic to embryonic sites. However, there are net falls in the lipid content, about 2.8 g, in protein, 0.5 g and in carbohydrate, about 175 mg.

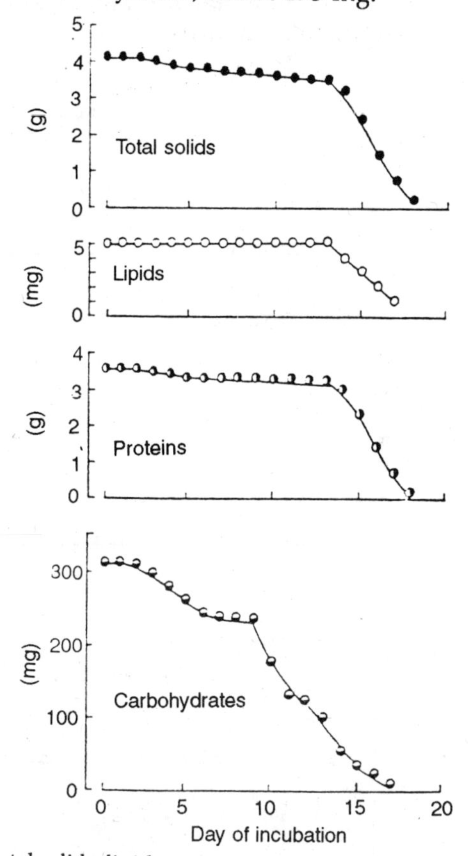

Fig. 11.2. Total solids, lipids, proteins and carbohydrates of the albumen during development of the fowl.

The Albumen and its Utilization
Composition

We have already seen that the major component of the albumen is water accounting for about 90% of its total weight.

Proteins form over 90% of the solids found in the albumen. By far the most abundant protein is ovalbumin (54%) followed by similar quantities of ovotrans ferrin (conalbumin) and ovomucoid. Some details of these and the minor protein components are detailed in Table 11.1: for detailed physical and chemical data of the albumen proteins, including their amino acids, of a wide range of birds.

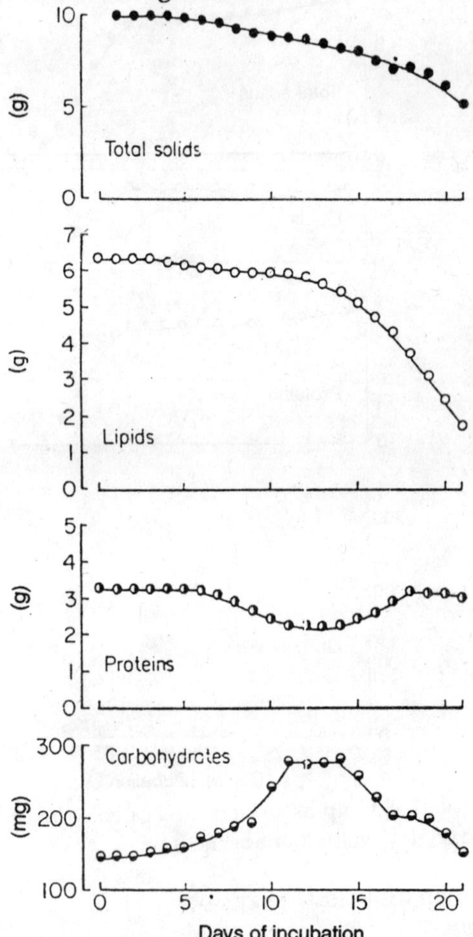

Fig. 11.3.   Total solids, lipids, proteins and carbohydrates of the yolk during development of the fowl.

Of the non-protein solids, approximately 500 mg in the domestic fowl, carbohydrate accounts for 300 mg while the remainder is inorganic ions, sulphur, potassium, sodium and chloride being the most abundant.

TABLE 11.1

The Percentage Composition of the Proteins of Egg Albumen

Species	Ovo-transf-errin	Ovo-mucoid	Ovo-inhib-itor	Lysoz-yme	Ovomacro-globul-in	Flavo-prote-in	Ovom-ucin	Avidin[1]
Domestic fowl	12	11	1.5	3.4	0.5	0.7	2.9	11.5
Turkey	11	15	0.5	3.1	0.0	0.4	—	16.2
Japanese quail	12	—	—	—	—	—	--	—
Duck	2	—	—	1.2	—	0.3	—	7.1
Goose	4	—	—	0.6	—	0.3	—	8.0
Ostrich	3	10	0.6	0.5	0.5	0.3	2.8	—
Adelie penguim	4.5	10	—	0.05	1.0	0.3	—	5.0

[1]Units g^{-1}.

Reference to Figure 11.1 will confirm that the albumen is not homogenous: rather it is composed of two types — thin and thick albumen. Their protein content is similar but the thick albumen contains about four times more ovomucin.

Absorption

While water is actively moved from the beginning of incubation there is little gross uptake of the solids until the thirteenth day. However, in the period up to the thirteenth day some limited movement of material does occur since egg white proteins have been found in the yolk, blood and amniotic fluid as early as the fifth day. There is evidence, moreover, of a preferential uptake, by an unknown mechanism, of certain proteins.

The continuous removal of water results in a reduction in the volume of the albumen, and as development proceeds the albumen becomes concentrated towards the narrow pole of the egg and is gradually surrounded by the chorio-allantois which finally forms a discrete albumen sac. Some of the histochemical changes occurring in this sac during the second week of incubation have been described by *Medda* and *Bose* (1967). The major movement of albumen begins on about the thirteenth day. It is at this time that the sero-amniotic connection becomes perforated allowing the albumen to flow into the amniotic cavity.

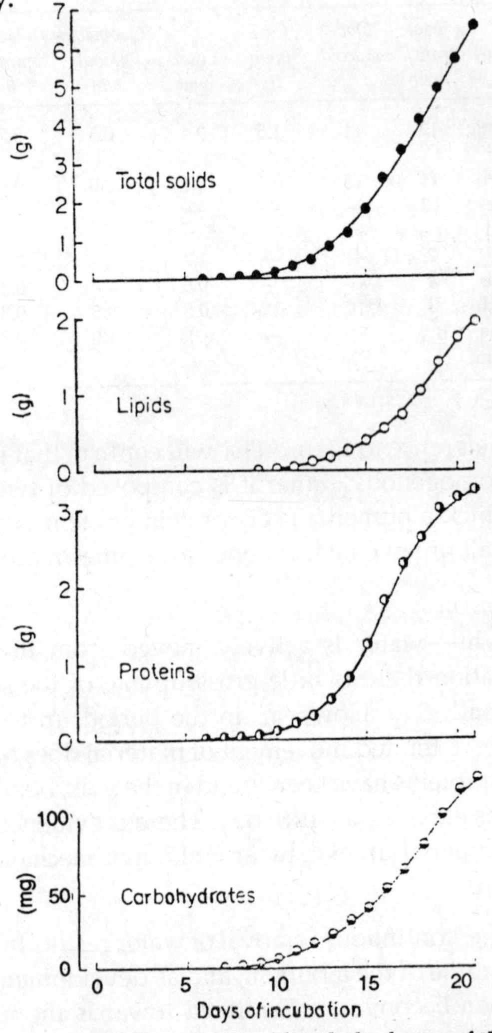

Fig. 11.4. Total solids, lipids, proteins and carbohydrates of the fowl embryo.

Transfer of the albumen is rapid - up to 70% of the protein being moved within a day. However, it is not complete and a little thick albumen may be left.

At the same time as part of the albumen is flowing into the amniotic cavity there is a simultaneous movement of albumen into the yolk sac probably through the yolk sac umbilicus and as a result there is a net increase in the amount of protein in the yolk. The albumen proteins present in the amniotic fluid finally become available to the embryo when this fluid is actively imbibed. This usually begins on about the twelfth day but becomes particularly active after the fourteenth day.

Functions

We now know much about the chemistry and physics of the proteins of albumen but very little about the functions of albumen. Suggested functions include:

(1) Provision of a suitable aqueous environment for development. (Anti-bacterial properties may be included in this function).

(2) A reservoir of water and protein.

(3) The supply of particular proteins — e.g., ovotransferrin for the blood (ovotransferrin and plasma transferrin are probably the same protein through the carbohydrate moieties differ — and calcium-binding proteins.

Feeney and *Allison* (1969) have pointed out that the high pH of the albumen (about pH 7.6 immediately after laying rising to pH 8.5 after a day or two) is itself a powerfully antibacterial property. At the same time various of the proteins have properties which will reduce the likelihood of bacterial growth. Thus lysozyme will lyse the cell walls of certain bacteria, ovotransferrin (conalbumin) chelates iron, zinc and copper thereby removing these elements from bacteria which are dependent on them, and ovomucoid inhibits trypsin.

The function of the albumen as a water reservoir is probably of paramount importance.

The Yolk and Its Utilization
Composition

Yolk is of two types—white and yellow. White yolk surrounds the blastoderm and extends down through the yellow yolk to the latebra.

It represents no more than 2% of the total yolk and contains less solid overall but a greater proportion of proteins to lipid than does the yellow yolk. Yellow yolk may be considered as an oil and water emulsion the continuous phase being an aqueous protein with fluid-filled yolk spheres, which in turn contain electron-dense sub-droplets.

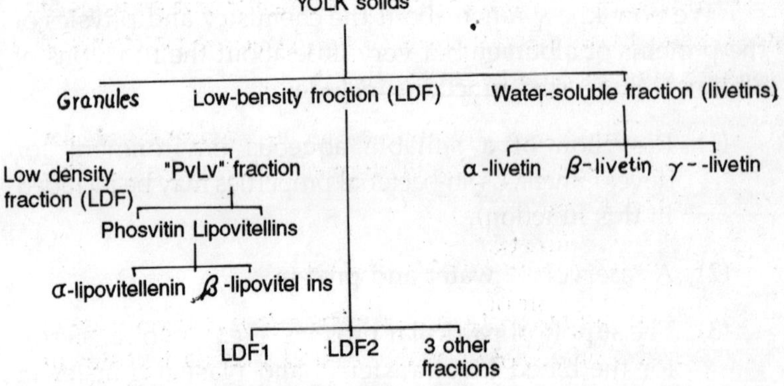

Fig. 11.5. The fractions of the yolk solids.

The yolk contains about 10 g of solids, almost all of which is protein-aceous. There is only a percentage ($< 1\%$) of free lipid, carbohydrate, etc. The yolk proteins are usually divided into three major fractions — granular, low-density and water-soluble. Further subdivisions are possible. The livetins (α-, β- and γ-livetin) have been identified as plasma albumin, plasma α_2-glycoprotein and plasma γ-globulin. The relative proportions of the various proteins are given in Table 11.2. It will be noted that the lipids are almost completely confined to the low-density fraction (LDF) and are associated with vitellin and vitellenin almost exclusively. It appears that the lipovitellenin fraction carries the major proportion of the total lipid and nearly all the triglycerides while lipovitellin fraction carries the remaining triglycerides and the phospholipids. About 70% of the phosphorus and 75% of the calcium of the yolk are found in the phosvitin fraction. The fatty acid composition of the yolk lipids is given in Table 11.3.

TABLE 11.2

Composition of the Yolk

Protein class	% of total yolk solids	% of total yolk proteins	% of total yolk lipids
Gramules			
Phosvitin	4	12	0
Lipovitellins	16	36	7
Low-density fraction	5	0	0
Low-density fraction	65	22	93
Water-soluble fraction	10	30	0

TABLE 11.3

Fatty and Composition of the Yolk Lipids

Fatty acid	Yolk total lipid	% of total fatty acid methyl-ester		
		Triglyceride	Lecithin	Cephalin
Palmitic (16:0)	23.5	22.5	37.0	21.6
Palmitoleic (16:1)	3.8	7.3	0.6	trace
Stearic (18:0)	14.0	7.5	12.4	32.5
Oleic (18:1)	38.4	44.7	31.4	17.3
Linolei (18:2)	16.4	15.4	12.0	7.0
Linolenic (18:3)	1.4	1.3	1.0	2.0
Arachidonic (20:4)	1.3	0.5	2.7	10.2
Eicosapentaenoic (20:5) } Docosapentaenoic (22:5)	0.4	0.2	0.8	3.0
Docosahexaenoic (22:6)	0.8	0.6	2.1	6.4

The yolk sac membrane

Williams (1967) has differentiated three stages in the utilization of yolk in the domestic fowl:

(1) All cleavage cells in the area pellucida are laden with yolk inclusions which are utilized up to the time of the formation of the primitive streak.

(2) As the primitive streak becomes visible the embryo is able to utilize the yolk in the area opaca, the yolk being actively phagocytized by the endodermal cells. Intracellular digestion occurs and with the formation of the blood vessels in the developing yolk sac, the products of intracellular digestion can be transported to the embryo.

(3) The third stage persists from the second day to the end of incubation. Here the yolk is subjected to extracellular digestion, the endoderm of the yolk sac secreting enzymes and absorbing the products of digestion, and also an amount of undigested, large molecules (e.g., complete proteins), and conveying them to the embryo via the vitelline circulation.

Here we shall be concerned with only the third stage but it will become evident there are several similarities between this and the second stage. The first two stages are fully described by *Williams* (1967).

The yolk sac membrane plays a key role in the nutrition of the developing bird, not only as a digestive and absorptive organ but also as a site for the synthesis of specific proteins amino acids and blood, as a source of lymphoid precursor cells for the thymus and bursa of Fabricius a a site for glycogen storage and as a temporary excretory organ.

Development. Initially the yolk is bounded by the vitelline membrane, a four-layered structure between 6 and 11 μm thick. The two inner layers, the true vitelline membrane and the perivitelline membrane, are laid down in the ovary and the outer two, the continuous membrane and the extra-vitelline membrane, in the oviduct. The true vitelline membrane becomes discontinuous even before ovulation and the remaining layers fall away as the yolk sac membrane spreads out over the yolk. Their rupture is perhaps the result of the uptake of water from the albumen with the attendant increase in volume. The membrane remains adhering to the yolk, until the latter part of incubation, as the yolk sac umbilicus.

Like the other extra-embryonic membranes, the yolk sac membrane is essentially two-layered, being composed of an outer mesoderm and an inner endoderm. It is continuous with the hind gut of the embryo. As it spreads, from the intestine, over the yolk two morphological regions can be differentiated. Peripherally there is the non-vascular area vitellina, while medially there is a highly vascular area, the area vasculosa. As development proceeds the blood vessels invade the area vitellina until the whole sac is vascularized. The yolk is surrounded by the end of the fifth day. From the fourth day the surface area of the membrane increases dramatically by the development, on the inner surface, of elaborate folds.

The blood circulation of the membrane passes through two distinct developmental phases. Briefly, the primary circulation is composed of an indifferent network of capillaries bounded peripherally by the sinus terminals which returns blood to the two anterior vitelline veins and thence to the heart. There are no arterial trunks. After two days' incubation, only some 12 h or so after the membrane begins to differentiate, the secondary or definitive circulation begins to develop, though the primary circulation is still not fully differentiated. Firstly, the vitelline (omphalo-mesenteric) arteries begin to differentiate while the right anterior vitelline vein of the primary circulation degenerates. This is followed by the development of the intermediate and posterior vitelline veins and a reduction in the size of the sinus terminalis (now termed the marginal vein). Next the collateral veins emerge and the marginal vein degenerates. Finally, the omphalomesenteric arteries branch richly to give a large capillary network, the blood returning to the embryo, though to lessening extent from the tenth day, by way of the single anterior and posterior vitelline veins and also by the paired omphalomesenteric veins.

In spite of the numerous functions ascribed to the yolk sac membrane, little morphological variation has been detected in the structure of the endoderm. The endodermal cells are columnar and form a single layer. Mitochondria, rough-endoplasmic reticulum and glycogen granules are commonly found in the cytoplasm together with numerous yolk inclusions and intracellular laminated materials. The apical surfaces of these cells are studded with short microvilli. A feature of the endoderm during the last week of incubation is the appearance of numerous lipid spherules, 50 to 150 mm in diameter, which *Lambson* (1970) considers as indicative of the large amounts of lipid being transported at this time.

Absorption of yolk. We have already seen the gross changes in the composition of the yolk as development proceeds. Now we can consider these changes in detail. Although the yolk sac membrane is essentially an extension of the gut little or no yolk material passes directly into it. Rather all the material required by the embryo is actively taken up by the endoderm, transferred to the mesodermal blood vessels and then transported to the embryo. The enzyme trimetaphosphatase is thought to be concerned in this process and is very active in the yolk sac membrane. All materials take this route including lipids for there is no lymphatic system.

Proteins The question of whether proteins are absorbed intact or only as amino acids is still not entirely resolved, though *Klein* (1968) has shown that complete proteins are necessary for normal development. Certainly some of the livetins may be transferred without alteration to the embryo to act as plasma proteins: α-livetin is identical with plasma albumin, β-livetin with plasma α_2-glycoprotein and γ-livetin with plasma γ-globulin. While some proteins may be transported via the yolk sac membrane others may follow a more circuitous route — from yolk to albumen to amniotic fluid and finally to digestive tract and blood stream. However, the amount of protein that is transported intact is probably small and the main absorption product may be considered to be amino acids.

There are at least three proteolytic enzymes involved in the hydrolysis of proteins and these may be pre-formed for their presence has been demonstrated in the unincubated egg. Other proteinase activity can be detected in the yolk sac from the fourth day and this activity increases during incubation. As a result of the activity of these enzymes, free amino acids in the yolk increase both quantitatively and qualitatively during the first 9 days. However, while there is considerable extracellular digestion of yolk proteins there is morphological evidence that intracellular digestion may also take place.

An active amino acid transport system exists in the yolk sac membrane by the sixth day. Removal of protein nitrogen from the yolk to the embryo can be seen to proceed at a steady rate for about 13 days, the proteins of the granular fraction being preferentially degraded. After 13 days there is a large influx of albumen proteins, through the yolk sac umbilicus—which causes a marked rise in the water-soluble protein content of the yolk. The uptake of protein nitrogen by the embryo from the granular fraction continues at the same rate until hatching while the uptake from the low-density fraction increases during the third week.

It is perhaps not surprising that the granular fraction should be selectively absorbed during the first fortnight for the PvLv fraction contains 85% of the iron and calcium stores and most of the protein phosphorus. How this selection is effected, however, is unknown. Furthermore, there is a more rapid removal of phosphorus from the phosphoproteins of this subfraction than from the phospholipids suggesting that the phosphorus in this form is more readily available to the embryo.

The uptake of phosphoprotein is maximal during the last week of incubation. The importance of these proteins is probably as a source of inorganic phosphate which is required for the production of ATP. Phosphoprotein phosphatase is concerned with the liberation of the phosphate groups and its presence has been demonstrated in both the yolk sac endoderm. However, *McIndoe* (1960) suggests that the phosphate is not in fact made immediately available but is first incorporated into an acid-soluble organic fraction and then converted to inorganic phosphate.

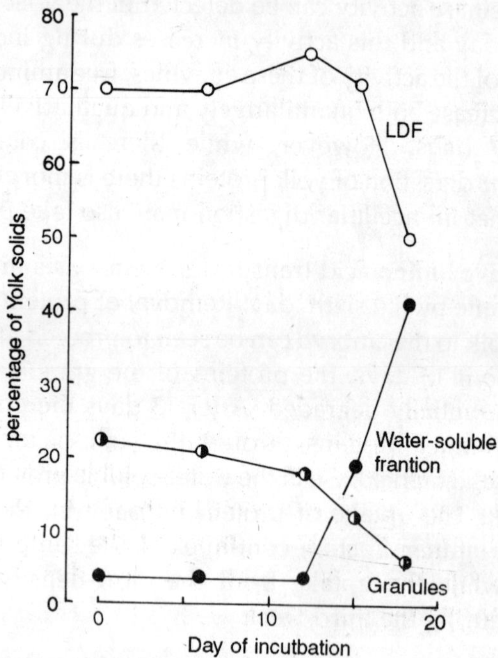

Fig. 11.6.   Percentage changes in the yolk solids during incubation.

Lipids.   The mechanism of lipid absorption is poorly understood. The amount of lipid transported in the first fortnight of incubation is probably about 350 mg a considerably lesser amount than that determined by *Saito et al.* (1965). In the last week of incubation the uptake of lipid accelerates dramatically to reach about 1 g per day and as a result the yolk sac membrane contains as much lipid as the yolk by the seventeenth day.

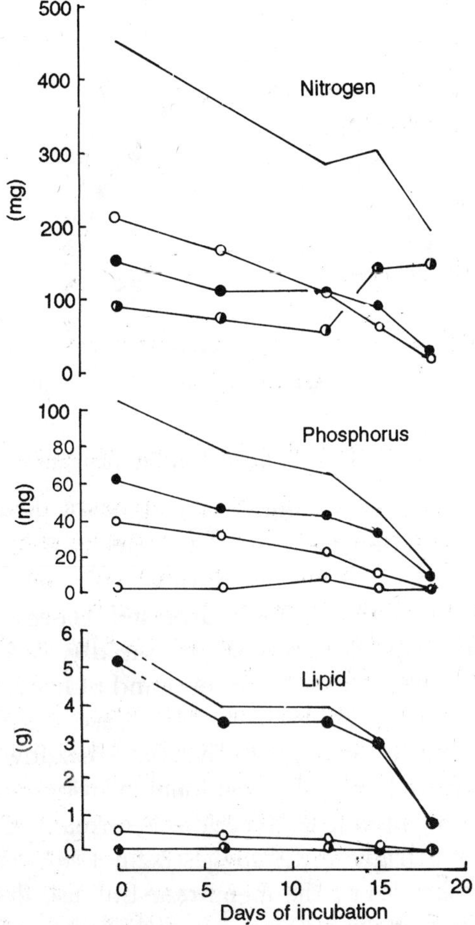

Fig. 11.7.   Amounts of nitrogen, phosphorus and lipid in the yolk and yolk fractions during incubation. Solid line = amount in whole yolk ; O = amount in granules; O = amount in low-density fraction; O = amount in water-soluble fraction.

With no lymphatic system the absorption of lipids poses an intriguing problem. It might be reasonably expected that absorption is preceded by digestion. However, mono- and

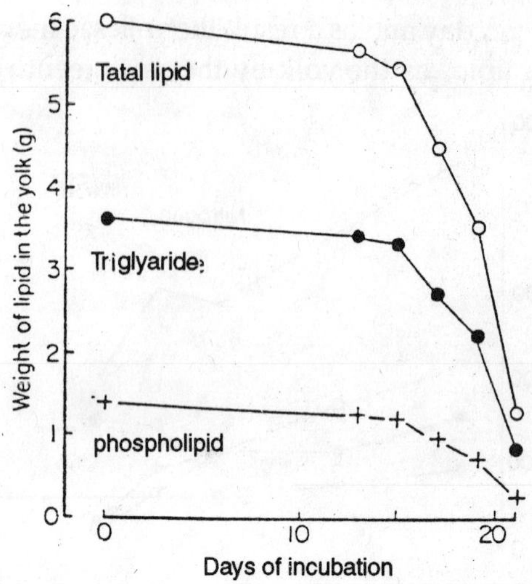

Fig. 11.8. Change in the yolk lipids during incubation.

diglycerides, free fatty acids and lyso-phosphatides are present in only trace amounts, about 0.1% of the total lipid, and their concentration does not increase during incubation. This suggests that either the lipids are not hydrolysed before absorption or that the products of hydrolysis are rapidly, and completely, absorbed by the yolk sac endoderm and utilized there for the synthesis of new lipids. Certainly lipase is present in the membrane but it seems likely that the breakdown of lipid is basically intracellular. The positional distribution of the fatty acids in the membrane triglycerides is virtually identical with that of the yolk triglycerides suggesting not only the absorption of intact molecules by the membrane but also that this is the major method of absorption. The earlier work of *Budowski et al.* (1961) supports this conclusion. the uptake of the phospholipids, the second most abundant class of yolk lipid, would also seem not to require hydrolysis and there is evidence that the uptake of the intact molecules is enhanced as incubation proceeds.

While the uptakes of triglyceride and phospholipid proceed at similar rates during incubation it is noteworthy that the proportion of esterified to free cholesterol in the yolk increases progressively from the ninth day. It seems likely that this esterification takes place in the membrane, since the ratio of esterified to free cholesterol is even higher there, and that a proportion of the cholesterol ester is transported back to the yolk. Nearly all this cholesterol ester is cholesterol oleate, the oleic acid possibly coming from the abundant phospholipid, phosphatidyl choline. The purpose of this esterification is unclear but the relatively high proportions of cholesterol esters in the lipids of the chylomicrons and lipoprotein in embryonic plasma (Schjeide, 1963) suggest that they play an essential role in the assembly of some lipoprotein complex in which from the lipids are transported from the yolk sac membrane to the embryo.

There does not appear to be any preferential absorption of a triglyceride during incubation. With the exception of the C_{18} polyunsaturated acids, the fatty acid composition of the yolk triglycerides remain fairly constant. *Issacks et al.* (1964) suggested a preferential absorption of C_{20} polyunsaturated acids but this has not been confirmed. There is evidence, however; of a preferential absorption of phospholipids containing docosahexaenoic acid.

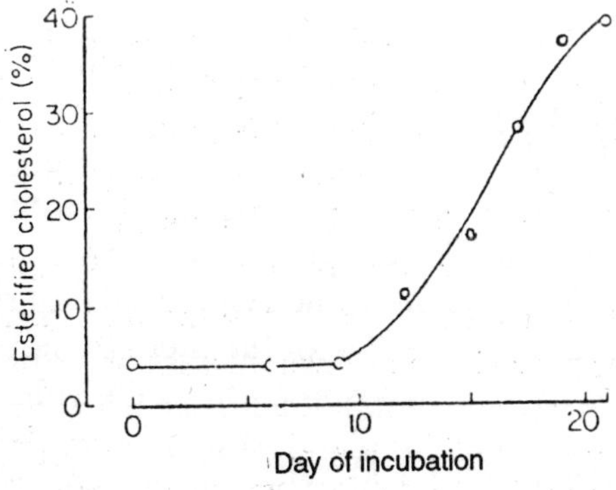

Fig. 11.9. Percentage esterification of the cholesterol in the yolk during incubation.

The principal phospholipids of yolk are phosphatidyl choline (about 75%) and phosphatidyl ethanolamine (20%)

The latter is selectively absorbed during incubation especially those molecules in which stearic acid occupies the α-position and arachiodonic acid or docosa-hexaenoic acid in the β-position.

Carbohydrates. We have already seen that the carbohydrate content of the yolk is never great, about 150 mg at the start of development rising to a maximum of 275 mg on the tenth day. It is not surprising,therefore, that active transport of carbohydrate does not occur in the membrane for the greater part of incubation. It is only in the last 3 days that the ability to concentrate sugars is developed.

Utilization of yolk

The proteins, amino acids and lipids absorbed by the yolk sac membrane are transported by the blood stream to the embryo where they can be utilized for growth or the production of energy.

Proteins and *amino acids*. We have already noted that some intact proteins are apparently necessary for normal development, that there is a limited transport of antibodies to the embryo and that some of the livetins perhaps act as plasma proteins.

Details of the rates accumulation of amino acids by the embryo are given by *Rupe* and *Farmer* (1955). These authors have shown that the amino acids taken up during the first 10 days or so of incubation come exclusively from the yolk. Furthermore, the uptake of the various amino acids is not uniform; after 16 days 77% of the glycine, 50% of the aspartic acid, 50% of the histidine, 31% of the methionine and 11% of the tryptophan have been absorbed from the yolk.

Most of the amino acids absorbed from the yolk and albumen are used to synthesize proteins or new amino acids while a proportion is used for the synthesis of glucose *de novo*. Several enzymes are present within the yolk sac itself that are concerned in protein metabolism: these include cystein lyase serine hydrolase cysteine desulphydrase and glutamotranferase.

Gluconeogenesis. The newly laid egg of the domestic fowl contains about 500 mg of carbohydrate of which approximately 70% i present as glucose. By the tenth day of incubation this glucose is virtually exhausted. It is not surprising therefore that the embryo should have an active gluconeogenic system from an early point in development. The key enzymes concerned are present in embryo homogenates or liver homogenates from about the seventh day. These include phosphoenol pyruvate carboxykinase, pyruvate carboxylase, glucose-6-phosphatase and fructose-1, 6-diphosphatase. The activities of those enzymes generally increase to optima on the sixteenth or seventeenth days and then decline thereafter.

The preferred amino acid substrate is as yet unknown. Both alanine and glutamate have been shown to be incorporated into glucose suggests that: serine may be more glucogenic in the embryo. Certainly it is an abundant amino acid, there being 330 mg in every gram of phosvitin. There is, furthermore, a preferential uptake of this amino acid during development and it can easily be converted to pyruvate.

Much of the gluconeogenic effort is directed towards the production of glycogen which can be made available to the embryo as and when necessary. While glucose is incorporated into glycogen pyruvate is probably the substrate of choice for it has been found that the embryo preferentially uses three carbon molecules rather than glucose for glycogenesis Uridine-diphosphate glucose glucagon synthetase is present in the liver from the seventh day have shown that the glycogenic enzymes increase in activity during incubation.

The liver is an important site for the synthesis and storage of glycogen; in the bird these functions are assumed as soon as the liver begins to differentiate on the sixth or seventh day though it has recently been suggested that by using more

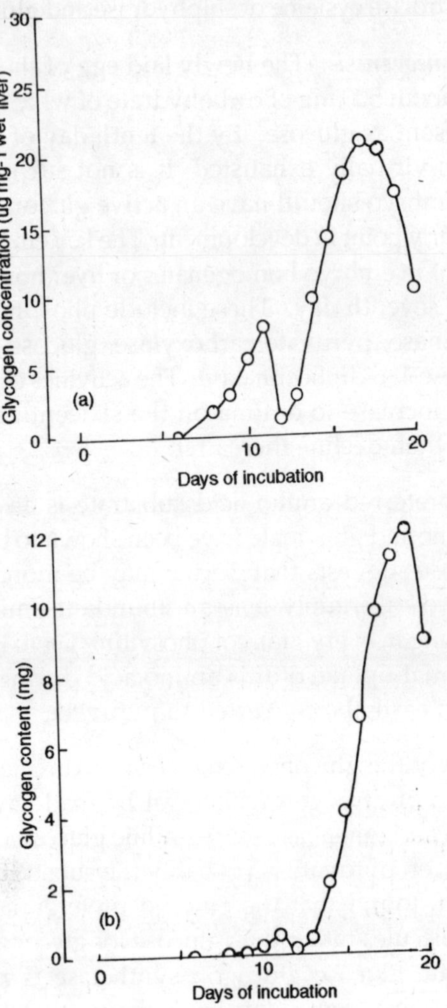

Fig. 11.10. Changes in hepatic glycogen during development: (a) concentration (b) total amount.

sensitive techniques glycogen may be detected as early as the fourth day. These authors suggest that gluconeogenesis has therefore begun by this time. This is difficult to reconcile with the observations on the appearance of the gluconeogenic enzymes. The glycogen concentration and content of the liver during development have been determined on many occasions The results of *Daugeras* (1968) are typical and are shown in Figure 11.12. However, the amount of glycogen stored at this site is less than that stored in the yolk sac membrane. While the membrane has been recognized as a site of glycogen storage for 100 years less work has been done on it than on the liver. It was believed at first that the yolk sac membrane acted as a transitory liver, the function of glycogen synthesis and storage being 'lost' to the liver as it (the liver) became functional. However, modern research has shown that the glycogenetic activity of the membrane is retained throughout incubation.

Lipids. As we have already seen the absorption of lipids proceeds with a variable amount of modification to the native molecules. Once the substances are made available to the embryo further modifications may occur and new classes of substances synthesized.

Much new phospholipid is synthesized by the embryo which may require the extensive breakdown and resynthesis of the absorbed phospholipids. Thus, the fatty acid compositions of the phosphatidyl choline, phosphatidyl ethanolamine, phosphatidyl serine and diphosphatidyl glycerol in the liver differ markedly from the corresponding fatty acid compositions of the same phospholipids in the yolk. At the same time phosphatidyl inositol, absent from the yolk, is synthesized in the embryo. However, not all the phospholipids modification. The sphingomyelin of the yolk and embryo are similar suggesting little modification at any stage.

Triglycerides are little altered during absorption and deposition in the liver but may be subjected to considerable modification thereafter. *Noble & Moore* (1964) has noted important differences in their free fatty acid composition. The preferential uptake of docosahexaenoic acid has already been noted. While this could indicate a special requirement for this acid much of it is found in the triglycerides rather than the phospholipids suggesting that the requirement is not for the acid *per se* but for some other constituent of the phospholipid molecule which incidentally contains a high proportion of the acid.

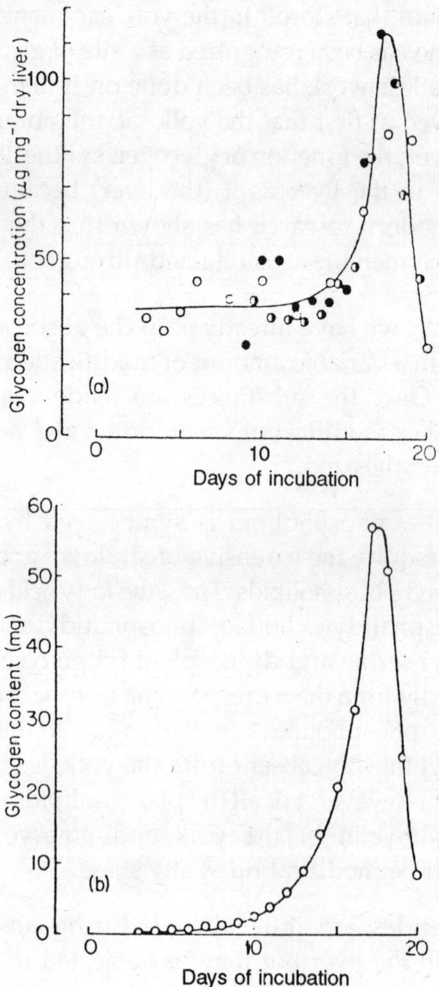

Fig. 11.11. Glycogen stores in the yolk sac membrane: (a) concentration.

The major lipid stored in the liver is cholesterol. After the fifteenth day the increase in these stores is brought about almost completely by cholesterol esters, 80% of which is in the form cholesterol oleate. In the extra-hepatic tissues triglyceride is the main lipid and accounts for 75% of the total extra hepatic stores, reflecting, no doubt, its importance as an energy source during development.

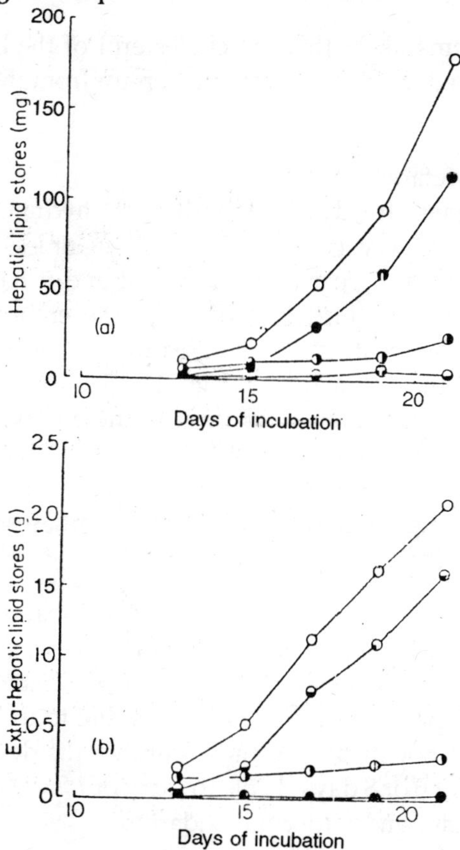

Fig. 11.12. Lipid stores of the developing embryo. (a) hepatic stores; (b) extra-hepatic stores; O = total lipid; ● = cholesterol; O = phospholipid; O = triglyceride.

An amount of sphingomyelin is utilized for the myelination of the nervous system but the low concentration of this substance in the brain - about 3% suggests that myelination is incomplete.

Indeed the process does not begin until the tenth to thirteenth day of incubation 11% while the adult level of sphingomyelin (115) is achieved only after 5 months. significant changes in the lipid composition of the myelin have been noted during the last 3 days of incubation.

It seems likely that the cholesterol of the brain is newly synthesized and is not simply taken up from the yolk or liver stores.

Energy Metabolism

We have already seen that the carbohydrate content of the incubated egg is very small, about 500 mg or less. This material is utilized during early development and it has been shown that during the third to sixth days the embryo utilizes the pentose phosphate shunt. The pathway is again particularly active in the brain tissue from 12-15 days. This metabolic pathway is extremely important to the embryo in providing both the pentoses necessary for nuclei acid biosynthesis and the reduced nicotinamide adenine dinucleotide phosphate (NADH) required for reductive syntheses involved in development.

For about half the period of incubation the overall respiratory quotient is about 0.7, indicating a fat metabolism. Indeed it has been calculated that approximately 94% of the total energy expenditure (24 kcal) is due to the catabolism of lipid. It is not surprising therefore that the developing chick from about 6 or 8 days is able to carry out fatty acid activation, acyl transfer and fatty acid oxidation.

It might be expected that with such an intense level of fat catabolism there would be a significant accumulation of ketone bodies. All the early studies, however, failed to demonstrate any but recent work has shown significant concentrations during development. The latter authors found the concentration rose from about 120 μg ml^{-1} at 14 days to a peak of 177 μg ml$^-$

[1] at 17 days and then fell to about 10 μg ml^{-1}, a concentration then maintained until after hatching. *Best* (1966) reported values as high as 270 μg ml^{-1} just before hatching. A high activity of βs-hydroxybutyrase has been reported in the kidney (mesonephros).

While lipid catabolism is undoubtedly the most important method of producing energy for development, the pathways of carbohydrate metabolism show a general increase in activity with time. Thus there are increases in the glycolytic flow rate tricarboxylic acid cycle activity gluconeogenesis and glycogen metabolism.

The Sub-embryonic Fluid

This liquid arises and fills the sub-embryonic (sub-germinal, sub-blastodermic) cavity. It is actively secreted by the blastodermal endoderm from the second day of incubation and persists until the fifteenth day though its volume is greatest by the end of the first week. Both the yolk and albumen contribute to its formation, though the majority of the water comes from the albumen. It is an important fluid for its mineral composition is such as to support normal development of the embryo.

The ratio of Na$^+$ to K$^+$ in the fluid falls from about 8 at 2 days to 1.5 at 11 days. These ratios are considerably lower than those usually encountered in blood plasma (34) or in media used to culture cells (25). *Grau et al.* (1962) have shown that these low values are necessary for the survival of the embryo. Since the actual concentration of Na+ in the sub-embryonic fluid (>90 mEq 1^{-1}) exceeds those of the yolk and albumen (39 and 86 mEq 1^{-1}, respectively) it seems likely that the blastodermic endoderm actively secretes Na-. At the same time a few K$^+$ are transported or diffuse across the gradient, <20 mEq 1^{-1} in the fluid as compared with 43 mEq 1^{-1} in the albumen yand about 70 mEq 1^{-1} in the yolk, thus achieving the optimum ratio of Na$^+$ to K$^+$. The fluid is also deficient in Ca^{2+} but again an excess of these has been shown to have a deleterious effect on development. While the sub-embryonic fluid may provide the proper   physiological medium for development it does not act as a source of nutrients.

12

Physiology of Hatching

Terminating the embryonic existence, a process usually described by the deceptively simple word 'hatching', is in fact an extremely complex procedure and it is not surprising, therefore, that there is a peak of mortality at this time. Unlike the mammal, essentially all the stimuli are generated by the embryo itself, though it is able to respond to stimuli from both the parents and the other members of the clutch.

Superficially there are three major events in the sequence of hatching: the onset of pulmonary respiration, pipping (that is the single point fracture of the shell, usually in that overlying the air space), and the emergence of the hatchling. However, underlying these more obvious events are many others, e.g. maturation of muscles. changes in the circulation, withdrawal of the yolk sac, all following a sequence which will ensure a successful outcome of incubation.

Pulmonary Respiration
Introduction
Birds and reptiles pass through a stage of development unknown to the mammal — a period when the embryonic and adult respiratory surfaces, the chorio-allantois and the lungs, function side by side. This period is conveniently described as the parafetal period and the animal, the parafetus. These terms are due to *Romijn* (1948) and were originally used to describe the period from the onset of breathing to pipping. Here we shall use the term to cover the period when the two respiratory systems function simultaneously.

The Respiratory System

The embryology of the system is fully described by *Hamilton* (1952) and *Romanoff* (1960). In brief the system develops from two distinct sites: the larynx and trachea are derived from the median laryngotracheal grooves; the lungs, bronchi and air sacs develop from paired endodermal diverticula of the embryonic fore-gut. The air sacs, which appear on the sixth day, are initially found as six paired structures but two pairs fuse to form the median clavicular sac and, in some species including the fowl, another pair fuse to form the median cervical sac. Thus, there are normally eight or nine sacs: the median (or paired) cervical sac(s), the median clavicular and the paired cranial thoracic, caudal thoracic and abdominal sacs. These sacs are connected, via the ostia, to the bronchi and are concerned, in the hatched bird, with the movement of air within the system. Their innervation, in the fowl and mute swan, is described by *Groth* (1972).

The architecture of the bronchi is complex. Each lung is supplied with one primary bronchus (the mesobronchus) which gives rise to the secondary bronchi (ento-, ecto-, latero- and dorsobronchi) and these in turn give rise to the tertiary bronchi (parabronchi). No tertiary bronchus ends blindly and many join to form long, curved circuits. It is these tertiary bronchi that are concerned with gaseous exchange. Each is pierced by numerous openings which lead in turn to the atria and then the infundibula. The infundibulum leads to a network of fine anastomosing air capillaries which are in intimate contact with the blood capillaries.

The flow of air within the respiratory system is complex and the air sacs, far from being evolutionary relics or buoyancy bags, have an important function in maintaining the correct pattern of air flow. They are not concerned with gaseous exchange *per se*, however. There is no true diaphragm. It will be sufficient to point out here that both inspiration and expiration are active processes and, because of the flow pattern, fresh air is probably passing unidirectionally through the tertiary bronchi continuously. The avian respiratory system is, then, an extremely specialized system, designed to meet the heavy energy requirements of flight, and, as a result, is able to supply the bird with more oxygen per unit time than any other system.

The onset of breathing : the Palmonary stimulus

A prerequisite for breathing is the removal of the fluid invading the whole of the respiratory tract. Undoubtedly a major component of this fluid is of amniotic origin, but within the lung itself the fluid may contain a large proportion of an ultrafiltrate of the embryo's blood. The uptake of amniotic fluid, by active imbibition, is begun, in all birds examined, when 70% of the incubation period is complete and is virtually complete just prior to the initiation of breathing. Thus in the fowl imbibition is completed about a day before the chick hatches. While some of the fluid remaining in the respiratory tract proper may flow into the pharynx and then be swallowed, some remains and presumably is taken up by the lung tissue itself.

Early attempts to identify the pulmonary stimulus though not completely successful, served to indicate that it is probably gaseous and that it is possibly a high partial pressure of carbon dioxide in the arterial blood (Pa_{co2}). Thus *Windle* and *Barcroft* (1938) with the fowl, and *Windle* and *Nelson* (1938) with the duck, found that clamping the umbilical vessels of the mature embryo stimulated deep rhythmic respiratory movements almost immediately. That carbon dioxide was the more likely stimulus was suggested by the observation that respiratory movements could be initiated by raising the Pco_2 in the atmosphere surrounding the egg by only 12 mmHg while a reduction of 68 mmHg in Po_2 was required.

It is only comparatively recently that there has been further interest shown in the nature of the pulmonary stimulus. *Visschedijk* (1962, 1968b) found that increasing or decreasing the permeability of the shell over the air space did not lead to any change in the timing of pulmonary respiration. However,

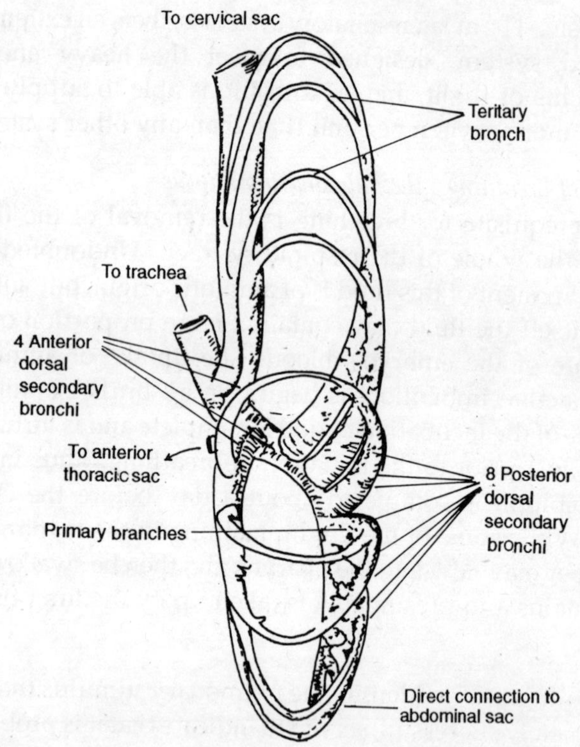

Fig. 12.1. Bronchi of the goose lung, Dorsal view of the right lung: the ventral series of posterior scondary bronchi has been omitted. (Reproduced, with permissiion, from Brackenbury, 1971).

Freeman (1964a) using paraffin was instead of liquid paraffin for reducing permeability found a significant advance of five hours. He confirmed that perforating the shell was without effect as was ventilating the air space.

At first sight it would seem that either the enhanced P_{CO_2}, reduced P_{O_2}, or both in the air space was responsible for stimulating breathing. There is an important objection to this interpretation, however: perforating or ventilating the air space failed to delay breathing.

TABLE 12.1

Effects of Sealing, Perforating or Ventilating the Air Space on Hatching

Lung ventilation			*Pipping*			*Hatching*			
Sea-led	*Per-forated*	*Venti-lated*	*Sea-led*	*Per-forated*	*Venti-lated*	*Sea-led*	*Perfo-rated*	*Venti-lated*	*Ref-erence*
+0.2	-0.2	—	+5.3	-5.9	—	-0.8	+1.3	—	Visschedijk (1968b)
+5.3	-1.4	-1.0	+13.7	-7.7	-9.8	+1.6	+0.9	-1.2	Freeman (1964a)

Each figure is the time in hours that the event is advanced (+) or retarded (-) as compared with normal eggs.

Another approach to the problem has been made possible by the recent development of micro-techniques suitable for analysis of gases in the small volumes of blood available from the avian embryo. Towards the end of the embryonic period the Pa_{CO_2} may reach the very high level of 60 mmHg — but not exceptionally so. The likelihood of a high Pa_{CO_2} being the stimulus is strengthened therefore by these direct observations, though the further observation that there is a significant *fall* in Pa_{CO_2} before breathing commences weakens the hypothesis.

In an attempt to resolve this paradox we have examined a modification of the hypothesis suggested by the above results— that the stimulus is a particular ratio of Pa_{CO_2} to Pa_{CO_2} as it is in the sheep—but without success (Vince *et al.*, 1974).

A different approach has been suggested by the demonstration that all aspects of the hatching process can be substantially advanced by sound — 'clicking.' This suggests that under certain circumstances the gaseous stimulus can either be initiated earlier or overridden by the sound stimulus. An earlier initiation of pulmonary respiration has been observed in the fowl, duck, goose and Japanese quail but not in the bobwhite quail.

The technique of sound stimulation has been used in conjunction with blood gas analysis in an attempt to determine whether or not there is a unique change in the partial pressures of the blood gases prior to breathing. Such a change might be interpreted as the pulmonary stimulus. However, while sound stimulation led to an advance in the initiation of breathing no clear differences in $Paco_2$ or $Paco_2$ could be identified. The positive identification of the pulmonary stimulus remains an intractable problem therefore and it seems very likely that it is gaseous but whether it is a high $Paco_2$, a low $Paco_2$ or a critical ratio of the two still remains to be elucidated.

The development of the breathing pattern

Breathing probably begins about the time the beak penetrates the inner shell membrane though the embryo is capable of showing respiratory movements many hours before The lungs neither inflate immediately nor completely. Rather the tertiary bronchi are aerated first in limited areas. These areas can be recognized by the change in colour of the lung from dark to bright red. *M.A. Vince* and *B. Tollhurst* (personal communication) have found that inflation is complete in both lungs after about 6 h (the fowl), 4 h (duck) or 9 h bobwhite quail). The role of surfactant in facilitating breathing has not been completely elucidated.

Initially breathing is both irregular and intermittent. Gradually regular patterns of medium amplitude breathing emerge and come to dominate the pattern. In the third phase both frequency and amplitude increase and each breath is often accompanied by a 'click sound.' This pattern has been observed in the several species examined though the length of any one phase may vary from species to species. The rate of breathing in the hatching fowl reaches about 90 to 100 breaths per minute while in the duck it is a little higher at 110 breaths per minute.

Recent work by *Dawes* (1973) indicates that the control mechanisms of respiration are at least relatively mature at hatching. Thus during experimental anoxia he found that breathing amplitude rose quickly indicating some peripheral stimulation. Thereafter, the parafetus began to gasp, breathing rate slowed and finally ceased illustrating the excitatory and depressant effects respectively of anoxia on the respiratory centres of the brain.

Circulatory and Associated Changes

The circulation of the embryo is characterized by two short circuits — the ductus arteriosus between the pulmonary and aortic arches and the intra-atrial foramina, the avian equivalent of the foramen ovale. The ductus venosus, unlike its mammalian counterpart, is lost on the seventh day of incubation. When the bird commences breathing these short circuits, together with the chorio-allantoic circulation, have to be closed off in order that a double circulation can be formed.

It has been found that the walls of the ductus arteriosus have well-developed muscle layers particularly at the proximal end. As breathing begins these muscles contract to restrict and finally prevent blood flow. This is a gradual process, and is complete in 91% of the chicks at hatching. The physiological mechanism is uncertain, neither acetycholine nor catecholamines have marked stimulatory properties on the muscle cells.

The intra-atrial foramina are so constructed that they normally act as valves. During diastole of the right atrium the blood flows into the chamber and pushes the flaccid inter-atrial septum towards the left thereby opening the formina and allowing a proportion of the blood to flow directly into the left atrium During systole the pressures in the atria become equalized, thus allowing the septum to take medial position and in so doing the foramina are closed. As breathing becomes established more blood flows into the left atrium from the now-functioning pulmonary vein and as a result the pressures within the atria tend to be equal. This closes the formina and their permanent closure follows, usually within four or five days.

The mechanism by which the blood flow, via the umbilical

vessels, to and from the chorio-allantois, is progressively restricted is uncertain though the closing of the umbilicus itself may be involved. Clearly this is a gradual process, for gaseous exchange continues for more than 20 h. But it should be noted that the volume of blood flowing through the chorio-allantois begins to decline some four days or so earlier. Why this is so is uncertain and in some ways paradoxical. It is precisely at this period of development that the problems of oxygenating the embryo are probably at their greatest

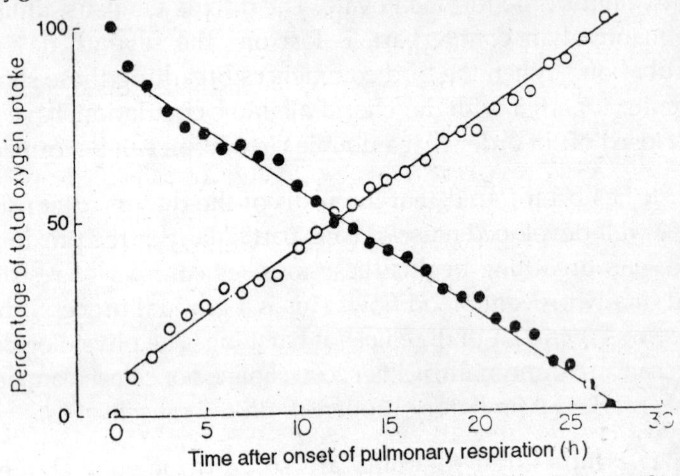

Fig. 12.2.   The percentage contributions of the chorio-allantois (O) and the lungs (O) to the oxygen requirements of the hatching chick.

Blood pressure rises markedly during hatching to at least 43/23 mmHg and continues to increase in the immediate post-hatching period and the heart rate increases from about 260 beats per minute to 295 beats per minute or more

Pipping

Some 8 or 9 h after breathing is established in the fowl the shell over the air space is usually fractured at one point — 'pipped.' The timing of this event can be advanced by waxing or oiling this part of the shell, delayed by perforating it or virtually abolished by ventilating the air space with atmospheric air.

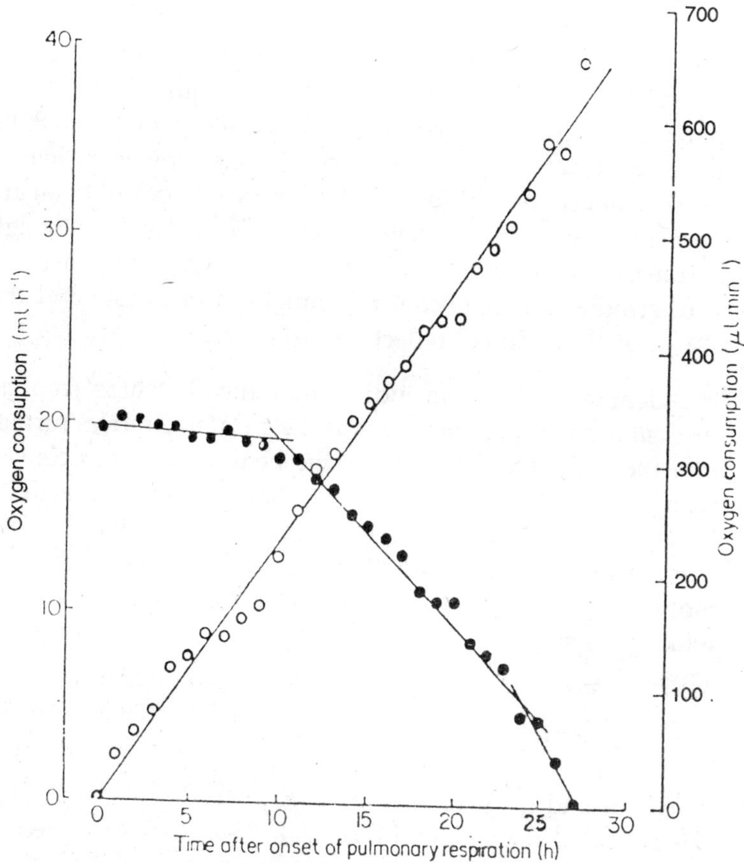

Fig. 12.3. The contributions of the chorio-allantois (O) and the lungs (O)
to meeting the oxygen requirements of the hatching fowl.
The degeneration of the chorio-allantois as a site for gaseous
exchange does not begin until some 10 hours after breathing
commences and thus coincides with the time of piping.

This pipping, like breathing, would seem to result from a
gaseous stimulus. In a series of elegant experiments *Visschedijk*
(1968c) has established that either a reduction in the P_{O_2} or a
rise in the P_{CO_2} in the air space advances the time of pipping
and that for a given change, carbon dioxide is twice as effective
as oxygen.

Just prior to pipping the Po_2 in the air space of the domestic fowl may fall to as little at 60 mmHg while the Pco_2 may reach 60 mmHg. At this point in the hatching process the lungs are meeting about 40% of the total oxygen requirements of the parafetus and it might be expected that the parafetus will be faced with increasingly hypoxic conditions. However, there is little evidence that this occurs: cardiac stores of glycogen are normally low, suggesting few problems during hatching though a transient depletion can be detected at pipping There is, furthermore, no significant accumulation of lactate and the Pao_2 and $Paco_2$ do not reflect hypoxia.

General activity, including that the 'hatching' muscle (*musculus complexus*), increases as the parafetus moves further into the air space. Eventually the beak comes into contact with the calcareous shell and the enhanced muscular activity leads to the egg tooth the reinforced tip of the upper beak — penetrating it. Once the shell is pipped the parafetus enters a more quiescent period, at least in terms of physical activity, which persists until 3 or 4 h before the chick emerges from the shell.

Active Hatching

The final activity of the parafetus is directed towards cutting round the shell and the emergence of the chick. That this period is independently controlled was first suggested by *Freeman* (1962). He found that the oxygen consumption of the parafetus began to rise an hour or two before active hatching began. It was suggested that this increase in metabolism resulted from the release of a hormonal stimulus which leads to the active cutting round of the shell.

It is now generally agreed that the hatching stimulus is not gaseous. *Visschedijk* (1962, 1968b) and *Freeman* (1964a) found that altering the permeability of the air space shell did not have any effect on the time of hatching. The suggestion that it is hormonal in nature was strengthened by the demonstration that the thyroid hormones accelerate hatching and that they stimulate active hatching specifically. However, it has been suggested that progesterone is the hormone responsible though.

An important role has been ascribed to the muscle, *musculus complexus,* in the hatching processes. This muscle shows marked development just before hatching begins and acts to raise the head of the parafetus and is presumed to provide most of the force necessary for breaking down the shell. The timing of its maturation seems to be unique but attempts to explain this have largely failed.

Withdrawal and Fate of the Yolk Sac

As incubation draws to a close there remains outside the embryo a variable amount of yolk in the yolk sac. This is progressively withdrawn into the abdominal cavity, from the nineteenth day in the fowl, probably as a result of the activity of the abdominal musculature, the process in the fowl being completed some 14 h before emerging from the shell. The control of the withdrawal activity is uncertain, though there is limited evidence that the thyroid or adrenal or both may be involved.

Although the yolk sac is connected directly with the duodenum there is little or no movement of material via this route even after hatching. Instead it is absorbed through the yolk sac membrane and transported to the chick by the omphalomesenteric vessels. There are approximately 5 g of yolk in the yolk sac of the neonate fowl but this is rapidly utilized by the bird and the yolk sac is usually vestigial by the fifth day.

Oxygenation and Energy Metabolism During Hatching

With the onset of breathing the oxygen requirements of the parafetus are met by the combined activities of the chorio-allantois and lungs. It is possible to determine their relative contributions during the parafetal period from the data of *Visschedijk* (1962). Given that the reduction in the chorio-allantoic component is similar over its whole surface, we can that after the initial 3 h or so the change-over from chorio-allantoic to pulmonary respiration is linear and is not completed, at least in the fowl, until the very end of the hatching period. However, the degeneration of the chorio-allantois would not seem to be initiated until 10 h after the initiation of breathing for it continues to provide the same amount of oxygen up to that point. Thereafter its importance as a respiratory surface declines rapidly.

The oxygen requirements of the parafetus, not unexpectedly, rise during hatching. Typically the fowl consumes 25 ml h^{-1} prior to breathing ml h^{-1} prior to active hatching and about 40 ml h^{-1} at the moment of escape from the shell. After this, at least in the homeothermic species, a further substantial rise occurs.

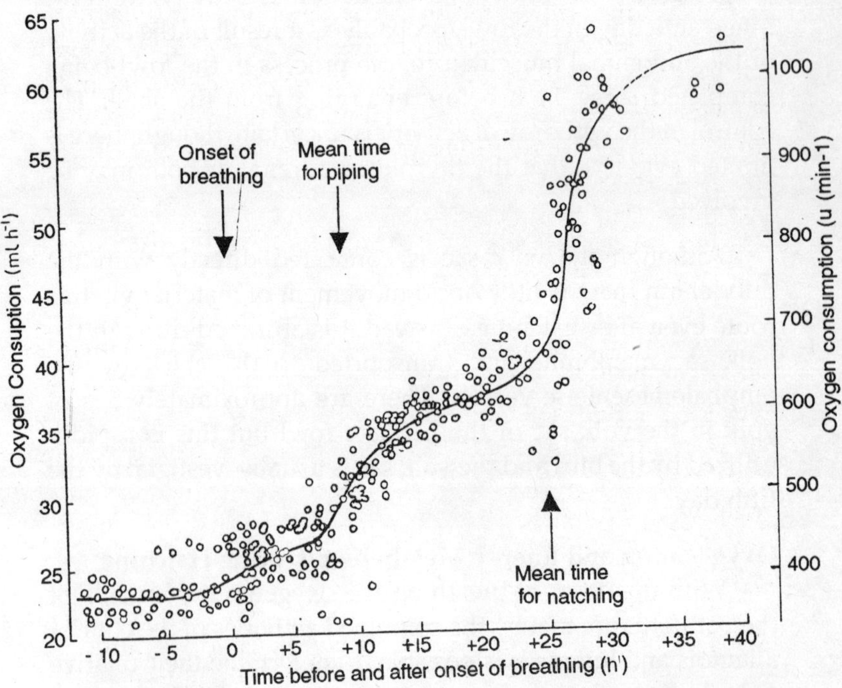

Fig. 12.4. oxygen consumption of the flow during the hatching period. The data have been taken from Visschedijk (1962) and Freeman (1962 and unpublished observations).

It would seem that the bird's requirement for carbohydrate increases markedly during hatching for there is a dramatic mobilization of glycogen stores a   little before breathing commences. It is likely that the glucose that is made available is necessary for the proper functioning of the central nervous system. Indeed a marked rise in the glycogen stores of the brain have been noted as this time. The overall respiratory quotient does not show any increase during this period, however, remaining at about 0.7. A transient period of mild hypoxia is indicated by the fall in cardiac glycogen stores at the time of pipping.

13

The Neonate

Birds show a wide range in their degree of functional maturity at hatching. Some species, the duck for instance, are relatively mature, in that they have sight, are able to feed themselves and generally show a marked degree of independence. Others, the passerines for instance, are distinctly immature being blind and naked and totally dependent upon the parent birds for their survival. But whether the hatchling is mature or immature all neonates are confronted by the same hostile environment. In this chapter we shall discuss three of the many problems: changing environmental temperature, digestion and absorption of food and the problem of bacterial and viral diseases.

Thermoregulation
Ontogeny of thermoregulation
Birds and mammals are the only two classes able to produce heat for the maintenance of their body temperature. They may therefore be termed endothermic homeotherms and are thus distinguishable from the ectothermic homeotherms which, while able to control their body temperature, do so by deriving heat from outside the body. While the adults of the two classes are homeotherms their embryos are poikilothermic. Homeothermy, furthermore, does not necessarily develop at birth: when it does the species is said to be precocious. Those species which become homeotherms later are described as altricial. This concept, while useful, is too rigid and a complete range of types may be distinguished. While they may show marked metabolic responses to cold immediately after hatching

not all the precious species are complete homeotherms even within a day of hatching. Of those species studied the Anatidae are probably the best able to maintain their body temperature similarly the slender-billed shearwater and the western gull are quite good while the Galliformes are less good and the Japanese quail is very poor. Most precocious species have very good thermoregulatory ability by the end of the first week of post-embryonic life. The ability to regulate heat production in the altricial species develops a variable, though constant for each species, time after hatching and may not be complete until the end of the nestling period.

Mechanisms of Thermoregulation

The attainment of homeothermy in both altricial and precocious species has often been correlated with the development of insulation, in particular the feathering (including the down feathers). There is no doubt that this is a factor but it is not the most important one. Perhaps the rapid increase in the ratio of mass to surface area is more important *Freeman* (1965b) has shown that part of this increase in mass is brought about with no change in surface area by the replacement of the relatively inert yolk by metabolically active tissue. At the same time *Wekstein* and *Zolman* (1968, 1969, 1970) have shown that the capacity to thermoregulate depends on age rather than the degree of feathering.

Chemical regulation of heat production has been studied mostly in the fowl. While the neonate is capable of a limited amount of shivering it seems likely that a proportion of thermoregulatory heat is produced by non-shivering means though not through the activity of brown adipose tissue for the fowl at least does not possess this tissue. The thermogenic stimulant may well be thyroxine. The ability to shiver develops rapidly after hatching and is probably responsible for most of the thermoregulatory heat produced after a week. A similar situation exists in other avian species both precocial and altricial.

Precocious species are able to pant or utilize gular fluttering but the altricial species show little ability to increase heat loss by these methods.

Most neonate birds, whether altricial or precocious show behavioural responses to temperature. They will huddle and reduce their activity to a minimum in order to conserve energy. *Kleiber* and *Winchester* (1933) have shown that such actions lead to a 15% reduction in energy expenditure in the domestic fowl.

The Alimentary Tract

In general the smaller the bird the smaller, relatively, are its food reserves at hatching. During incubation the domestic fowl utilizes 42% of its lipid stores whereas the eastern house wren uses 99%. It is not surprising therefore that the alimentary tract is usually mature at hatching. However, as we shall see, some digestion probably occurs before hatching.

The maturation of the tract has been best studied in the domestic fowl where it has been shown that there is a marked increase in development during the final 6 days or so of incubation. Thus the oesophageal mucus glands are mature at hatching the gland cells of the proventriculus commence secretion on the thirteenth day the innervation of the gizzard becomes operative at hatching the gizzard itself is mature at hatchingthe duodenum and small intestine enter their final phase of differentiation three or four days before hatching the gut transport systems for amino acids and sugars become functional, and for the latter become markedly more efficient the activities of pancreatic chymotrypsin, carboxypeptidase, amylase, maltase, sucrase and lipase increase dramatically from the seventeenth day. The changes in invertase activity do not parallel the changes shown by these enzymes, however. Activity increases progressively from 13 days, falls significantly at 20 days and then increases again Brown

A limited amount of digestion begins before hatching for the embryo actively imbibes the amniotic fluid from the thirteenth day. This fluid is particularly rich in protein for the albumen floods through the seroamniotic connection into the amniotic cavity from the twelfth day. It is of interest, therefore, that the amino acid transport system of the small intestine should be mature by the sixteenth day. Recent work has shown that the major molecular species of chymotrypsin in the duodenum at this time is chymotrypsin 3. Since this species is rapidly replaced by chymotrypsins 1 and 2 after hatching Cohen and Kulka suggest that this species is particularly active towards the albumen proteins, one of the most prevalent amino acids being leucine. At the same time, however, it seems likely that intact protein can be transported across the gut epithelium for antibodies of maternal origin can be detected in the embryo. This ability is lost at hatching.

Finally it should be pointed out that the residual yolk is not transferred through the yolk stalk directly into the small intestine for digestion. *Fritz* (1961) has shown that the yolk material is absorbed exclusively via the omphalo-mesenteric blood vessels.

Immunological Competence

It is generally assumed that the avian embryo develops in a sterile environment. By and large this is true but it is not always the case for some diseases, for instance, infectious avian encephalomtyelitis, are transmitted from the mother to her progeny via the egg. The cuticle of the egg is an important barrier to the penetration of the shell by bacteria during incubation and even should some organisms pass through the pores into the egg the anti-bacterial properties of the albumen often prevent the establishment of an infection. Once the shell is fractured during the hatching process the bird becomes exposed to a contaminated environment and the survival of the bird will largely depend upon its ability to react immunologically to the antigenic stimuli.

The neonate usually possesses a variety of circulating antibodies but all these have been absorbed, probably intact, from the yolk having been secreted by the hen. Thus the chick is passively protected for up to four weeks after hatching by these maternal antibodies. The ability of the neonate to respond to novel antigenic stimuli has been shown to exist at hatching, though not before and then only to a very limited degree. This primary antigenic response develops rapidly and progressively after hatching and is mature after three weeks.

The importance of the bursa of Fabricius in conferring immunological competence upon the chick is now recognized.

Concluding Remarks

In the life-cycle of an animal any stage is the product of earlier ones and at the same time it must necessarily influence future stages. Thus, the embryo is affected by the availability of nutrients laid down by the hen, the newly-hatched chick is the product of incubation and the adult is, in turn, the product of that newly-hatched chick as further influenced by the environment. We have, of course, dealt here with only part of the life-cycle — that part conveniently identified with the egg. It will be clear from both sections of the book that the concept of the cleidoic egg being complete and isolated is too simple. The embryo is not only affected by many environmental factors but also reacts to them. Thus there is a constant interaction between embryo and environment. Hatching should not therefore be considered as an end-point for it is simultaneously a beginning. It is perhaps nothing more than a convenient reference point.

14

Hormones in Development

In the course of elucidating the endocrine status of an organ it is usual to remove it surgically from the animal and then, if obvious changes result, to attempt replacement therapy either by means of grafting or by injecting simple extracts of the organ. If this course of action succeeds then the active principle or principles will be isolated and identified and their precise function determined by further experimentation. One difficulty in this approach is that the amount of substance required to elicit a response is not likely to be known, and the observed response might therefore not be the normal response. The use of specific inhibitors overcomes to a large degree these problems and, as we shall see below the results of experiments using these substances can produce more satisfactory results than those using the purified hormones.

There is also another complication. The fact that a response may be elicited following the application of a particular stimulus which may mimic a normal response of the developing animal, does not necessarily mean that this is the normal stimulus.

The avian egg, free from material influences, offers the endocrinologist an excellent experimental system, yet its potential has, surprisingly, not been fully exploited and large areas remain to be explored. Moreover, while it has been known for many years that certain endocrine glands become functional during incubation, it is only relatively recently that the extent of hormonal influences in development has been appreciated.

Thyroid Hormones
Development of the thyroid glands

The thyroid of the domestic fowl arises as a midventral diverticulum of the pharynx on the second or third day of incubation. By the fifth day the anlage has become two-lobed and the cells have migrated towards the third aortic arch as double sheets surrounded by connective tissue. These sheets separate into clusters of cells and colloid droplets form intracellularly by the eighth day. The droplets coalesce in the clump of cells to produce first two large extracellular droplets then a single follicular space. The follicles increase in size by the fusion of adjacent follicles and by the division of the surrounding cells. The apical surface of the cells is characterized by cilia. The Golgi bodies are well-developed by the eighth day and rough endoplasmic reticulum increases from about the eleventh day to achieve the appearance of the adult thyroid by the fourteenth day.

Synthesis and secretion of hormones

There is little evidence of diversity in the biosynthetic mechanisms of the thyroid hormones in the different classes of animals and it seems likely that the avian embryo conforms to this pattern. Iodide ions are oxidized to iodine by a peroxidase system and the iodine reacts with the phenolic ring of free tyrosine or of tyrosine in proteins to form mono-and diiodotyrosines. Thyroglobulin is synthesized as polypeptide sub-units containing uniodinated amino acids, which are then modified by the addition of substituted hexosamines to form a glycoprotein. Iodination occurs from the completion of the glycoprotein subunit until the molecule of thyroglobulin is secreted into the follicular colloid. Thyroxine is then synthesized from one free iodotyrosine molecule and one iodotyrosine bound into the thyroglobulin molecule. The thyroglobulin molecules thus contain thyroxine, triiodothy-ronine, mono and diiodotyrosine. The thyroxine and triiodothyronine are liberated by the hydrolysis of the thyroglobulin — the other amino acids are re-utilized by the thyroid cells.

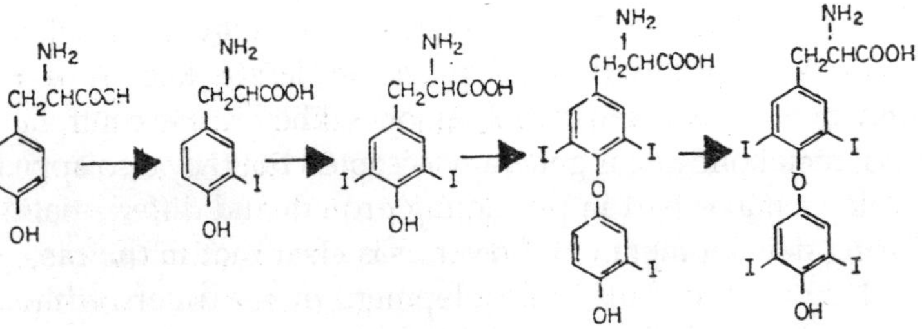

Fig. 14.1. Metabolic pathway for the synthesis of the thyroid hormones.

In mammals triiodothyronine is between five and seven times more potent than thyroxine but in the hatched bird at least, and probably the embryo, these hormones have a similar potency. This is due to similar degrees of binding by plasma protein.

The age at which iodine is concentrated by the thyroid has been variously estimated as $5\text{-}1/_2$ days, 7 days, 8 days, 9 days, 10 days, and 11 days, Since monoiodotyrosine can be detected in the thyroid by $8\text{-}1/_2$ days, diiodotyrosine by the ninth day and thyroxine by the tenth day it seems likely that the concentration of iodine begin no later than the eighth day. It is generally agreed, however, that secretion of thyroxine and presumably, triiodothyronine, begins on the tenth day *Trunnell* and *Brayer* (1953) having shown that this is the time when the thyroid itself matures.

Secretion of the thyroid hormones (it is not known in what proportions thyroxine and triiodothyronine appear in the blood of the embryo) is controlled by the hypophyseal hormone, thyroid-stimulating hormone (thyrotrophin). The development of the thyroid and its ability to concentrate iodine are influenced by vitamin B_{12} insulin (maraud *et al.*, and also by the bursa of Fabricius.

It is probable that the rate of secretion of the hormones is low until the fifteenth day. There is a marked increase in the protein bound iodine content of the blood during hatching.

Metabolic effects of the thyroid hormones

Although the age at which secretion of the thyroid hormones begins is known with some precision, less is known of the fundamental actions of these hormones. Their role in controlling basal metabolic rate is generally accepted but they also appear to play a major part in promoting growth and differentiation during development. However, it is clear that in the case of the bird perhaps half of development in *ovo* occurs without the modifying influences of thyroxine and triiodothyronine. Indeed, while the tissues of the young embryo are probably sensitive to their action there is evidence that even low concentrations — 100 ng per egg — are teratogenic.

The injection of thyroid hormones at about the time the thyroid begins its own secretory activity has been found to produce variable results and has thus failed to provide much information on the significance of the hormones in the second half of incubation. *Romijn et al.* (1952) found thyroxine stimulated metabolic rate but no increase in growth rate resulted. *Portet* (1966), on the other hand, could not detect any change in either parameter, while *Beyer* (1952) found similar increases in growth rate and oxygen consumption with overall increases in metabolic rate.

The increase in oxygen uptake growth rate and the significant, though temporary fall in the hepatic stores of glycogen are all thought to result from the onset of thyroidal secretion. Attempts to delay the depletion of hepatic glycogen by treating the embryo with the goitrogen, thiourea, have been been completely successful though partial decapitation ('hypophysectomy') prevents completely this fall.

While exposure of the embryo to enhanced concentrations of the thyroid hormones has not produced consistent results, reduction or suppression of the natural secretion of the thyroid with substances such as thiouracil or thiourea or by hypophysectomy is generally agreed to result in a decreased growth rate, decreased metabolic rate, hypertrophy of the thyroid and a delay in the time of hatching though these effects do not become manifest before the tenth day when the thyroid normally begins active secretion. Glycogenesis is reduced in both liver and yolk sac membrane following treatment of the embryo with thiourea. Again these effects are not seen before the tenth day.

The thyroid hormones influence bone development. *Adamson* and *Ingbar* (1967a, b) have shown that triiodothyronine, but not thyroxine, stimulates the uptake of neutral amino acids by developing bone. Indeed there is some evidence to suggest that thyroxine may inhibit growth. The morphogenesis and keratinization of the scales of the leg are also influenced by the thyroid for thiourea delays development while exogenous thyroxine advances it. Similarly the development and pigmentation of the down feathers are affected by the thyroidal status of the developing chick.

The importance of the thyroid hormones in the morphogenesis and histogenesis of the duodenum is now well-established. Development of the duodenum ceases at $17^1/_2$ days in hypophysectomized embryos and is retarded in embryos treated with thiourea but continues if thyroxine is made available. The thyroid also apparently influences adrenal development though it is uncertain whether this relationship is direct or whether it involves the pituitary. *Bargman* and *Gardner* (1967) have implicated the thyroid gland in the development of the auditory system.

The thyroid and hatching

In the course of studies on hatching chicks of the domestic fowl *Freeman* (1962) noted that oxygen consumption began to rise about 2 h before the chick began the final breaking down of the shell. Indeed since there was no concomitant increase in overt activity it seemed possible that this rise was hormonally induced and that the hormone might therefore be considered to be the hatching stimulus. In view of their effects on metabolic activity, the thyroid hormones were soon considered for this role.

As we have already seen from the results of work using goitrogens, the thyroid is capable of influencing the rate of development and the time of hatching. However, this action may be seen as a general response by the tissues and it does not necessarily follow that the thyroid produces a specific stimulus which initiates active hatching. Treating the embryo at a relatively late stage of development with thiouracil (17 days — $18^3/_4$ days — was found to have little effect on development but a delaying action on the time of hatching. The relationship between thyroid hormones and the stimulation of active hatching was further emphasized by Freeman (1964) who showed that by treating the embryo with either thiouracil, triiodothyronine, thyroxine or thyrotrophin at $18\text{-}1/_2$ to 19 days the time of hatching could be greatly influenced without, significantly, affecting the time at which pulmonary respiration was initiated. Unfortunately neither *Balaban* and *Hill* (1971) nor *Oppenheim* (1973) have been able to confirm that part of the work employing thiouracil. Both groups have found that the thyroid hormones accelerate hatching, though when the treatment was given at 17 days, pulmonary respiration was also significantly advanced suggesting that at this earlier time the embryonic response is general rather than specific.

TABLE 14.1

Effects of Changing Thyroidal Status on the Hatching Processes

| Treatment | Dose | n | *Houri advance (+) or retardation (-) of* | | Percentage advance or record, in period from |
			Pulmimary respiration	*Hatching*	*pulm, resp. to hatch*
Thiouracil	2 mg	78	-1.0	-10.7	-39
	4 mg	78	-0.3	-17.3	-69
Triiodothyronine[1]	1μg	144	+0.7	+5.3	+22
Thyroxine	1 g	144	+0.5	+4.0	+15
T4 + Thiour	2 mg+ 2 μg	58	+0.5	+0.5	0
Thyrotrophin	1 IU	40	+0.5	+3.5	+12
	2 IU	40	+0.5	+4.0	+14

[1] Sodium salt.

Thus while the thyroid is implicated in stimulating active hatching specifically the hypothesis remains unproven. Protein-bound iodine concentrations in the plasma have been shown to double during the last 30 h of incubation, from 6.3 to 13.2 μg I $^3/_4$100 ml.$^{-1}$ However, further work on this aspect of thyroidal physiology is needed.

The thyroidal status of the embryo appears to affect the rate at which yolk material is absorbed by the embryo during the latter stages of incubation. Clearly this is important in relation to the movement of the yolk sac into the abdominal cavity prior to hatching. Finally, it should be noted that the development of the hatching muscle *(musculus complexus)* may be influenced by the thyroid hormones.

Parathyroid Hormone and Calcitonin
Introduction
It will be convenient to consider these two hormones together for though they are secreted by different glands - parathyroid hormone (parathormone) by the parathyroid glands, calcitonin by the ultimobranchial bodies - both are concerned with the control of calcium metabolism. Their functions both in the embryo and in the adult have not yet been fully elucidated, though it seems probable that they have antagonistic effects.

Development of the parathyroids and ultimobranchial bodies
The parathyroid glands probably arise from the endoderm from the third and fourth visceral pouches. The pair on each side generally fuse to give one glandular mass which is close to, or even attached to, the thyroid gland of that side. Accessory parathyroid tissue often occurs in the thymus and in the ultimobranchial bodies, while active ultimobranchial C-cells are often found in the parathyroids.

The ultimobranchial bodies are derived from the endoderm of the sixth visceral pouch. The cells migrate to a position below the parathyroids and thyroids. Details of their development may be found in *Dudley* (1942). Their structure and ultrastructure are described by *Stoeckel* and *Porte* (1967a, b; 1969a, b) and *Hodges* (1970). The endocrine cells of the ultimobranchial bodies are distinct and are termed C-cells.

Secretion of parathormone and calcitonin

Sun (1932b), on histological grounds, suggested that there was no secretion of parathormone during development in *ovo*. However, later works has generally indicated that synthesis of the hormone begins on about the eighth day and that there may be actual secretion from about 10-1/2 days. The rate of synthesis seems likely to increase sufficiently by the thirteenth day for storage vesicles to be formed in the cells. Bone resorption through parathormone-induced osteoclastic activity can be stimulated by the eleventh day though *Jones* (1970) believes the action of parathormone in ovo is to inhibit ossification rather than stimulate resorption.

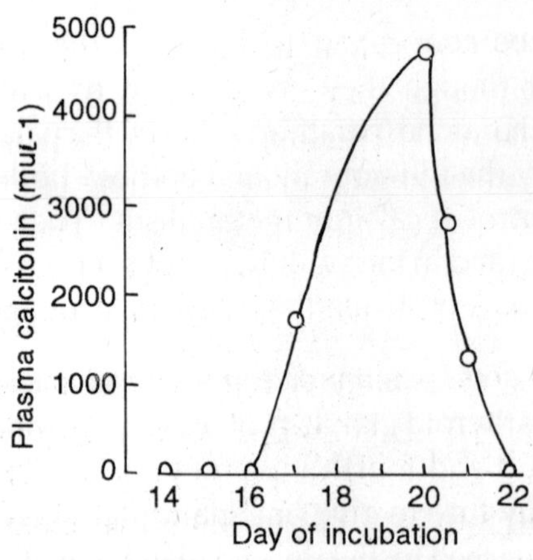

Fig. 14.2.   Plasma calcitonin concentration during the development of fowl.

Calcitonin contains about 32 amino acids and has a molecular weight of the order of 4500. *Stoeckel* and *Porte* (1969a) presented histological evidence of secretion from the eleventh day though direct measurements show no appreciable amounts in the plasma before the seventeenth day. The concentration of calcitonin the ultimobranchial bodies rises rapidly during hatching but the peak plasma concentration occurs when pulmonary respiration is established. The concentration of calcitonin in terms of body weight is very high in the embryo 2500 mU kg^{-1} rising to 5200 mU kg^{-1} 3 days after hatching. It reduces thereafter to reach 600 mU kg^{-1} in the adult.

Significance in development

The role of these two hormones in calcium metabolism of the adult is still equivocal. The situation with regard to the embryo is even more problematical.

In the adult parathormone stimulates a rise in plasma calcium through bone resorption while calcitonin probably has the reverse effect. The importance of controlling plasma calcium levels within fine limits for the proper functioning of muscles and nerves cannot be over-emphasized. One might reasonably expect the parathyroids and ultimobranchial bodies to be active, particularly during the period of active calcium resorption from the shell, that is from 14-19 days. However, no evidence has been presented to indicate any parathyroid involvement in this process. On the other hand the excretion of phosphorus by the developing embryo may be partly controlled by parathormone.

The secretion of calcitonin is probably maximal immediately before hatching and coincides with a lowered plasma calcium concentration. The significance of this is unknown.

Adrenal Hormones
Development of the adrenal glands

The primordial adrenal cortical cells arise from the mesoderm medial to the mesonephros and can be distinguished by the fourth day in the domestic fowl. The cortical cells develop rapidly and by the seventh day they are arranged characteristically in cords. Medullary cells begin to be associated with these cortical cells between the fourth and fifth days. The glands become encapsulated by the tenth day.

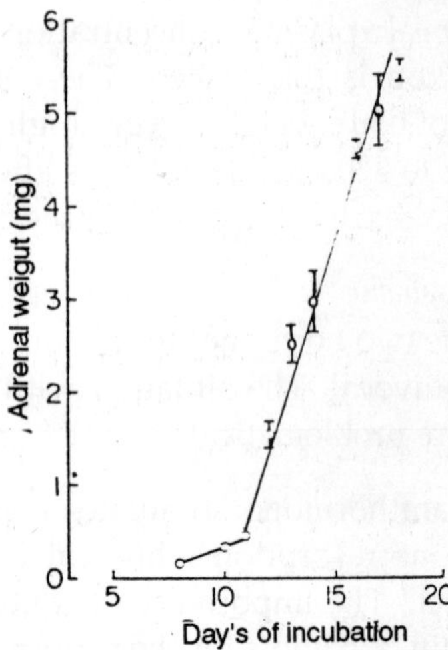

Fig. 14.3. Growth of the adrenal glands of the fowl during incubation. Each point is the mean net weight of the paired glands. Standard errors are indicated.

The growth curve of the adrenal is clearly biphasic. The first period, from the fourth to the eleventh day, is one of slow growth but during the second period, from the eleventh day to hatching, the growth rate increases markedly. The second period is controlled by the secretion of ACTH by the adenohypophysis and it is not surprising, therefore, that the greater proportion of the adrenal should be cortical in nature. *Hall* (1970) has shown that from the thirteenth to eighteenth day the proportion of cortex increases from 65% to 77% though *Payne* (1955) noted up to 90% in the newly hatched chick. (In the adult cortex and medulla are in equal proportions. There is some evidence that the growth rate of the adrenal of the male is greater than that of the female.

Synthesis and secretion of adrenal hormones

Adrenal steroids. The biosynthetic pathways have not been elucidated completely in either the embryo or the adult but they almost certainly follow the general scheme set out. The ability of the 14-day-old embryo to synthesize 11-deoxycorticosterone, corticosterone and aldosterone from acetate

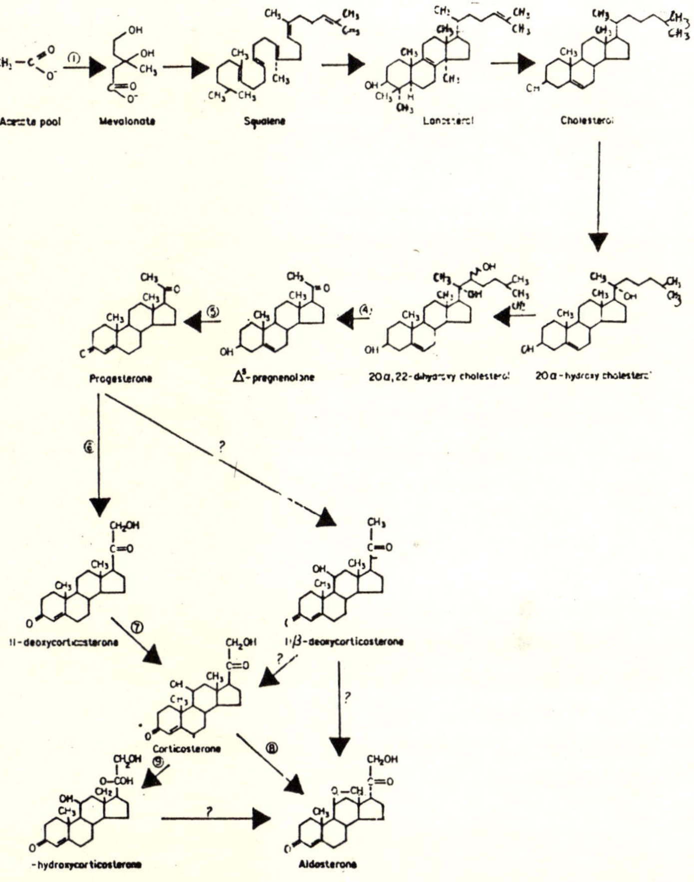

Fig. 14.4. Metabolic pathways for the synthesis of adrenal cortical steroids. Notes on enzymes involved : 1. The ability to synthesize cholesterol from acetate is possessed by virtually all cells. 2. Cholesterol 20x-hydroxylase. 3. Cholesterol 22-hydroxylase. 4. 20-22 desmolase. 5. $\Delta^{5-3}\beta$-ol dehydrogenase and Δ^5 isomerases. 6. Steroid 21-hydroxylase. 7. Steroid 11β-hydroxylase. 8. Steroid 18-hydroxylase. 9. Steroid 18-hydroxylase.

has been demonstrated by *Bonhommet* and *Weniger* (1967a, b) and semi-quantitated by *Pedernera* and *Lantos* (1973). It is possible that, as in the adult, the main secretory product is corticosterone. It is not certain whether hydrocortisone (cortisol) and cortisone are synthesized by the bird, though the former steroid has been shown to have considerable activity in the embryo.

It has been shown both histochemically and by bioassay that the adrenal cortical cells are able to synthesize corticosteroids from the fifth day of incubation — only 24 h after their differentiation. However, natural secretion *in vivo* does not begin until the eighth day when the endogenous secretion of ACTH begins. The establishment of the pituitary-adrenal axis is thus achieved at an early stage of development. Early work had suggested that the adrenal cortex functions, at least to a degree, autonomously but this has now been disproved. *Woods et al.*, (1971) found the axis established by 14-$\frac{1}{2}$ days, *Adjovi* and *Idelman* (1969) by the twelfth day but more sensitive technique have shown it to be established as early as 8 days.

While the amount of circulating corticosterone rises progressively during development *Piddington* (1970) and *Woods et al.*, (1971) have found that the secretion rate undergoes a marked rise at about 14-$\frac{1}{2}$ days indicating an increase in the rate of turnover.

It is generally agreed that the hypothalamus is not involved in the control of the normal secretory pattern of the adrenal cortex, at least during incubation though is probably required in the mediation of the stress response.

Catecholamines. The biosynthetic pathways of adrenaline and noradrenaline are outlined. Catechol substances can be detected in the adrenal at 4 days — a time coincident with the differentiation of the adrenal medullary cells. However, it does not seem likely that the secretary rate is very high before the tenth.

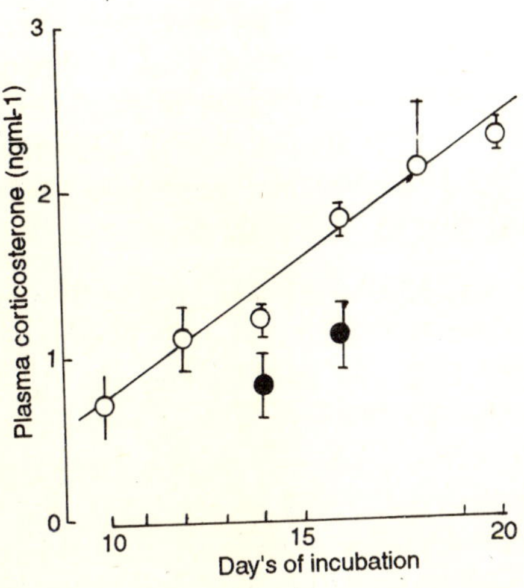

Fig. 14.5. Plasma corticosterone concentrations during development of normal (O) and hypophysectomized (●) fowl.

Wassermann and *Bernard* (1970) have determined the adrenaline content of the adrenals during development. Similar estimates of the noradrenaline have not been made though a semi-quantitative study shows that noradrenaline predominates throughout development. Estimates made on the adrenals of the neonate show that from between 60% and 80% of the total medullary catecholamine stores are noradrenaline.

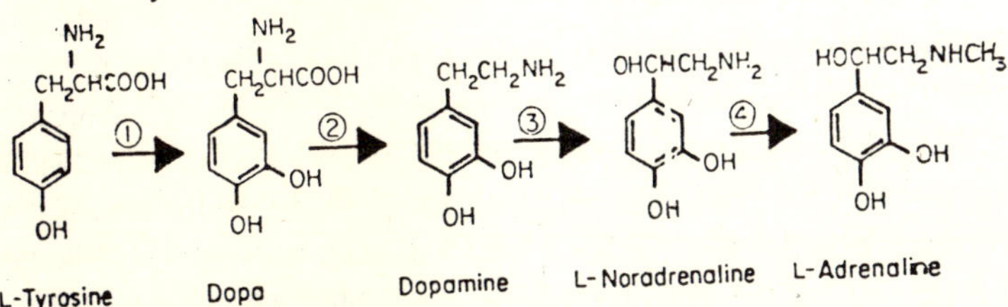

Fig. 14.6. Metabolic pathway for the synthesis of catecholamines. The enzymes involved are as follows :

(1) Tyrosine 3-hydroxylase

(2) Dopa decarboxylase

(3) Dopamine hydroxylase

(4) Phenylethanolamine *N*-methltransferase.

Significance of adrenal cortical hormones in development

As we have already seen the developing fowl synthesizes at least two cortical hormones, corticosterone and aldosterone. While their roles in adult life are reasonably well understood, their function in the developmental processes remain, with one important exception, to be elucidated.

The exception concerns the differentiation and maturation of the duodenum in the last few days before hatching. Several authors believe that both the adrenal and the thyroid are concerned in this though *Pedernera* (1971) suggests that the adrenal alone is involved. The effects of corticosteroids on alkaline phosphatase activity may be relatively non-specific for there is indirect evidence that they act on that enzyme in bone and thus influence bone formation There is, furthermore, a suggestion that alkaline phosphatase may be stimulated by another, as yet unknown, mechanism for the activity of this enzyme is elevated in mature embryos treated with a corticosteroid inhibitor.

There is some evidence that the adrenal corticosteroids are involved in the maturation of the exocrine pancreas. Exogenous hydrocortisone stimulates the accumulation of analyse, procarboxypeptidase A, chymotrypsinogen and to a less extent endonuclease. This maturation of the pancreas is initiated on the fourteenth day, a time, as we have seen, when adrenal cortical activity is increasing. *Hijmans* and *McCarty* (1966), *Dautlick* and *Strittmatter* (1970) and *Strittmatter* (1972) have also found that the maturation of the small intestine, and the activities of the maltase and sucrase that it secretes, are influenced by the adrenal corticosteroids. The prevention of splenomegaly is also a function of the adrenal hormones.

Other functions that have been ascribed to the corticosteroids include the maturation of both ovary and testes stimulation of (Na + K)-stimulated ATPase influence over cholesterol metabolism and the stimulation of phenyletholamine-N-methyl transferase activity in the adrenal

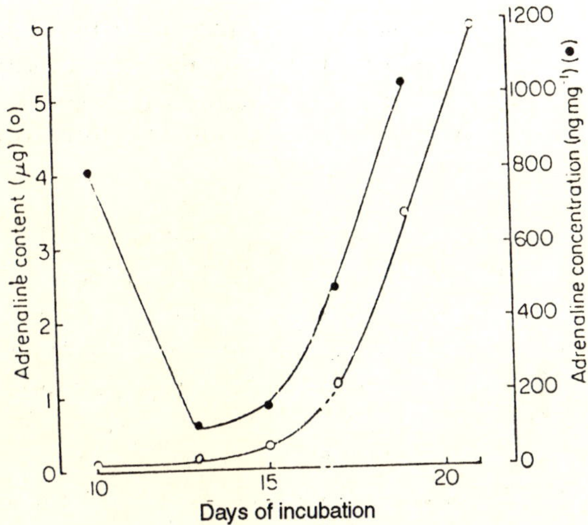

Fig. 14.7.   Adrenaline concentration (●) and content (O) of the adrenaline during development of the fowl.

medulla. This enzyme is concerned in the conversion of noradrenaline to adrenaline and an increase in its activity might explain the increased concentration of adrenaline in the adrenal from the fifteenth day.

Significance of adrenal medullary hormones in development

The catecholamines are the secretions of the sympathetic nervous system of which the adrenal medulla is but part. Thus this tissue differs from other endocrine tissue in being part of a larger functional unit which is under the direct control of the central nervous system. While the sympathetic nerve cells have long axons and secrete catecholamines locally at the surface of the cells they innervate, the medullary cells lack axons and secrete their hormones into the circulation thereby causing general reactions throughout the body. Because of this dual system it is necessarily difficult to subscribe with certainty any catecholamine-induced reaction to the adrenal medulla. Indeed the role of catecholamines in the development processes generally remain largely speculative, though there has recently been speculation that they play an important role in early morphogenesis.

It is possible that the progressive increase in blood pressure during development may be due to circulating catecholamines for *Girard* (1967, 1973b) has shown that the circulatory system is sensitive to both adrenaline and noradrenaline from the sixth day.

Immediately prior to the initiation of hatching (i.e., before pulmonary respiration is established) there is a profound mobilization of glycogen stores of the embryo. This certainly involves the liver probably the yolk sac membrane but, significantly, neither the heart nor the brain. The mechanism of this mobilization has been studied only in the liver. Adrenaline stimulates glycogenolysis though norardrenaline is ineffective. It seems possible, therefore, that the adrenal medulla is concerned though the data of *Freeman* and *Manning* (1971) show that, on a molar basis, glucagon is perhaps twenty times more potent than adrenaline in mobilizing glycogen. This might be thought to indicate that glycagon is the more likely stimulus. However, the natural mobilization of hepatic glycogen is impaired by pretreatment of the embryo with α- or β-receptor blocking agents indicating a role for the catecholamines rather than glucagon. At the same time it has been shown, albeit in rabbits, that the mobilization of hepatic glycogen is most rapidly achieved by stimulation of the splanchnic nerves innervating the liver cells. Thus, paradoxically, while catecholamines may well be implicated, it seems more likely that they emanate from the autonomic nervous system rather than the adrenal medulla.

Pancreatic Hormones
Introduction

In addition to the two well-known hormones, insulin and glucagon, the avian pancreas secretes a physiologically-active substance, at present known simply as avian polypeptide (APP: Its significance remains to be elucidated.

The role of insulin in carbohydrate metabolism in the adult remains problematical indeed its significance has occasionally been questioned. However, evidence is now being collected to confirm its role in carbohydrate metabolism. As we shall see, its role in embryonic development has been partly elucidated.

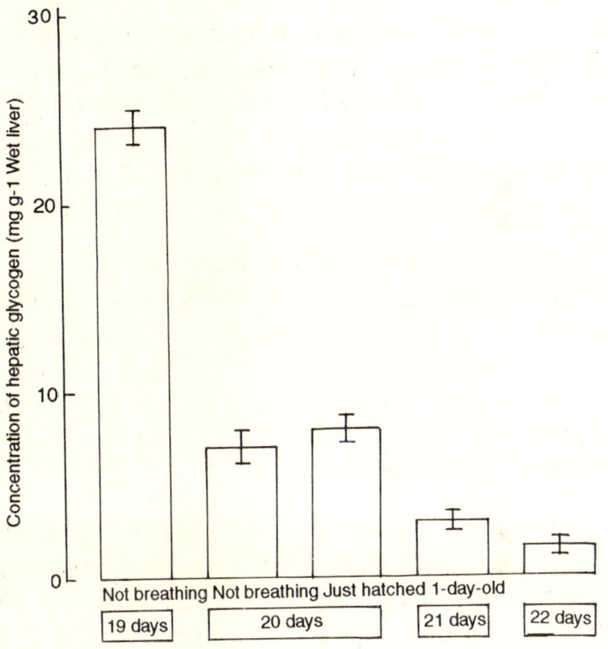

Fi. 14.8.   Hepatic glycogen stores during hatching.

Glucagon, a powerful glycogenolytic hormone in mammals and birds, is a particularly potent lipolytic agent in the latter. Recent research has demonstrated that a glucagon-like substance is also secreted by the small intestine. It seems likely that the pancreatic and gut glucagons have different roles in metabolism though that of the gut factor has yet to be determined.

Development of the endocrine pancreas

The pancreas arises as three separate primordia: one is a dorsal evagination of the gut while the other two are lateral evaginations. These anlagen arise, in the fowl, after three days and eventually fuse. The various cell-types of the islets of

Langerhans begin to differentiate early. Their distribution is not random; there is a predominance in the splenic lobe which is derived from the dorsal anlage. Studies with the light microscope suggested α- and β-cell differentiation on about the eighth day but with the electron microscope these cell types can be recognized on the third day — that is shortly after evagination. A third cells type, the s-cell, can also be differentiated by the tenth day. Its significance remains to be elucidated: some suggest it is either equivalent to the α-cell or is a precursor of the β-cell; others recognize it as a distinct cell type and suggest it is concerned with gastrin secretion.

Structure and secretion of insulin and glucagon

Insulin. Avian insulin, like its mammalian counterpart, is composed of two chains of amino acids. There are six different amino acids in the sequence of fowl insulin as compared with porcine and ox insulins fowl and turkey insulins are identical but duck insulin has three different amino acids in the sequence as compared with fowl insulin.

Insulin is secreted by the β-cells, from the fourth of fifth day though not to any significant extent before the twelfth or thirteenth day. The importance of the splenic lobe as a source of insulin is shown by the data of *Kedinger et al.*, (1972): 66% of the total pancreatic insulin is found in this tissue which itself represents 6% of the pancreatic tissue.

Attempts to measure the concentration of circulating insulin during development have been unsuccessful, probably because the concentrations are below the sensitivity of the methods currently available. At hatching the concentration is about 6 $\mu U\ ml^{-1}$.

*Glucagon.*Two molecular speciès of pancreatic glucagon have been isolated from the duck, one with a molecular weight of 3000 the other of 6000. The jejunum and ileum, but not the duodenum, secrete only the 'large' molecular from In the adult, however, the circulating molecule, whether from pancreas or gut, is the large form. No information regarding other avian species is available.

The pancreatic x-cells begin to secrete glucagon on the third fifth eighth or tenth day. There are, as yet, no data on circulating glucagon concentrations.

Role in insulin in development

Zwilling (1951) was the first to show that injected insulin causes hypo-glycaemia and enhanced the concentration of glycogen in the yolk sac membrane. The latter observation was confirmed by *Thommes* and *Mathew* (1969) who further established that the response could be elicited as early as the fourth day of incubation. This response is probably mediated through UDPG-glycogen synthetase. The increase in the concentrations of glycogen in the yolk sac membrane and liver during the third week of incubation may therefore be controlled by the natural secretion of insulin stimulating the activity of this enzyme.

Insulin also affects the maturation of the hepatocytes. *Benzo* & de la *Haba* (1972) have shown that insulin is needed for the development of smooth endoplasmic reticulum in the hepatocytes. This insulin, moreover, has to be complexed to zinc.

The lipogenic mechanisms of the heart tissues are stimulated by insulin as is glucose uptake by the same tissue though only from the seventh day. It should be remembered, however, that the rate of insulin secretion is probably low until the third week.

Insulin has also been implicated in the stimulation of the uptake of iodine by the thyroid and of the uptake of neutral amino acids by the nervous system.

Role of glucagon in development

While glucagon plays a very important role in lipid metabolism of the hatched bird, being a potent mobilizer of lipid from the adipose tissue it does not seem that this activity is present during incubation or at least only to a very restricted degree. This reduced activity is due partly to the relative immaturity of the adenyl cycalse-cyclic AMP system. Glucagon has activity in the carbohydrate metabolism of the embryo and may well have an important role in controlling circulating glucose concentrations.

Glucagon stimulates glycogenolysis in both the yolk sac membrane and the liver. This is brought about, at least in the yolk sac membrane, by an activation of phospharylase.

It has already been noted that there is a drastic mobilization of glycogen just prior to the onset of breathing. Given the glycaemic activity of glucagon it would seem possible that this hormone is responsible.

On balance, however, the evidence currently available suggests that this mobilization is mediated by the adrenal medulla or the autonomic nervous system, i.e., by cazecholamines. Perhaps the most important piece of evidence for this interpretation is that α- and β-adrenergic receptor blockers impair the natural mobilization of glycogen.

During hatching the adenyl cyclase-cyclic AMP system of adipose tissue matures so that glucagon is able to stimulate both lipolysis and glycogenolysis. However, by reducing the dose rate to 1 μg kg^{-1} found that only the lipolytic mechanism was stimulated suggesting that this becomes the more important role of glucagon after hatching.

Early reports that glucagon has growth-promoting activity have not been subsantiated.

Gonadal Hormones
Development of the genital system

Initially the genital system of the avian embryo develops similarly in both males and females — for about 5 days in the house sparrow, 7 days in the domestic fowl and 7-$^1/_2$ days in the duck. Sexual dimorphism commences thereafter.

In the female the Mullerian ducts, derivatives of the urogenital ridges, and the gonadal tissues develop asymmetrically : the left ovary and duct develop normally while the right ovary and Mullerian duct develop indifferently at first and then degenerate. Both mesonephric (Wolffian) ducts degenerate though they usually persist into post-embryonic life.

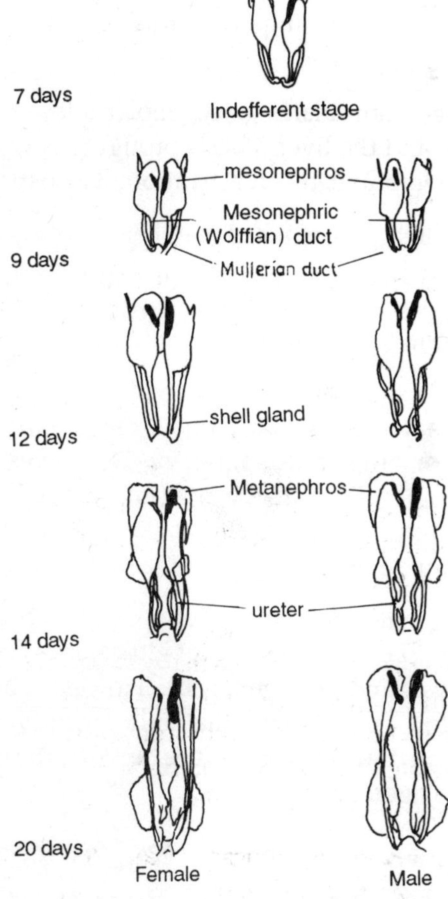

Fig. 14.9. Development of the male and female genital tracts of the fowl.

Role of glucagon in development

While glucagon plays a very important role in lipid metabolism of the hatched bird, being a potent mobilizer of lipid from the adipose tissue it does not seem that this activity is present during incubation or at least only to a very restricted degree. This reduced activity is due partly to the relative immaturity of the adneyl cycalse-cyclic AMP system. Glucagon has activity in the carbohydrate metabolism of the embryo and may well have an important role in controlling circulating glucose concentrations.

Glucagon stimulates glycogenolysis in both the yolk sac membrane and the liver. This is brought about, at least in the yolk sac membrane, by an activation of phospharylase.

It has already been noted that there is a drastic mobilization of glycogen just prior to the onset of breathing. Given the glycaemic activity of glucagon it would seem possible that this hormone is responsible.

On balance, however, the evidence currently available suggests that this mobilization is mediated by the adrenal medulla or the autonomic nervous system, i.e., by catecholamines. Perhaps the most important piece of evidence for this interpretation is that α- and β-adrenergic receptor blockers impair the natural mobilization of glycogen.

During hatching the adenyl cyclase -cyclic AMP system of adipose tissue matures so that glucagon is able to stimulate both lipolysis and glycogenolysis. However, by reducing the dose rate to 1 μg kg^{-1} found that only the lipolytic mechanism was stimulated suggesting that this becomes the more important role of glucagon after hatching.

Early reports that glucagon has growth-promoting activity have not been subsantiated.

Gonadal Hormones
Development of the genital system

Initially the genital system of the avian embryo develops

similarly in both males and females — for about 5 days in the house sparrow, 7 days in the domestic fowl and 7-$\frac{1}{2}$ days in the duck. Sexual dimorphism commences thereafter.

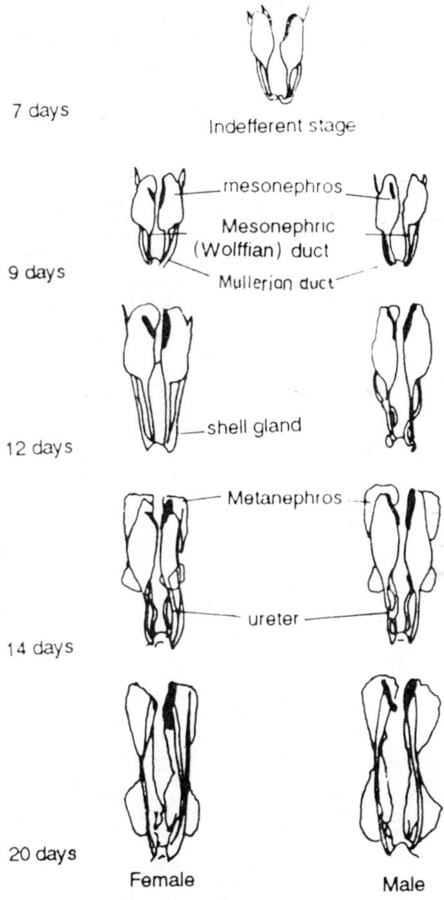

Fig. 14..10. Development of the male and female genital tracts of the fowl.

In the female the Mullerian ducts, derivatives of the urogenital ridges, and the gonadal tissues develop asymmetrically : the left ovary and duct develop normally while the right ovary and Mullerian duct develop indifferently at first and then degenerate. Both mesonephric (*Wolffian*) ducts degenerate though they usually persist into post-embryonic life.

Both gonads are Wolffian ducts develop in the male to give the testes and vasa deferentia, whereas the Mullerian ducts regress. In the duck and fowl regression begins on the ninth day and is complete by the fourteenth.

Ovarian oestrogenic hormones are synthesized and secreted by medullary interstital cells, for those cells have high concentrations of lipids and cholesterol Δ^5-33-hydroxysteroid dehydrogenase. Ovarian androgenic secretion may be a property of the cells of the primary albuginea and stroma.

In the male early steroid secretion is found in the cord cells but is progressively lost during development. At about $8\text{-}^1/_2$ days in the fowl ($13\text{-}^1/_2$ days in the Japanese quail) the interstitial cells of the differentiating tests show Δ^5-3β-hydroxysteroid dehydrogenase activity.

Synthesis and secretion of gonadal hormones

The general metabolic pathways of the biosynthesis of gonadal steroids in mammals are illustrated. Studies on the bird have generally confirmed the similarities between the two classes.

Fig. 14.11. Metabolic pathways for the synthesis of gonadal steroid hormones.

The enzymes involved are as follows :

(1) Δ^5-33-ol-dehydrogenase and -isomerases

(2) C_{21} steroid 17α-hydroxylase

(3) 20α-dehydroxy-dehydrogenase

(4) 20b-dehydroxy-dehydrogenase

(5) 16a-hydroxylase

(6) 17-20-desmolase

(7) 17b-hydroxysteroid dehydrogenase

(8) Aromatizing enzymes

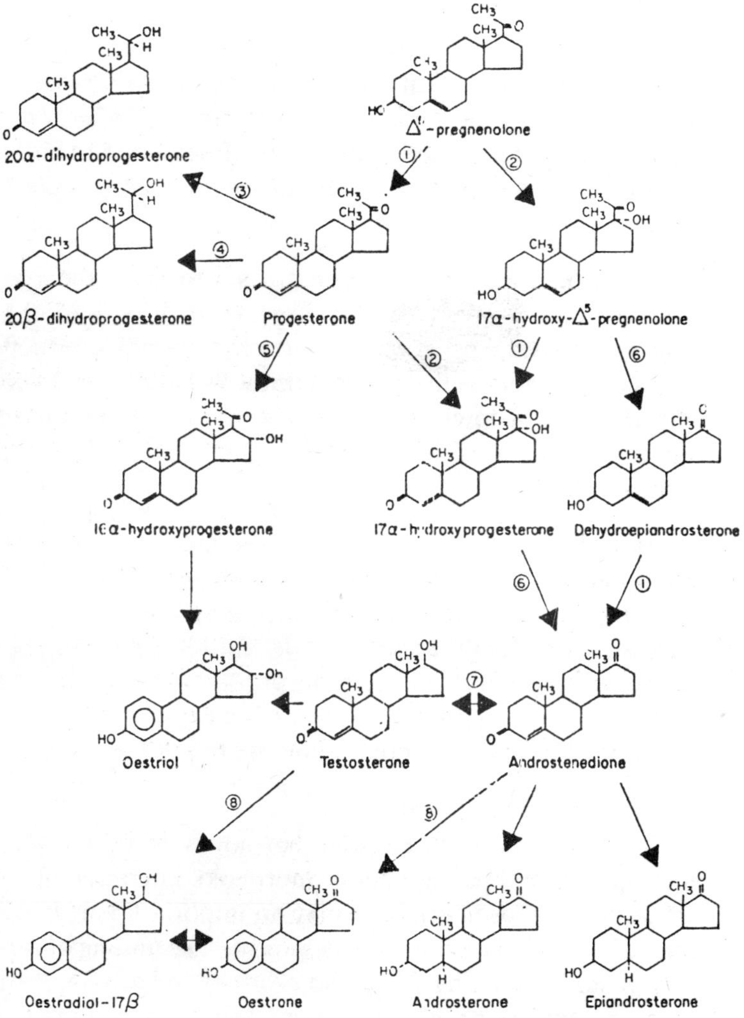
Δ⁵-pregnenolone
20α-dihydroprogesterone
20β-dihydroprogesterone
Progesterone
17α-hydroxy-Δ⁵-pregnenolone
16α-hydroxyprogesterone
17α-hydroxyprogesterone
Dehydroepiandrosterone
Oestriol
Testosterone
Androstenedione
Oestradiol-17β
Oestrone
Androsterone
Epiandrosterone
① ② ③ ④ ⑤ ⑥ ⑦ ⑧

Synthesis of hormones certainly begins shortly after the gonadal anlagen appear on the fourth day and there is some evidence that this begins earlier and is more active in the female. Between the fourth and sixth days, while the gonadal tissue is in the indifferent state, oestrogen and oestradiol-17β are synthesized though *Galli* and *Wassermann* (1973) suggest that these are synthesized only by gonads destined to become ovarise. On the seventh day, still in the indifferent state, the gonads synthesize testosterone.

As differentiation of the gonads proceeds the ovary produces progressively more oestrogens with oestradiol-17β always exceeding oestrone production. Oestriol production begins on the fourteenth day. Similar results have been obtained with the guinea fowl and duck. The testes do not produce significant amounts of oestrogens but increasingly more testosterone.

Initially gonadal hormone secretion is probably autonomous. The pituitary-gonadal axis is established from about the thirteenth day. Gonadotrophins have been found to increase markedly the secretion of oestradiol-17β in both sexes but have little effect on oestrone production. *Akram et al.,* (1973), however, found that the secretion of oestrone and oestradiol-17β was unaltered following hypophysectomy.

Significance in development

That secretion of gonadal hormones begins early in development and perhaps before morphological differentiation suggests that these hormones play an important part in early gonadal differentiation and development. The finding of earlier and greater activity by the gonad destined to become ovarian in nature conforms with this notion. However, with perhaps two exceptions, the roles of the various hormones remain to be properly elucidated.

The mechanism of regression of the Mullerian ducts differs between sexes. Both regress in the male but only the might one in the female. In the male regression is controlled by the testicular secretions. The active principle is not testosterone but remains to be positively identified though there is some evidence that it may be an oestrogen-like substance . It is known that the maximum activity of the 'regression' hormone is from 11-13 days and it is suggested that the subsequent fall in activity is related to the onset of the secretion of thyrotrophin, follicle-stimulating hormone and growth hormone by the pituitary.

The regression of only the right Mullerian duct in the female is probably stimulated by the oestrogens though it is likely that there is also a genetic component at work. Studies on the hatched bird have indicated that oestrogen is important in stimulating development and that progrestrone generally accentuates this action. Thus it is the right duct that regresses even in embryos where the left ovary has been ablated.

The androgens, particularly testosterone, are required for the transformation of the Wolffian ducts into the vasa deferentia.

Some instances of the gonadal hormones influencing an extra-gonadal metabolic pathway have been tentatively identified. Young embryos have been found to be able to degrade testosterone only to 5β-reduced metabolites. These 5β-androstanes induce -aminolevulinic acid synthetase, the rate-limiting enzyme in porphyrin synthesis. Thus the embryo may well be able to influence erythropoietic activity. There is also some evidence that testosterone may stimulate chondroitin sulphate synthesis in bone while oestradiol-17β may reduce it. There are, furthermore, the necessary enzymes present in the embryonic cartilage to catabolize the sex hormones. Finally, and perhaps not unexpectedly in view of the marked sexual dimorphism in the plumage of sexually active birds, the gonadal hormones affect the pigmentation of the down feathers.

Brooks and *Ungar* (1967) have reported that progesterone, 16- or 17-hydroxyprogesterone stimulates the maturation of the so-called 'hatching muscle' *(musculus complexus)* which is concerned in the pipping and hatching processes. Progresterone not only stimulated the uptake of water, a process that is probably unique to this muscle, but also advanced hatching. Progesterone might therefore be considered for the role of a hatching hormone. However, attempts to repeat this work have not been successful.

Hypophyseal Hormones
Introduction

The pituitary gland or the hypophysis consists of two distinct parts, the adenohypophysis and the neurohypophysis: further subdivisions are possible. Its endocrine functions are of two basic types: the control of the activity of other endocrine glands, such as the thyroids, adrenals and the gonads and the production of hormones which have a general metabolic effect. But while the hypophysis may be seen as the co-ordinator of other endocrine glands it is itself, at least in the hatched bird, influenced by higher centres of activity, notably the hypothalamus. However, as we shall see its activities in the embryo are independent of hypothalamic control.

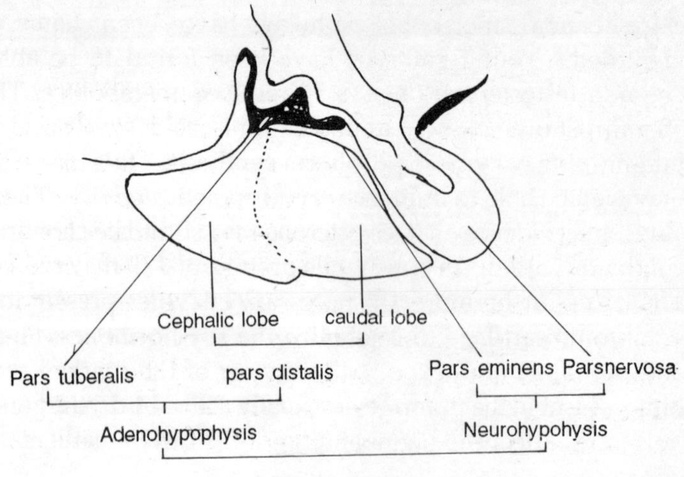

Fig. 14.12. A generalized scheme of the avian hypophysis showing the main divisions.

Development of the hypophysis

The adenohypophysis is derived from the oral epithelium of Rathke's pouch which extends upwards, from the second day, towards the developing infundibulum. Concurrently there is an outgrowing from the infundibulim to form the neurohypophysis. Together these two diverticuli become associated to form the hypophysis. Further details of their embryological development and cytodifferentiation may be found in *Wingstrand* (1951), *Romanoff* (1960), *Doskocil* (1965, 1966) and *Mikami et al.,* (1973).

The adenophypophysis is made up of the pars distalis and the pars tuberalis : there is no pars intermedia in the birds as there is in mammals. The pars distalis, perhaps the most studied portion of the hypophysis, is usually subdivided further into the cephalic and caudal lobes. The cephalic lobe probably secretes adrenocorticotrophic hormone (ACTH), thyroid-stimulating hormone (thyrotrophin: TSH) and follicle-stimulating hormone (FSH) exclusively; the caudal lobe probably secretes luteinizing hormone (LH: also known as interstitial cell-stimulating hormone, ICSH) while growth hormone (somatotrophin : GH) may be produced by either only the caudal lobe or both lobes. It is uncertain whether melanocyte-stimulating hormone (MSH) or prolactin are secreted by the pars distalis. Species differences may well occur, however. TSH may be secreted by both lobes in the duck the pigeon by only the cephalic lobe in the white-crowned sparrow and the Japanese quail.

Little is known of the pars tuberalis. It usually covers part of the diencephalon and encircles the infundibular stem.

The neurohypophysis remains distinct from the adenohypophysis, there being a sheath of connective tissue between them. This neural lobe, the equivalent of the posterior lobe of the mammal, contains neurosecretory fibres which extend back to the median eminence (pars eminens) and the hypothalamus. Neurosecretory material becomes visible in the neurohypophysis on about the thirteenth day. Thereafter the amount of material increases rapidly.

Secretion of hormones by the pars distalis

As we have already noted little or no information on the secretions of the pars tuberalis of the adenohypophysis or the neurohypophysis is available. We shall therefore confine this section to a discussion of the hormones of the pars distails of the adenohypophysis.

It is clear from the literature that the various cell-types of the pars distalis mature at different but precise times in development. ACTH is generally considered to be the first trophic hormone to be secreted : in the domestic fowl it probably appears on the eighth day. On the tenth day TSH secretion begins on the thirteenth day the secretion of gonadotrophins (LH and FSH) and on the fifteenth day the secretion of GH.

Significance of the hypophyseal hormones in development

Hypophyseal hormones are essential for normal development. Only some 6% or so of hypophysectomized embryo survive to 21 days and of these none hatches whereas hypophysectomized embryos with a par distalis grafted on to their chorio-allantoic membrane show an almost complete restoration of their viability and physiology. It is therefore not surprising that hypophysectomized embryos are dwarfed (lack of GH), have small, non-functioning thyroids (lack of TSH), adrenals (lack of ACTH) and gonads (lack of gonadotrophins). It is perhaps a little surprising, however, that all these hormones are secreted by the pars distalis acting autonomously, that is without the need for the so-called 'releasing factors', thyrotrophin releasing factor, adrenocorticotrophin releasing factor, etc., secreted by the hypothalamus. The successful restoration of a normal physiological state by grafting the pars distalis on to the chorio-allantoic membrane of the hypophysectomized embryo confirms this to be the case.

The evidence for a role by GH in the differentiation of the

duodenum remains equivocal : *Bellware* and *Betz* (1970) considered it necessary but more recent work by *Hart* and *Betz* (1972) has thrown doubt on this. *Betz* (1967) found that the extra-embryonic membranes are poorly developed in the hypophysectomized embryo and suggested that a hormone from the pars distalis stimulates their development. It is conjectural that this hormone might be GH.

The progressive inhibition of the Mullerian duct regression hormone is probably accomplished by the increasing activity of FSH, TSH and GH. On the other hand, the hypophyseal gonadotrophins are not required for the transformation of the female left Mullerian duct into the oviduct.

It has recently been suggested that the pituitary might influence directly bone differentiation and mineralization. The evidence is admitted to be slender in view of the known effects of the trophic hormones on other endocrine glands which in turn affect bone metabolism and much more work will be needed before a direct role can be assigned to the adenohypophysis.

Several authors have noted that the hypophysectomized embryos suffer from a disturbed water balance. This disturbance can be partially relieved by a pars distalis graft, perhaps through a restored secretion of aldosterone, though it seems reasonable to assume that the neurohypophysis plays some part in this aspect of the physiology of the embryo.

Glands of Uncertain Endocrine Status
Introduction

In the preceding sections of this chapter we have considered the known endocrine glands in some detail. There are, in addition, several glands whose endocrinological status is uncertain. Inevitably the evidence for considering some of these as endocrine glands is stronger than for others. Thus, the kidney (mesonephros, metanephros) and liver may secrete erythropoietin the thymus may secrete a hormone concerned in the maturation of the lymphoid system; the bursa of Fabricius may secrete a hormone with similar activity to that of the thymus and the pineal gland may produce a hormone which acts upon the hypophysis to enhance growth and maturation of the embryo. In this section we shall consider two of these: the *bursa* of *Fabricius* and the pineal gland.

Bursa of Fabricius

This gland arises as a diverticulum of the proctodaeum of the cloaca. Its function has yet to be precisely defined though there is now sufficient evidence to sustain the concept of its being endocrine. The evidence that it produces secretions concerned with immunological competence is particularly strong. Thus in view of *Thorbecke et al.*, (1968) secretion of a bursal hormone begins in the period 16-18 days, the hormone being transported either from cell to cell or by the migration of hormone-producing cells to other peripheral lymphoid sites. They consider that the hormone stimulates differentiation and maturation of the lymphocytes.

The pineal gland

The demonstration that embryos exposed to light during incubation tend to hatch earlier than those incubated in total darkness but of indicates that the embryo is photosensitive. There are two known discrete photosensitive organs in the bird, the eye and the pineal gland or epiphysis cerebri. At present it is not known which if either of these two organs might be concerned with the reception of the photostimulus and its transmission to the hypophysis though the late maturation of the eye makes it a doubtful candidate. Possible functions of the pineal gland have been fairly exhaustively explored by *Stalsberg* (1965) who found that the incubation period was unaffected by pinealectomy in *ovo* though regrettably the influence of light *per se* was not studied. The involvement of the pineal gland in determining the rate of development thus remains conjectural.